MW01630256

HINTS & KINKS

for the Radio Amateur

Edited by Steve Ford, WB8IMY

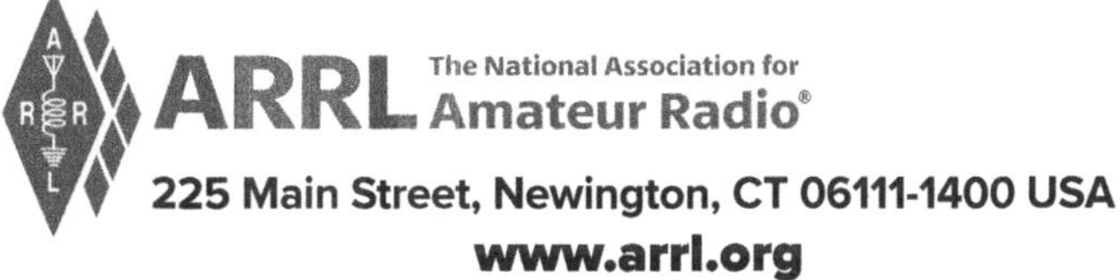

ARRL The National Association for **Amateur Radio®**

225 Main Street, Newington, CT 06111-1400 USA
www.arrl.org

Copyright © 2024 by
The American Radio Relay League, Inc.

*Copyright secured under the
Pan-American Convention*

International Copyright secured

All rights reserved. No part of this work may be
reproduced in any form except by written per-
mission of the publisher. All rights of translation
are reserved.

Printed in the USA

Quedan reservados todos los derechos

ISBN: 978-1-62595-197-7

Nineteenth Edition

Contents

Preface

Welcome to the 19th edition of *Hints and Kinks for the Radio Amateur*. This book is a compilation of material published in *QST* magazine's popular "Hints and Kinks" column between January 2012 and December 2016.

Between these covers you'll find handy advice on a broad range of topics from antennas to vintage radio restoration. The great thing about "Hints and Kinks" is that all these tips come from your fellow amateurs, people just like you who've discovered creative solutions to everyday problems. That's why "Hints and Kinks" has held an honored position for decades as one of the most-read sections of *QST*.

If you have hint or kink to contribute, by all means send it to our Hints and Kinks editor. You can reach the editor by e-mail at **h&k@arrl.org**.

Please note: The information in this book was current when it was originally published in *QST*. Call signs, postal addresses, e-mail addresses and websites cited in articles may now be out of date.

US Customary to Metric Conversions

International System of Units (SI)—Metric Prefixes

Prefix	Symbol			Multiplication Factor
exa	E	10^{18}	=	1 000 000 000 000 000 000
peta	P	10^{15}	=	1 000 000 000 000 000
tera	T	10^{12}	=	1 000 000 000 000
giga	G	10^{9}	=	1 000 000 000
mega	M	10^{6}	=	1 000 000
kilo	k	10^{3}	=	1 000
hecto	h	10^{2}	=	100
deca	da	10^{1}	=	10
(unit)		10^{0}	=	1
deci	d	10^{-1}	=	0.1
centi	c	10^{-2}	=	0.01
milli	m	10^{-3}	=	0.001
micro	μ	10^{-6}	=	0.000001
nano	n	10^{-9}	=	0.000000001
pico	p	10^{-12}	=	0.000000000001
femto	f	10^{-15}	=	0.000000000000001
atto	a	10^{-18}	=	0.000000000000000001

Linear
1 metre (m) = 100 centimetres (cm) = 1000 millimetres (mm)

Area
$1 \text{ m}^2 = 1 \times 10^4 \text{ cm}^2 = 1 \times 10^6 \text{ mm}^2$

Volume
$1 \text{ m}^3 = 1 \times 10^6 \text{ cm}^3 = 1 \times 10^9 \text{ mm}^3$
$1 \text{ litre (l)} = 1000 \text{ cm}^3 = 1 \times 10^6 \text{ mm}^3$

Mass
1 kilogram (kg) = 1 000 grams (g)
  (Approximately the mass of 1 litre of water)
1 metric ton (or tonne) = 1 000 kg

US Customary Units

Linear Units
12 inches (in) = 1 foot (ft)
36 inches = 3 feet = 1 yard (yd)
1 rod = 5½ yards = 16½ feet
1 statute mile = 1 760 yards = 5 280 feet
1 nautical mile = 6 076.11549 feet

Area
$1 \text{ ft}^2 = 144 \text{ in}^2$
$1 \text{ yd}^2 = 9 \text{ ft}^2 = 1\ 296 \text{ in}^2$
$1 \text{ rod}^2 = 30\frac{1}{4} \text{ yd}^2$
$1 \text{ acre} = 4840 \text{ yd}^2 = 43\ 560 \text{ ft}^2$
$1 \text{ acre} = 160 \text{ rod}^2$
$1 \text{ mile}^2 = 640 \text{ acres}$

Volume
$1 \text{ ft}^3 = 1\ 728 \text{ in}^3$
$1 \text{ yd}^3 = 27 \text{ ft}^3$

Liquid Volume Measure
$1 \text{ fluid ounce (fl oz)} = 8 \text{ fluidrams} = 1.804 \text{ in}^3$
1 pint (pt) = 16 fl oz
$1 \text{ quart (qt)} = 2 \text{ pt} = 32 \text{ fl oz} = 57\frac{3}{4} \text{ in}^3$
$1 \text{ gallon (gal)} = 4 \text{ qt} = 231 \text{ in}^3$
1 barrel = 31½ gal

Dry Volume Measure
$1 \text{ quart (qt)} = 2 \text{ pints (pt)} = 67.2 \text{ in}^3$
1 peck = 8 qt
$1 \text{ bushel} = 4 \text{ pecks} = 2\ 150.42 \text{ in}^3$

Avoirdupois Weight
1 dram (dr) = 27.343 grains (gr) or (gr a)
1 ounce (oz) = 437.5 gr
1 pound (lb) = 16 oz = 7 000 gr
1 short ton = 2 000 lb, 1 long ton = 2 240 lb

Troy Weight
1 grain troy (gr t) = 1 grain avoirdupois
1 pennyweight (dwt) or (pwt) = 24 gr t
1 ounce troy (oz t) = 480 grains
1 lb t = 12 oz t = 5 760 grains

Apothecaries' Weight
1 grain apothecaries' (gr ap) = 1 gr t = 1 gr a
1 dram ap (dr ap) = 60 gr
1 oz ap = 1 oz t = 8 dr ap = 480 fr
1 lb ap = 1 lb t = 12 oz ap = 5 760 gr

Multiply $\rightarrow$

Metric Unit = Conversion Factor × US Customary Unit

$\leftarrow$ **Divide**

Metric Unit ÷ Conversion Factor = US Customary Unit

Metric Unit =	Conversion Factor ×	US Unit
(Length)		
mm	25.4	inch
cm	2.54	inch
cm	30.48	foot
m	0.3048	foot
m	0.9144	yard
km	1.609	mile
km	1.852	nautical mile
(Area)		
mm²	645.16	inch²
cm²	6.4516	in²
cm²	929.03	ft²
m²	0.0929	ft²
cm²	8361.3	yd²
m²	0.83613	yd²
m²	4047	acre
km²	2.59	mi²
(Mass)	(Avoirdupois Weight)	
grams	0.0648	grains
g	28.349	oz
g	453.59	lb
kg	0.45359	lb
tonne	0.907	short ton
tonne	1.016	long ton

Metric Unit =	Conversion Factor ×	US Unit
(Volume)		
mm³	16387.064	in³
cm³	16.387	in³
m³	0.028316	ft³
m³	0.764555	yd³
ml	16.387	in³
ml	29.57	fl oz
ml	473	pint
ml	946.333	quart
l	28.32	ft³
l	0.9463	quart
l	3.785	gallon
l	1.101	dry quart
l	8.809	peck
l	35.238	bushel
(Mass)	(Troy Weight)	
g	31.103	oz t
g	373.248	lb t
(Mass)	(Apothecaries' Weight)	
g	3.387	dr ap
g	31.103	oz ap
g	373.248	lb ap

Common Schematic Symbols Used in Circuit Diagrams

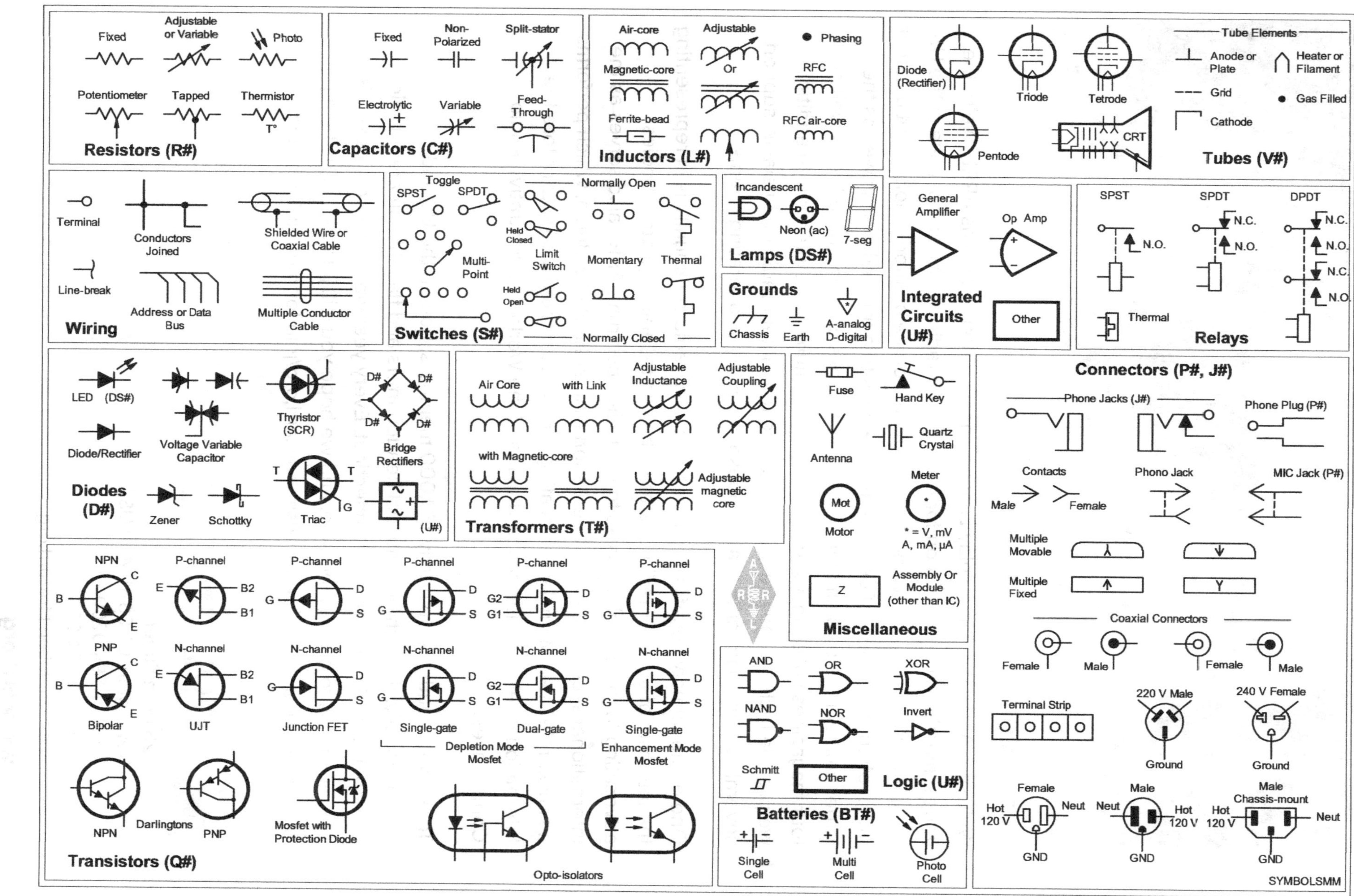

About ARRL

We're the American Radio Relay League, Inc. — better known as ARRL. We're the largest membership association for the amateur radio hobby and service in the US. For over 100 years, we have been the primary source of information about amateur radio, offering a variety of benefits and services to our members, as well as the larger amateur radio community. We publish books on amateur radio, as well as four magazines covering a variety of radio communication interests. In addition, we provide technical advice and assistance to amateur radio enthusiasts, support several education programs, and sponsor a variety of operating events.

One of the primary benefits we offer to the ham radio community is in representing the interests of amateur radio operators before federal regulatory bodies advocating for meaningful access to the radio spectrum. ARRL also serves as the international secretariat of the International Amateur Radio Union, which performs a similar role internationally, advocating for amateur radio interests before the International Telecommunication Union and the World Administrative Radio Conferences.

Today, we proudly serve nearly 150,000 members, both in the US and internationally, through our national headquarters and flagship amateur radio station, W1AW, in Newington, Connecticut. Every year we welcome thousands of new licensees to our membership, and we hope you will join us. Let us be a part of your amateur radio journey. Visit www.arrl.org/join for more information.

225 Main Street
Newington, CT 06111-1400 USA
Tel: 860-594-0200
FAX: 860-594-0259
Email: membership@arrl.org

www.arrl.org

ARRL MEMBERSHIP

Membership in ARRL offers unique opportunities to advance and share your knowledge of amateur radio. For over 100 years, advancing the art, science, and enjoyment of amateur radio has been our mission. Your membership helps to ensure that new generations of hams continue to reap the benefits of the amateur radio community.

Here are just a few of the benefits you will receive with your annual membership. For a complete list visit, **www.arrl.org/membership**.

KNOWLEDGE

ARRL offers knowledge to advance your skills with lifelong learning courses, local clubs, and publications to help you get and stay radio active.

ADVOCACY

ARRL is a strong national voice for preserving and protecting access to Amateur Radio Service frequencies.

SERVICES

ARRL offers a range of member services including the Technical Information Service (TIS), the ARRL Learning Center, and FCC license renewals.

RESOURCES

Digital resources including product review archives, webinars, training courses, e-newsletters, and more.

PUBLICATIONS

Members receive access to four digital magazines — *QST*, *On the Air*, *QEX*, and the *National Contest Journal (NCJ)*.

Two Easy Ways to Join

CALL Member Services toll-free at **1-860-594-0200**

ONLINE Go to our secure website at www.arrl.org/join

Antennas, Tips, and Tools

DIY Open-Wire Line

Antenna experts prefer open wire transmission line to coax. At one extreme is the 6-inch-wide line using 6-inch-long wooden dowels boiled in paraffin as spreaders for 600 Ω line for a doublet.[1] It was low loss but pesky to make and rather ugly. At the other extreme William Parmley, KR6L, used #22 AWG zip cord as a close-together pair.[2] Very portable, but 198 Ω in impedance and higher in loss because of small conductors and between-wire insulation. The de facto open-wire standard of today is window line with a nominal impedance of 450 Ω. Not bad and acceptable looking but you can do better at less expense. Here are three such lines that differ only in the spacers used.

All lines use #16 AWG insulated stranded speaker wire (RadioShack P/N 278-1267). This wire is all copper and more flexible than steel core ladder line wire. It will not break under moderate flexing. The wire's soft and the thick insulation is the key to the spacers staying in place.

Line A in **Figure 1** uses squares cut-from plastic milk containers. A half gallon container yields up to twenty 1⅜ × 1⅜ inch squares. Make holes with an ice pick, going first down and then back up. Space holes 20 mm apart both ways. Taper the wire end to

[1]Shackleford, W6YE "Custom Open Wire Line — It's a Snap," *QST*, July 2011, pp 33-36.
[2]Parmley, KR8L, "Zip Cord Antennas and Feed Lines For Portable Applications," *QST*, Mar 2009, pp 34-36.

make insertion easier. Set the squares about 6 inches apart. The spacer will not move along the wire by itself but you can move it by squeezing it then sliding it into position.

Line B in Figure 1 uses Styrofoam plastic as spacers. You need a thickness greater than the wire diameter. Many grocery meat trays have extra thick sides or bottoms. Cut the Styrofoam spacers 1⅟₁₆ inch wide × 1 inch long. Wrap a length of ¾ inch vinyl electrical tape around with the wire and insulator to hold them in place. [This method should be limited to short-term portable operations. — *Ed.*]

For Line C in Figure 1, the spacers are rectangular pieces cut from Coroplast, a plastic sign board material that is made with rectangular channels running through it (**coroplast.com**). Try your local sign company and purchase a piece that is 2 × 3 square feet. The side-to-side thickness is 4 mm and the core-to-core spacing is 5 mm. It slits easily lengthwise with an X-acto knife and cuts easily in width with scissors. Slit every sixth core to leave five full cores and cut that strip into ¾ inch lengths. The square core has only a loose grip on the wire so you will need a filler made of vinyl tape cut into ¾ inch squares and wrapped around each wire. [In the sample Willard provided, there was some slippage of the wire using the Coroplast insulator and tape filler. The use of glue is recommended. — *Ed.*]

Neither impedance, velocity factor, nor loss was measured for these lines, but here are some estimates. The *ARRL Antenna Book* calculated impedance using 0.0508

inch wire size at 20 mm spacing as 412 Ω. Since there is so little insulation between the wires, it should be very nearly that. If you need a different impedance, change spacing as calculated by the *Antenna Book* equation.[3] Velocity factor should be 0.91 or greater. Loss should be less than window line since there is less insulation between wires and near the 0.08 dB/100 ft @ 10 MHz as given for 600 Ω line.

Running any open line around and over supports and changing direction is always a bit of a problem, but my previous Hints & Kinks hint about using PVC end caps as standoffs will be of help.[4] Standoffs essentially eliminate losses. Of course, the Coroplast can be extended sideways to make a standoff of any shape or length. — *73, Willard Monahan, K6KH, 817 Pacific Ave, Manhattan Beach, CA 90266-5849, k6kh@aol.com*

Measuring Unknown Coaxial Traps

I found a couple of coaxial traps I had built some time ago and thought I might use them to build an 80-40 meter dipole. In order to model the antenna to determine the dimensions necessary and the expected performance, I needed to know the L/C ratio, as this will affect the length of wire beyond the trap for resonance on 80 meters. It is possible to determine these values with two relatively simple measurements.

I have an old grid-dip meter, to which I have added a BNC connection to the oscillator tank to connect to my homebrew counter for a more accurate frequency readout. If I made one measurement of the trap's resonant frequency, then a second measurement with a known capacitor soldered in parallel, I should be able to calculate the trap capacitance, and

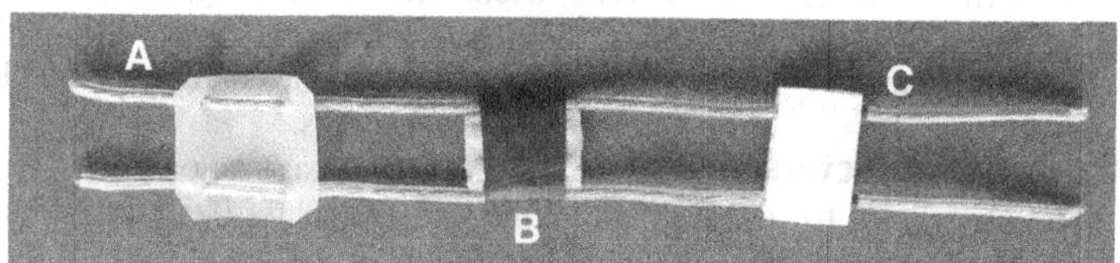

Figure 1 — Here are all three methods on one section of ladder line. At A is the "milk-container" spacer; at B is the "grocery tray" spacer, and at C is the Coroplast spacer. [Willard Monahan, K6KH]

[3]Silver, NØAX, Ed., *The ARRL Antenna Book*, 22nd Edition, pp 23-17–23-18.

from that, the inductance.

The familiar formula for a resonant LC circuit is:

$$f_1 = \frac{1}{2\pi\sqrt{LC}} \qquad \text{[Eq 1]}$$

If we add a capacitor (C_P) in parallel, the formula becomes:

$$f_2 = \frac{1}{2\pi\sqrt{L(C + C_P)}} \qquad \text{[Eq 2]}$$

After dividing one formula by the other, performing a bit of algebra and dividing out like terms, we obtain the following relationship:

$$C = \frac{C_P}{\left(\dfrac{f_1}{f_2}\right)^2 - 1} \qquad \text{[Eq 3]}$$

where C = trap capacitance, C_P = parallel capacitance, f_1 = trap resonant frequency without the added capacitor and f_2 = resonance with the added capacitor. Because of the ratios taken, if C_P is in pF, C will also be in pF.

With the value of C found, we can rearrange Eq 1 to find the value of L. The added factor 10^6 yields L in μH if C is in pF and f is in MHz:

$$L = \frac{10^6}{(2\pi f_1)^2 C} \qquad \text{[Eq 4]}$$

Using these three pieces of information, f_1, f_2 and C_p, and the formulas presented here you can learn the truth about those mystery traps lurking in your junk box. — *73, Ted Bergstrom, W1IQW, 50 Bayshore Dr, Mashpee, MA 02649-3967,* **w1iqw@arrl.net**

Ladder Line Standoff

I read a hint in the October 2010 Hints and Kinks regarding a coaxial cable standoff and would like to suggest a standoff for ladder lines.[5]

After erecting a new inverted **V** antenna up about 30 feet and using 450 Ω ladder line, I noticed that my SWR was poor for certain bands due to the proximity of the line to the vertical metal pole.

To remedy this, I glued a PVC **T** at the end of an 18 inch long PVC pipe and cut a thin slot big enough to insert the ladder line into the **T**. I then glued another ¾ inch **T** at the other end of the 18 inch tube, cut half of the perpendicular side to produce a semicircular seat that fits against the vertical pole (see **Figure 2**). I then used two hose clamps to attach the cut **T** to the pole. Following my standoff installation, my SWR dropped nicely.

[5]Kampe, KBØLSX, "Coax Cable Standoff," *QST*, Oct 2010, p 63.

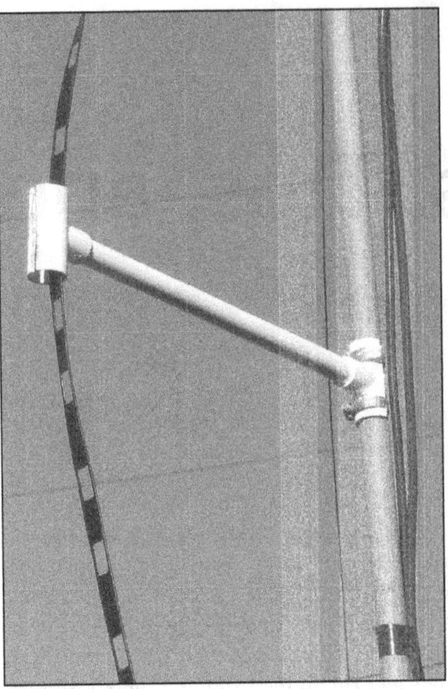

Figure 2 — Here is the PVC standoff holding the ladder line away from the metal mast. Note the modified **T** connector that is held to the mast using hose clamps. [Jim Wheeler, KJ6VX]

This is a very versatile and cheap method to make a standoff. The loose ladder line in the PVC **T** can move up and down with any gusts of wind. I also experimented with the length of the PVC and found 18 inches acceptable for both the SWR and the physical length of the standoff. — *73, Jim Wheeler, KJ6VX, PO Box 6564, San Diego, CA 92166-6564,* **wheeler513@cox.net**

Giving Your Radials an Edge

In our sandy soil here in Orlando, Florida, I needed to run at least eight radials in such a way that I could mow over them when the grass began to grow in earnest. I used an electric edger to cut slots into the lawn. I was then able to press #12 AWG insulated wire into the slots. The #12 wire was so stiff it would not lie in the slots so I used small hooks of wire to hold it in place and added a little topsoil to hold the wire and encourage the grass to grow over it.

I cut a wider slot to place an 8 foot length of PVC, sealed at both ends, that protects the RG-8 that is running to my Hustler 4BTV vertical antenna (**www.new-tronics. com**). — *73, Clinton Wills, WB4WMY, 17 Capehart Dr, Orlando, FL 32807,* **wb4wmy@bellsouth.net**

Fixing Shorted Connectors

There is nothing more frustrating than soldering a PL-259 connector and then finding that it is shorted. This has happened to me on many occasions when using an adapter to allow the use of smaller RG-58 type coax.

Figure 3 — Look closely inside this connector and you can see the small metal rim of the center pin. The RG-58 adapter can sometimes push the shield up against this rim, shorting the connector. [Jerry Semones, K4FJK]

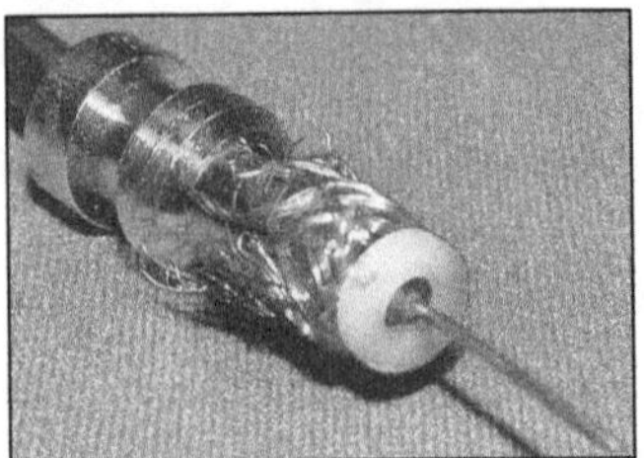

Figure 4 — Placing a plastic washer or other insulator on the end of the cable before inserting it into the connector body will prevent the shield from shorting against the center-pin rim. [Jerry Semones, K4FJK]

The reason for this has to do with the way many PL-259 connectors are made. Looking at **Figure 3** you can see that the center pin is swaged to the plastic insulator in a way that exposes a small ring. This ring can come into contact with the coax shield when the adapter is screwed in place.

To prevent this, add a small nylon washer (or other suitable insulator) over the center conductor before assembling the adapter. This added washer prevents the coax shield from contacting the center pin. The one shown in **Figure 4** is 5/16 inch diameter with a ⅛ inch hole and is 1/16 inch thick. I found them at a local hardware store. — *73, Jerry Semones, K4FJK, 9700 Forestwood Dr, Louisville, KY 40299,* **grandham@insightbb.com**

Tower Horses

In the middle of an antenna building project I found I had used all my saw horses in the process — some of the horses were holding completed antennas off the ground and others were holding up various "works-in-progress." As I was looking around for something else to use to support another project, I spied a couple of halves of a section of Rohn 45G. The tower sections proved to be the best antenna work platforms I have used. The vertical legs keep tubing from rolling off and one could work on different parts of an antenna on two or three levels of the tower pieces.

So, if you have an old tower section, or find one cheap at a flea market, buy it and cut it in half and it will work better for less money than even the least expensive commercial saw horses. — *73, Dave Patton, NN1N,* **nn1n@arrl.org**

Pipe Insulation Feed Through

There is a simple and inexpensive method to feed cables through a window. Just take a couple of pieces of soft water pipe insulation and cut the length to fit the inside of the window track. Place one on the window sill. Then lay down the coax or wire that you want to feed through. Finally, put the other piece of insulation on top to form a "wire sandwich" and close the sash. The insulation will compress and form a good weather seal. For the colder climates, take another piece of insulation and cut it lengthwise so that it is thin enough to fill the gap between the top and bottom sashes. This will prevent cold air from entering the shack. Make sure that you label each wire to eliminate confusion. Finally, don't forget to engage the window's security tabs to prevent someone from slipping in. — *73, Bob Steinberg, WA2KHR, 55 Daisy Farms Dr, New Rochelle, NY 10804,* **waveguide33@gmail.com**

Space Blanket Antenna

Ham radio operators often must set up and operate their equipment in remote areas with very limited resources. A necessity during such operations is a practical antenna. An operator can design a VHF or UHF vertical antenna on the spot but a counterpoise is needed.

This is where the space blanket antenna comes in. A space blanket is composed of a metal surface on a flexible Mylar base. It is easy to fold up and store and can fit into a shirt pocket or bug-out bag. It unfolds quickly and can be ready to provide an antenna counterpoise in only a couple of minutes. It can be used anywhere a vertical antenna is desired. Just lay the space blanket on a flat surface such as a table, the ground or sidewalk and it's operational in just minutes. **Figure 5** shows a mobile antenna with a physical support so it will stand upright sitting on top of a space blanket laid out on top of a folding card table.

How well does the space blanket antenna work? Without the space blanket, the small vertical antenna in Figure 5 has an SWR value above 2.0 to 1. By simply adding the space blanket beneath the antenna, the SWR is reduced to 1.2 to 1. The space blanket provides the necessary counterpoise for the vertical antenna very effectively.

The advantages of using the space blanket antenna in emergencies are obvious, but there may be other situations where it could be useful. One ham, while constructing a play house for his children was putting a wood cover on the structure, which had a roof height of about 12 feet. The roof was slanted a few inches. Tar paper was placed on the roof and then a space blanket was

[2]Larkin, W8RVT, "PA Module Repair," *QST*, Aug 2011, p 57.

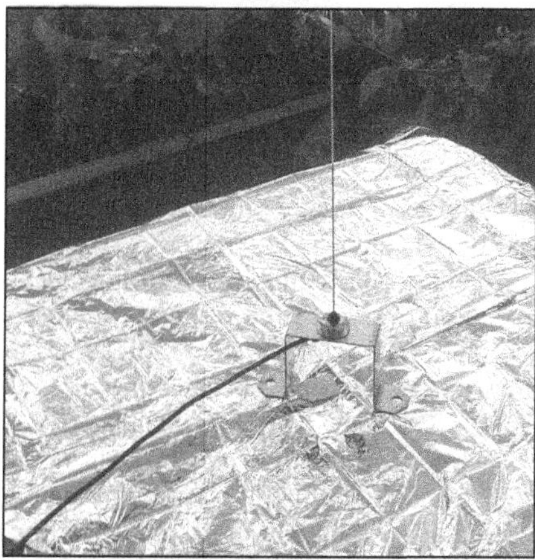

Figure 5 — A V/UHF antenna can be set up in a flash just about anywhere using a space blanket as a counterpoise. [Vernon Harris, W7GGM, photo]

lightly tacked down on top of the 12 × 12 foot roof. Next, rolled roofing was applied as the final roof material. A mobile antenna was anchored to the roof in the middle of the space blanket. It worked better than expected, providing an SWR of 1.25 to 1. A bit of adjusting the antenna whip length improved the SWR to 1.2 to 1.

One note: When the space blanket is laid out on top of a surface it would be wise to tack it down using thumb tacks or sticky tape of some kind to secure it. During testing of the antenna the SWR would vary as the breeze would move the edges of the space blanket and thus change the measured value. During the SWR testing small rocks placed on the edges of the table stabilized the blanket and kept the breeze from moving the edges.

The space blanket antenna concept is simple; the blanket is inexpensive and easy to store. It can be spread out and made operational in just minutes. The next time you have an urge to try a new antenna experiment, try out this simple concept. Happy antenna experimenting. — *73, Vernon Harris, W7GGM, 358 Meadow Brook Ln, Providence, UT 84332,* **bmsoffices@comcast.net**

Hole In One Antenna

Here's something I "invented" to help raise wire antennas. I cannot imagine it hasn't been done by others, but since I've never seen it written anywhere let me describe it here.

The parts needed are a golf ball, a small screw eye, a small fishing snap swivel and a 6-8 inch length of 8 pound monofilament line.

Drill an undersized hole about ¼ inch into the golf ball. Screw in the screw eye. Tie the monofilament to the screw eye with the standard fisherman's knot. Tie the other end of the monofilament line to the snap swivel. In the end of the antenna wire make a loop or solder a ring connector and then clip the snap swivel to the loop or ring (see **Figure 6**).

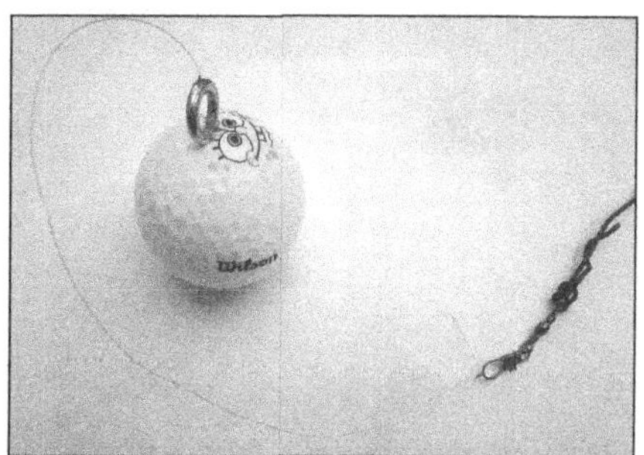

Figure 6 — This simple arrangement can be used for the quick deployment of a simple wire antenna for portable or emergency work. When the operation is over, a hard pull on the antenna breaks the fishing line and all the parts fall to the ground. [Stan Levandowski, WB2LQF, photo]

Now throw the golf ball up into the tree. Nine times out of 10 you can pull the wire out intact. If it manages to get hung up in a branch, the line will break before you exert enough force to break or damage the wire. The golf ball then falls to the ground. The antenna is easily retrieved and all you have to do is replace the short piece of line and you are back in business.

My antenna is 28 feet of #22 AWG stranded wire with a 33 foot counterpoise that I use with my Elecraft KX-1. It earned me my 1000 mile-per-watt award (900 mW on eight AA batteries for a New York to the Canary Islands contact with EA8BVP). — *73, Stan Levandowski, WB2LQF, 6 Chatham Ct, Fishkill, NY 12524-1530,* **wb2lqf@arrl. net**

Antenna Helper

The antenna helper idea started while attempting to make a 40 meter bobtail curtain antenna for Field Day. This is a large wire antenna and, working alone, I needed something

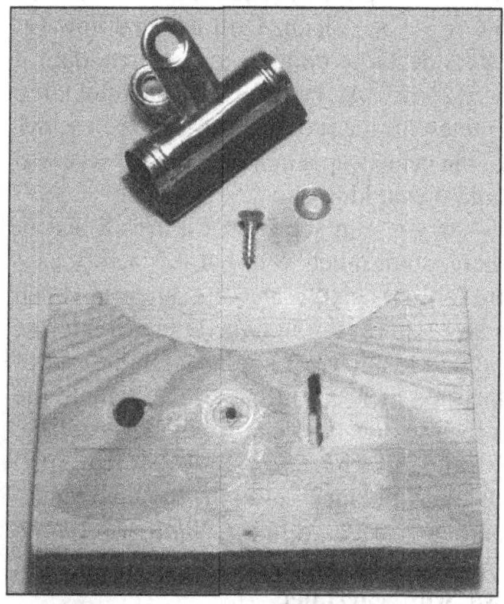

Figure 7 — To make the antenna helper you need a bulldog clip, wood screw and washer, and a wooden block. [Julian Blair, WA2WMJ, photo]

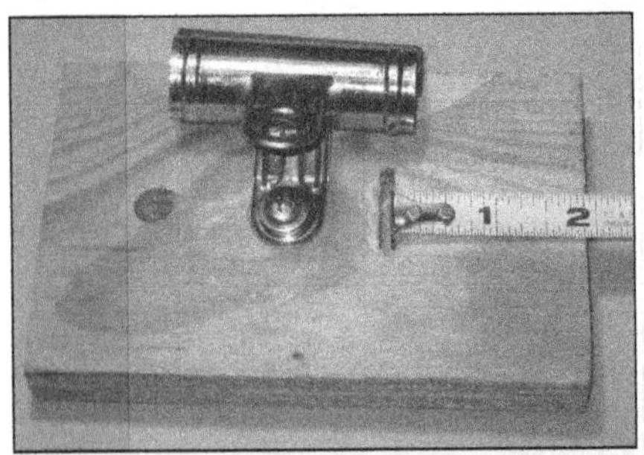

Figure 8 — The metal tab on the end of the tape measure is inserted into the slot on the board, then the clip is rotated over it to hold it in place. [Julian Blair, WA2WMJ, photo]

Figure 9 — Here is the antenna helper in use. The bulldog clip holds the end of the tape measure in place. The end of the antenna wire is held in the jaws of the clip. Pull them both back to the required length and cut. [Julian Blair, WA2WMJ, photo]

to hold the wire and tape measuring tool. So with that in mind, it was off to my junk box to see what I could come up with. Large bulldog clips came to mind and after about half an hour I had the idea and the parts to put this holder together.

All you need is a small piece of plywood to attach the clip to and hold the measuring tape (see **Figure 7**). The plywood measures 5½ × 3½ × ¾ inch. Drill a round hole on one side large enough to accommodate a screwdriver. At the other side cut a slot large enough for the metal tab on a tape measure. In the center tap a hole for a wood screw to hold the bulldog clip.

As you can see from **Figures 8** and **9**, there is not much to it, but it is a very useful tool for anyone making antennas. Push a screwdriver through the hole and into the ground to anchor it in place. Put the end of the tape measure in the slot and rotate the clip over the tape end to hold it in place. Next, put the wire in the clip, pull the wire and tape out to the required length and cut. — *73, Julian F. Blair, WA2WMJ, PO Box 602, Pine Bush, NY 12566-0602,* **wa2wmj@arrl.net**

Low Cost Radial Ring

I was given an older HF trap vertical antenna by a friend, along with some used coax. Knowing the importance of a good ground system, I scrounged enough house wire for 32 radials. But how to connect them at the base of the antenna?

Wandering the plumbing aisle of my local home improvement store I found a stainless steel sink strainer for less than $7 that looked promising. The inner diameter of the tailpiece connection matched the outer diameter of the 1½ inch galvanized water pipe I'd be using as the antenna mount and the outer rim would accommodate the connection points for the radials. I grabbed a stainless steel hose clamp on the way to the checkout and headed back to the shop.

Using a hacksaw I removed the "cross" in the bottom of the strainer, cut four slots in the tailpiece connection so it could be compressed around the mast and removed the burrs with a file. Using a drawing program, I created a paper template of 32 holes for attaching the radials. The holes were marked on the rim of the strainer and drilled.

The strainer was clamped to the antenna mounting pipe with the hose clamp. I inserted a short length of copper braid under the clamp and connected it to the coax shield at the base of the antenna (see **Figure 10**). Each radial wire was then attached with a stainless steel screw and nut, stretched into place and held down with a couple of landscape staples.

I assembled the antenna to the recommended starting dimensions and then tuned the traps to resonance using an antenna analyzer. On-air performance of this antenna system has been excellent, thanks in large part to the ground system. — *73, Ray DuFlon, KE6PXU, PO Box 188, Pescadero, CA 94060,* **ke6pxu@gmail.com**

Window-Mount Antenna

As you know, it is often impossible for amateurs living in an apartment to install a 2 meter antenna on the outside of their

Figure 10 — The sink strainer radial ring, hose clamp and braid installed at the antenna base. The ground braid connects to the coax shield. [Ray DuFlon, KE6PXU, photo]

Figure 11 — The homebrew window-mount antenna attaches to any window using suction cups. [Yvon Laplante, VE2AOW, photo]

building. Most of the time, they have to be satisfied with a handheld, a radio with a magnetic-mount antenna on top of the fridge or some other metallic object. Being in this situation, I made a small dipole antenna using telescoping antennas I took from old, broken FM radios (see **Figure 11**).

The antenna is mounted on a roughly 3 × 5 inch piece of Plexiglas with two suction cups. With my radio placed close to a window, I attached the antenna to the window in the vertical position and adjusted the two elements for 2 meters — about 19 inches. The antenna gives very good results. — *73, Yvon Laplante, VE2AOW, 304-1100 Curé Poirier Est, Longueuil, QC J4J 5M2, Canada,* **yvonlaplante2@videotron.ca**

Whipping Coax

A discussion I saw on a radio oriented forum focused on how to hang coax from the center insulator of a dipole antenna. A main concern was what would happen to the cable if it was bent too tightly over the support. The tight bend radius and tension involved could lead to shorting the center conductor to the shield.

The discussion caused me to remember a simpler method of hanging cable from a support using a technique called "whipping." This technique is normally used to keep twisted rope from fraying.

As shown in **Figure 12**, a loop of line, usually ³⁄₁₆ inch diameter for RG-8, is laid along the side of the cable. The free end "B" is then wrapped snugly many times around the cable. The length of the coiled turns should be roughly twice the diameter of the cable, then the end "B" is threaded through the loop and end "A" is pulled tight.

The end "A" of the support is held in place by the captured end "B" and by the same tension that holds the cable in place. One of the forum's correspondents commented that this is similar to a Kellem's grip, a woven steel cable support used in industry.

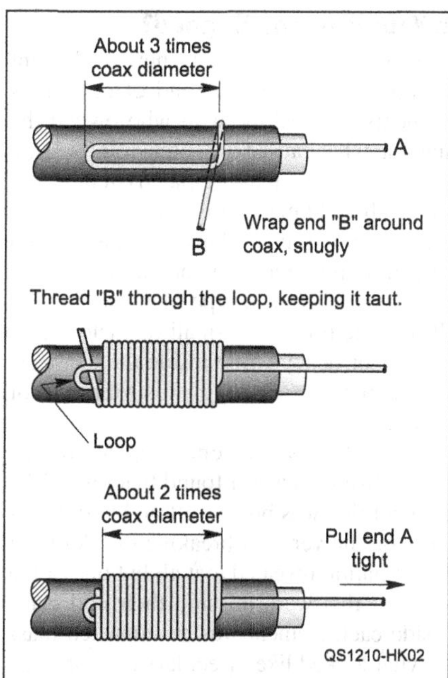

Figure 12 — This schematic diagram shows the process for tying a whipping knot on a piece of coax.

Figure 13 — To make a quick telescoping antenna, run a 6-32 tap through a PL-259 and add a piece of threaded rod to the base of any portable radio replacement antenna. [John Portune, W6NBC, photo]

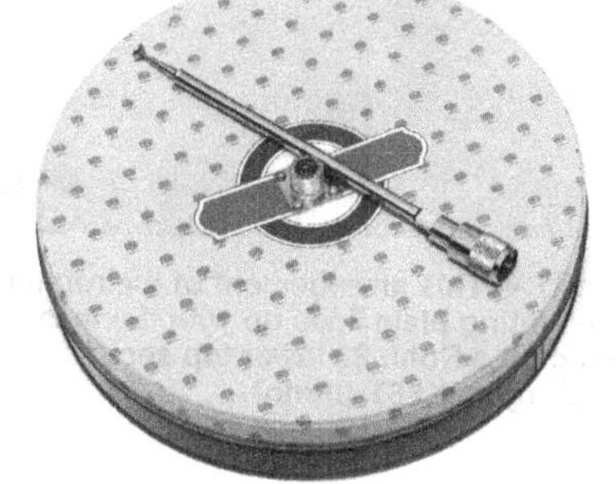

Figure 14 — Mount an SO-239 connector to the top of a cookie tin, add the quickie antenna and you're on the air. [John Portune, W6NBC, photo]

While whipping, when applied to the end of a rope, involves cutting the ends "A" and "B" close to the turns, if end "A" is left long, it can be used to tie the cable to the dipole antenna's center insulator, taking all the strain off the exposed wires and connections. The multiple turns, while only snug, hold the cable in a vise-like grip without deforming it, similar to how your hand would hold and turn a screwdriver, each finger contributing to the overall grip.

Since all you need to make this support is some small rope this technique lends itself to impromptu installations, like a Field Day outing. End "B" could be doubled over through the loop like a shoelace, which would allow the grip to be easily untied after the antenna is taken down. This knot is also suitable for more permanent installations if weatherproof rope is used. [For additional useful knots, take a look at the *QST* article, "The Knots of Ham Radio," available to members online through the Periodicals Archive. — *Ed.*] — *73, Stanley Labinsky, K2STN, 524 Plains Rd, Wallkill, NY 12589-4005,* **k2stn@arrl.net**

Quick Whip Antenna

A cookie tin can become a quick makeshift antenna with the addition of a vertical radiator. Many broadcast-band radio replacement whips from the local radio parts stores come with 6-32 threads for attachment. To mate one to a PL-259, just run a 6-32 tap into the end of the tip and all the way through the connector (see **Figure 13**). Then use a short length of brass 6-32 threaded rod or a cutoff screw for the connection. No soldering

required. Add a layer or two of heat-shrink tubing for insulation at the bottom of the whip. Add an SO-239 and a cookie tin and you're on the air (see **Figure 14**). — *73, John Portune, W6NBC, 1095 W McCoy Ln Spc 99, Santa Maria, CA 93455-1105,* **w6nbc@arrl.net**

Antenna Tube Talons

This method was inspired by Joel Hallas', W1ZR, article about adding 6 meters to a tribander.[1] Instead of the normal U-bolts and hardware to install the ½ inch element I used plastic copper pipe hangers called "Tube Talons." I found them at my local home improvement center for $1.98 a 10 pack.

The Talons come with a nail that must

[1]Hallas, W1ZR, "Add 6 Meters to Your Triband Trap Yagi," *QST*, Sep 2011, pp 40-42.

be removed. With the nail gone, drill out the nail hole with a ⁹⁄₆₄ inch drill bit (for #6 hardware). Cut off the bottom to a length that is short enough, when installed, to make a tight fit. Place the cut talon on the element and it should hang off to the side a little bit. That way when you tighten down the screw it will hold the element tightly. Install with a 6-32 × 1¼ inch screw. You will need four Tube Talons for each element (see **Figure 15**). The two opposing talons help to lock the element in place.

The Talons do not slide or move. You may have to experiment when you cut the Talon to get the correct length so it will tighten down and clamp the element in place.

Another Use for Tube Talons

I was working on my next antenna project, a dual-band wire beam.[2] Two of the parts needed were ½ inch and ¾ inch ID Teflon donuts. I wasn't about to order them from a specialty catalog so instead I used the Tube Talon clamps.

Both the ½ and ¾ inch Talons are cut the same way. A miter box used for cutting wood trim worked fine. You'll need two talons for one clamp, using only the tops. Cut the talon

[2]Baker, KG4JJH, "A Dual-Band Wire Beam for 17 and 12 Meters," *QST*, Aug 2005, pp 28-32.

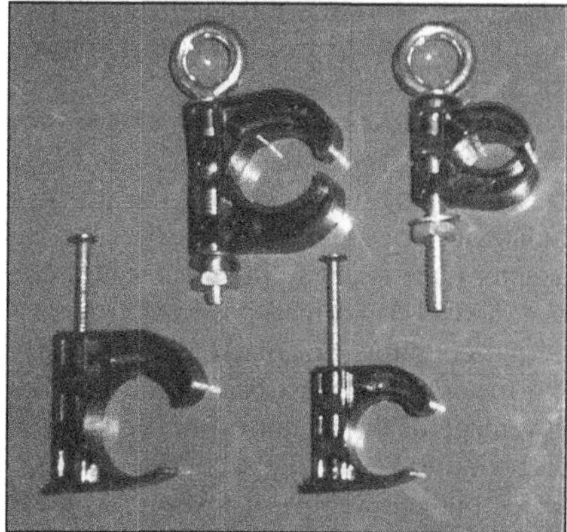

Figure 15 — Two Tube Talons, mounted in opposition make a secure anchor for an antenna element. [Walter Yatzook, W1ATV, photo]

Figure 16 — With their bottom fingers removed, two Talons mounted back-to-back can make a very effective clamp for tubing. [Walter Yatzook, W1ATV, photo]

square with the hook end. This is your cutting guide. After removing the nail, drill out to $3/16$ inch for a $3/16$ inch eye hook. Assemble the two top halves together with the eye hook using a flat washer on each end (see **Figure 16**).

The parts clamped well on the fiberglass tubes. The talons have a small dimple on the inside to help with clamping. — *73, Walter Yatzook, W1ATV, 77 Baker Ave, Meriden, CT 06451-5107,* **w1atv@arrl.net**

Coax Connector Covers

I discovered this simple trick on a camping trip. It helps prevent dirt and debris from fouling my PL-259 and SO-239 connectors. On our trip, we encountered heavy rain and had to stow the ham gear. We didn't want to roll up all of the coax and needed something to cover the ends. While looking for some sort of cover, I saw a spent 12 gauge shell casing on the table and tried it. It fit and was a cheap and effective solution. Although this won't make a waterproof seal, it will keep a lot of moisture out.

The 12 gauge shell casings fit snugly over PL-259s. Although they will fit over SO-239s with a little help, 16 gauge shell casings would be a better fit. — *73, Kevin Ballard, KB4RTR, 121 Jasmine Dr, Alabaster, AL 35007-5214,* **kb4rtr@arrl.net**

Tower Mount a TV Rotator

Many small VHF/UHF antennas don't require a heavy-duty ham antenna rotator. Tower mounted TV rotators may be adequate for the job and cost about $100 including an IR remote control!

Inline rotators like the Channel Master CM-9521A are designed to be clamped onto a stationary tubular mast with the antenna mounted on a short rotating top mast. The rotator becomes part of the antenna support and wind gusts can produce damaging side loads.

These types of rotators are not designed to mount inside a triangular tower like the Rohn 25 but using a homemade adapter plate makes such an installation possible. In addition, the bushing at the top of the tower section provides substantial lateral support for the system.

The adapter plate is shown in **Figure 17**. Remove and replace the rotator's four lower mast mounting studs with four stainless steel mounting bolts and lock washers. If the adapter plate is cut from ¼ inch aluminum plate and a 1¼ inch mast is used, the rotator's output coupling will match perfectly with the tower's top thrust bushing. This takes advantage of the rotator offset between top and bottom. Secure the adapter plate to the inside of the tower legs using four U-bolts.

You can use other rotators in place of the CM-9521A. The RadioShack 15-1245 is another inexpensive and popular rotator that

Figure 17 — This arrangement of the rotator and mounting plate lines up with the mast bushing on a Rohn 25 tower top section. [Jack Morgan, KF6T, photo]

will work interchangeably with this bracket. If you come across a rotator with a different offset or use a different mast diameter, you may have to make some adjustments.

A smaller diameter mast or a rotator with a smaller offset may require some spacer washers between the rotator body and adapter plate to get proper alignment with the tower top bushing. Alignment is not super critical — just make sure the rotator turns the mast 360 degrees without binding up. Be sure to use a little waterproof lube inside the bushing.

If you use a larger mast or rotator with an even greater offset, you could mount the plate on the outside of the tower and use spacers to achieve proper alignment. — *73, Jack Morgan, KF6T, 2040 Pheasant Hill Ln, Auburn, CA 95602-9673,* **kf6t@arrl.net**

Clip-N-Tune

While trying to load my 75-meter inverted L antenna on 160 meters, I discovered that my manual antenna tuner didn't have enough series inductance for resonance. Rather than place a coil in series with the antenna lead, I placed a small clip-on ferrite RFI choke over the wire where it connected to the antenna tuner output terminal.

The ferrite choke changed the antenna characteristics just enough to allow the tuner to match the antenna impedance. I did not determine the reasonable limits of this technique but it is incredibly easy to apply. A word of caution: The choke could saturate magnetically and produce unwanted harmonic signals if excessive transmitter power is used. — *73, Bruce L. Meyer, WØHZR, 9410 Blaisdell Ave, Bloomington, MN 55420,* **meyerbl@cpinternet.com**

Is Your Antenna Bugged?

I once saw a relatively new 3 element aluminum Yagi lose the end of one of its elements. I asked the ham who owned the antenna what caused it to break off. He said the element vibrated (oscillated) in the wind and he thought metal fatigue caused the failure. Since that time, I have had many beam antennas vibrate in the wind but have yet to experience the same type of metal fatigue. (It does help to place small diameter piece of rope or heavy twine inside the elements to dampen the effects of this vibration or oscillation).

Recently, while removing a Mosley TA-33 from a tower, I found that one of the element tips was broken off and two others were on the verge of breaking off. A closer investigation revealed that all but one of the element tips were missing their end caps. Inside each element was an impacted mass of what looked like insect larvae. It appears that hornets had laid eggs inside the element tips and, as the larvae matured, they secreted formic acid, which began to eat away the inside of the element tips.

Figure 18 shows formic acid's effect on a piece of aluminum tubing removed from the affected beam antenna. Next to the corroded aluminum tubing is a piece of tubing that was not affected so you can see the difference in the thickness of the element walls.

If you have a beam that is missing its end caps, lower the beam, clean out those elements thoroughly and replace the missing caps. This also applies to end caps on the boom. When I replace the end caps now, I usually give them a shot of super glue to help keep them in place. — *73, Lawrence W. Stark, K9ARZ, 1875 Chandler Ave, St Charles, IL 60174-4601,* **k9arz@arrl.net**

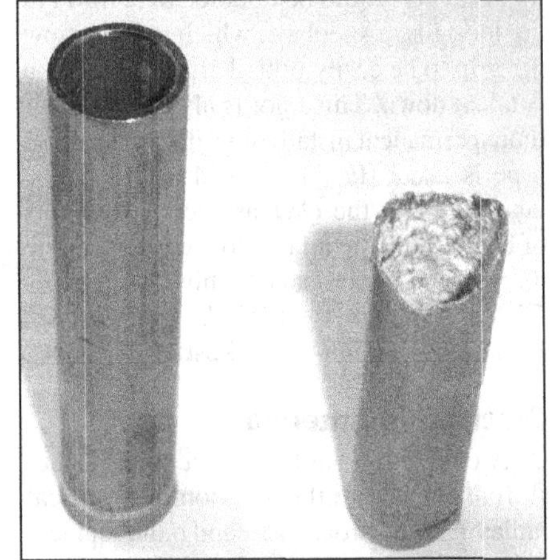

Figure 18 — Insect infestations in antenna elements can be much more than a simple annoyance. Many insects produce corrosive chemicals that can be very destructive to aluminum. Keep those end caps on. [Lawrence W. Stark, K9ARZ, photo]

Splicing Antenna Elements

For my V/UHF operations I frequently use an Arrow Antenna, model OSJ 146/440 (**www.arrowantennas.com**) solid aluminum rod J-pole, but I find the 5 foot length to be inconvenient when packing for travel to RACES or ARES® events. After some searching and minor fabrication with simple tools, I found a nice solution. If you have a similar antenna or just need to splice broken antenna elements, you may find this idea useful.

Arrow offers a version of the antenna with the longest element in two pieces that screw together, but that is not the version I have. So what to do? Since the antenna elements are $3/8$ inch diameter, I looked for a ⅜-inch shaft coupler for a reasonable price, say $10 or less, but wasn't able to find one. The reasonably priced shaft couplers I found online were flexible couplers. One type had insulators between the two shafts being spliced. Clearly the antenna coupler must be conductive. The other type of flexible coupler was conductive from one end to the other, but was spiral cut so it looked like a coil. I believed a coil at the center of the element would probably ruin the desired antenna characteristics.

I thought that the coil effect would be at a maximum if the element rod ends were inserted the minimum distance into the coupler and at a minimum if the elements were butted up against each other inside the coupler. I tried this flex coupler type anyway and, using a DG8SAQ Vector Network Analyzer, found no difference in VSWR in either configuration. That was a pleasant surprise, but the amount of flexing was a real concern to me. The top element swayed considerably. I was concerned that this might set up an undesirable mechanical resonance with the entire antenna. I started looking for a better, non-flexing solution and found it at a local hardware store.

The solution was a hollow round aluminum tube with an inside diameter of ⅜ inch — a perfect fit. The minimum purchase was 3 feet, so I had to buy a much longer length of tube than the 2 inches I needed, but it was inexpensive.

So the plan was to cut the longest element in two and use a short section of the $3/8$ inch tube as a collar to splice the pieces together. My feeling is that this makes for a stronger joint than the manufacturer's method since no material has to be removed from the element. The splicing tube is permanently attached to the upper element, allowing half the length of the splice to serve as a socket for the lower portion of the element. Some means of hand tightening the lower section to the tube is needed too.

There are many ways to solve these issues. My method is shown in **Figure 19**. I cut the longest element so the detached section was the same length as the remaining J-pole assembly: about 33 inches. I attached the tube to the upper element with four Allen screws (6 × 32) to discourage removal (see **Figure 20**). These bolts just happened to be in my junk box. It would be just as easy to use two longer screws that would pass through the splice and element, and secure them with a nut on the other side.

The lower section is held in place by an 8 × 32 eyebolt, which is easily turned by hand. I drilled through the splicing tube and element and then threaded the splicing tube for the eyebolt so no nut is needed. The eyebolt is attached to the upper section by a piece of flexible plastic cut from the lid of a food container. Now there is no loose hardware to drop and potentially lose.

The modified antenna received and transmitted successfully on both bands and an antenna analyzer sweep showed no meaningful difference in VSWR before and after the modification.

Another antenna system feature that I find very convenient is visible in **Figure 19**. The wooden dowel clamped at the bottom of the J-pole allows me to install the antenna into the mast sections I use for both portable and home operation. There's no need to tighten any nuts since gravity keeps the antenna securely in the mast. This makes for a true grab-and-go installation. There are wing nuts holding the dowel in place, so no tools are needed to remove the dowel if circumstances require. — *73, Doug Hart, AA3S, 6289 Beechwood Dr, Columbia, MD 21046,* **aa3s@arrl.net**

Quick Trim Your Dipole

After replacing the baluns on my loop antenna and dipole, I swept the band and found the lowest SWR on the dipole at the very low end of 80 meters, below where I wanted it to resonate.

I needed to know how much I should prune the antenna wires in order to achieve resonance. I came up with an idea but thought it was just too simple. However, I tried it anyway.

The principle was this: At 80 meters my minimum SWR was at 3.7 MHz; I wanted it to be at 3.8 MHz. I decided to calculate the wavelength for each frequency and simply subtract to determine what I need to cut.

I looked around the web for a basic frequency to wavelength calculator and found several. I settled on **www.wavelengthcalculator. com**. I entered 3.7 MHz and came up with a wavelength of 261.6 feet. I then entered the frequency of 3.8 MHz and came up with a wavelength of 254.7 feet. Next, I subtracted the two numbers and divided by four (because you have two quarter wave lengths for a dipole) the result was 261.6 – 254.7 = 6.9 feet. Divide by four and you get 1.73 feet.

I cut 1.73 feet (1 foot 9 inches) from both ends of the antenna. I swept the band again and voila! I was resonant at 3.8 MHz. The beauty of this method is that you already have the antenna up and all the various environmental conditions (height above ground, lengths of feed lines, wire size, surrounding structures, etc) affecting the antenna characteristics are present. Next time you put up a wire antenna follow this procedure: Cut the antenna to theoretical values. Run a sweep to find the frequency of minimum SWR. Do the simple math as above to find the difference in wavelength between the resonant frequency and the desired frequency, calculate the new antenna length, adjust the wires and you are done. [When adjusting the wire, don't cut the end off but fold it back onto the antenna. This will leave some excess for future adjustments. — *Ed.*] — *73, Gary Schmitz, KT7AZ, Oro Valley, AZ 85737,* **kt7az@arrl.net**

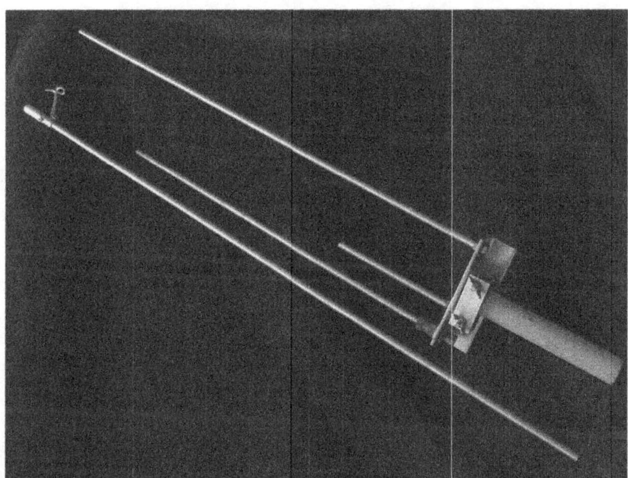

Figure 19 — **The disassembled antenna ready for compact storage. Also note the wooden dowel that can slip into a mast. The antenna will then be held in place by gravity with no nuts and bolts required, which helps you get on the air more quickly.** [Doug Hart, AA3S, photo]

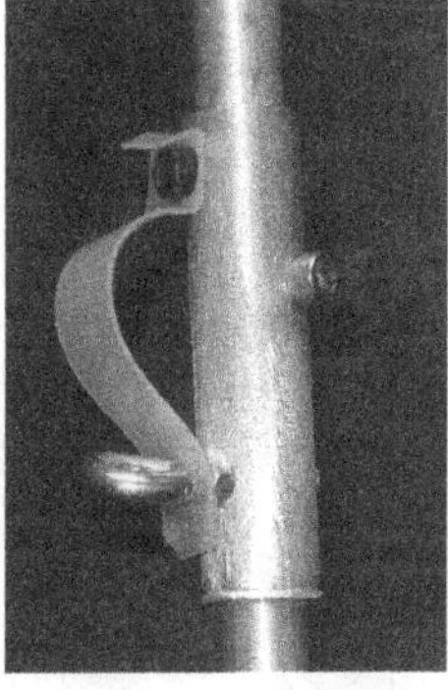

Figure 20 — **A close-up of the removable antenna element splice showing the eyebolt that serves as a thumb screw for easy removal of the upper element. Also shown are the bolts that permanently attach the upper element to the tube and the flexible plastic that keeps the eyebolt from getting lost.** [Doug Hart, AA3S, photo]

No-Scratch Mount

I recently replaced my car. I noticed that my 2 meter magnetic-mount antenna had scratched the paint on the roof of my old car. To keep this from happening on my new car, I used a piece of magnetic rubber sheet material commonly used for refrigerator magnets with advertising printed on them. If you don't have any on hand, a sign shop that makes magnetic signs for vehicles may give you a small scrap. The magnet material can easily be cut to the desired size. — *73, Al Forbes, KJ4YEV, 1908 Woodgreen Dr, Gastonia, NC 28052*, **alphaal@bellsouth.net**

Safer and Easier Hazer

I live in an area where "hurricanes happen." I have a Hazer on my tower to save the tower and antenna from wind. The Hazer (**glen martin.com**) is a triangular frame, about 3 feet tall, that fits around the tower and travels up and down on rollers. It has a lift cable running down the inside of the tower to raise and lower the antenna. When raised into position, it has a spring loaded latch that locks onto a tower rung and is released to lower the antenna. I wanted an easier and safer method for operating my Hazer and devised this technique.

I installed a post about 15 feet away from the antenna. The post is composed of two coaxial pipes, a 1½-inch OD pipe with a 1¼ inch OD inner pipe driven 10 feet into the ground. I installed an inexpensive all terrain vehicle winch at the top of the post with a watertight electrical box to hold the winch controller and pendant when not in use (see **Figure 21**). I threaded two brass ¼-20 screws through the bottom of the box. Inside, these connect to the winch controller (see **Figure 22**). I attach jumper cables from my car battery to the exterior ends of the brass screws to power the winch.

Next, I attached a pulley to the bottom of the tower (see **Figure 23**) away from the face with the Hazer latch and another at the bottom of the winch post. The lift cable runs about 2 inches above the grass when tight. The sideways force on the winch support post and tower leg is right at the ground level.

The cable is long enough so that, when the Hazer is on the ground for a storm, the cable runs from the Hazer to a pulley at the top of the tower. It does a 180° turn and then runs down the inside of the tower to a pulley at the bottom. At this bottom pulley it does a 90° turn and exits horizontally to the pulley at the bottom of the winch support pipes. From there it runs up the pipe to the winch.

To remove any tripping hazard when not in use, I put the cable into "clothesline mode" by loosening the winch cable enough to pull it up above head height where it is taped to the tower (see **Figure 24**). It is then wound around the winch at the post end (see **Figure 25**). This keeps the wire out of the way.

Figure 21 — The winch and control box mounted on the post about 15 feet from the tower. The steel cable runs down to a pulley at the base of the post. [Pat Hamel, W5THT, photo]

Figure 22 — The interior of the watertight box showing the winch control unit and pendant in their storage locations. At the bottom are the two brass screws used as 12 V connection points. [Pat Hamel, W5THT, photo]

Figure 23 — This pulley added to the bottom of the tower redirects the cable toward the post mounted winch. In this view the cable is not in use, runs out of the tower and turns upward and out of the way. [Pat Hamel, W5THT, photo]

Figure 24 —The cable, in "clothesline mode," is taped to the tower about 6 feet up. [Pat Hamel, W5THT, photo]

Figure 25 — In "clothesline mode" the cable runs from its taped point on the tower, over the top of the winch and down to the post pulley. This keeps the cable well above ground when not in use. [Pat Hamel, W5THT, photo]

To raise and lower the antenna, first, with the Hazer resting on its latch, the cable is brought down to the pulley level and the slack is taken out. Once the cable is under tension, the Hazer can be lowered.

Now I can raise and lower the antenna quickly and safely, by myself, without aching shoulder muscles. — *73, Pat Hamel, W5THT, 1157 E Old Pass Rd, Long Beach, MS 39560, w5tht@arrl.net*

Quick Disconnects for Window Line

Having used window line for feeding my horizontal loop antennas for over 20 years, I found that my rapid disconnects outside are great for reducing the chance of damage from lightning strikes. I realized that, for experimenting with antennas, feed lines, tuners and RF ammeters, rapid disconnects inside would save time, frustration and generally make life easier. Using screw terminals to connect my 300 Ω window line to my Drake MN-2700 tuners works until I want to disconnect/reconnect it. Dropped screws and lock washers only serve as an annoyance. My MFJ balanced line RF ammeter does use banana style jacks but not with banana plug spacing; a dual banana plug saves stress, flexing and breakage of the wires as opposed to individual banana plugs on each conductor. What was I to do?

After ruling out ideas like drilling holes and reconfiguring hardware, I decided to make some adaptors. With some offset brackets used to mount PC boards and banana

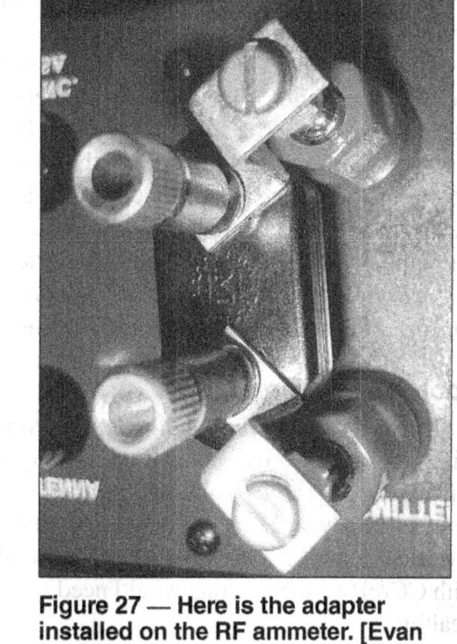

Figure 27 — Here is the adapter installed on the RF ammeter. [Evan Rolek, K9SQG, photo]

socket spacer plates, I constructed the two types of adaptors shown in **Figure 26**. The one with the yellow banana jacks is for my MN-2700 tuners that have screw terminals. The other adaptor is for my MFJ RF ammeter since its connectors don't have banana plug spacing (see **Figure 27**). Not high tech, but the convenience is helpful whenever I reconfigure things for experimentation purposes. I'm sure readers can greatly improve upon my initial design. Who says window line is hard to deal with? — *73, Evan Rolek, K9SQG, 1295 Oakleaf Dr, Beavercreek, OH 45434, k9sqg@arrl.net*

Just Add Wire

Every ham needs at least one dipole antenna and every dipole needs a center insulator, two end insulators and a feed line support. You can make these items in many ways, for example, by sawing some Plexiglas as shown in Figure 21.15 of the 2013 *ARRL Handbook* (see **Figure 28**).

What if you don't have the Plexiglas? Another alternative is to cut and drill a plumbing PVC 1 inch Tee fitting to the dimensions shown in Figure 28. The top half of the lateral tube is cut away. Cut this hemispheric section into three parts. Cut two end insulators, each $^{10}/_{16}$ inch wide, from the sides of the section and drill holes in each end. This leaves a 1 inch wide top section, which can then be drilled as needed to support either coax or twinlead. You get all the parts needed for just a small effort (see **Figure 29**).

The three holes in the center tube allow for either halyard mounting or pole mounting. If using coax, use Nylon clamps of proper diameter for support. You might find Tees whose lateral tube is either 3⅝ or 3⅛ inches long. If so, opt for the longer length because the shorter length requires rasping out the radius material between tube and the Tee as indicated.

For hardware, brass bolts rather than steel will give you longevity without corrosion. Solder lugs should be heavy duty to withstand both pulling and flexing. Three possible choices, all using material available at major hardware stores are:

- Gardner-Bender P/N 15-105 Ring Terminals, come in a package of 15. They have

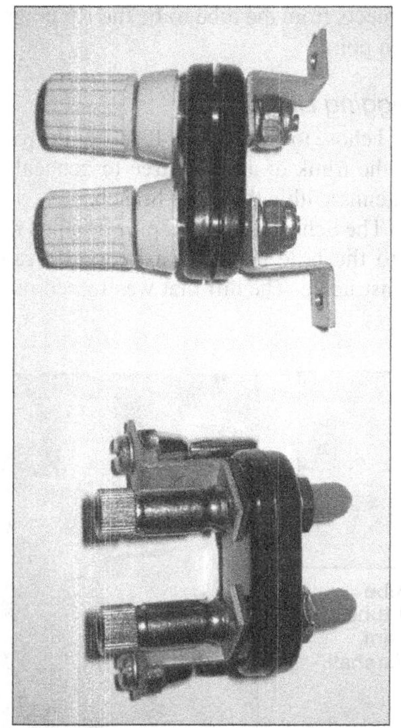

Figure 26 — Here are the two window line adapters, one for an MN-2700 tuner and the other for an MFJ RF ammeter. [Evan Rolek, K9SQG, photo]

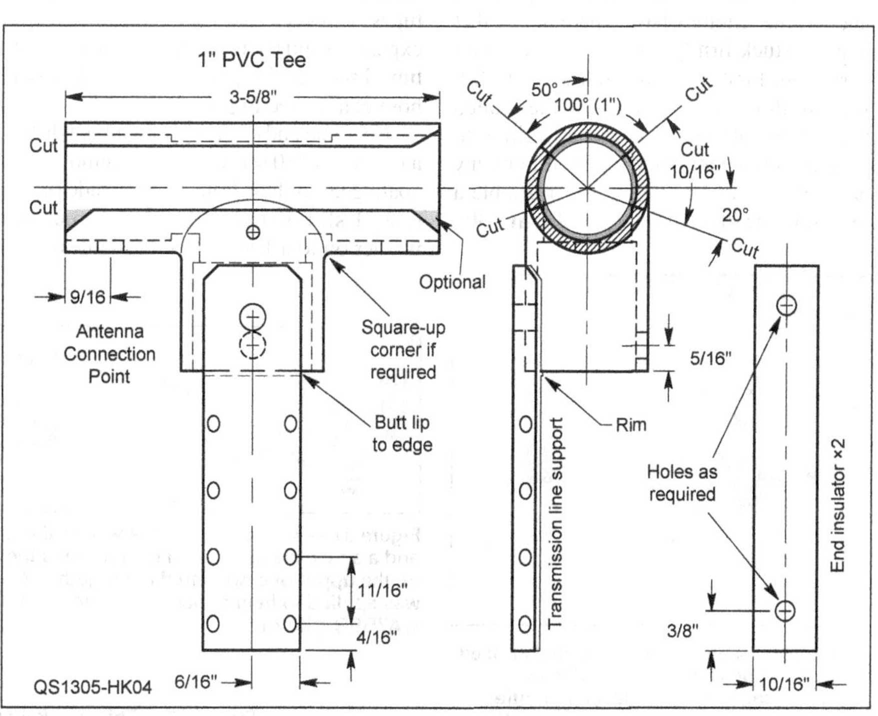

Figure 28 — This layout drawing shows how to cut the PVC Tee to create a complete dipole insulator set.

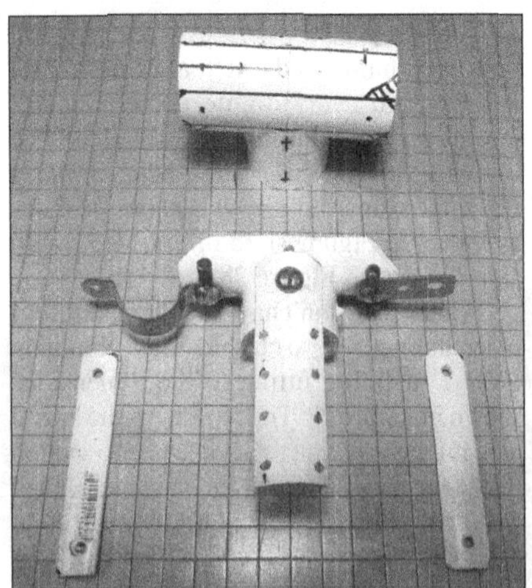

Figure 29 — In a portable or emergency situation, a PVC Tee can be cut into center and end insulators for a makeshift dipole. Just add wire. [Steve Sant Andrea, AG1YK, photo]

a ¼ inch hole, which accepts up to #14 AWG wire. These are plated steel, not copper, normally used for interior wiring.

■ Plumber's Tape, also called Hanger Strap, is ⅝ inch wide, predrilled, plated steel strap used for hanging galvanized pipes.

■ Conduit or Tube Strap is a ½ inch wide metal, omega-shaped support used for hanging ¾ inch copper pipe. When flattened, cut in half, and drilled, it will make two pieces.
— 73, Spud Monahan, K6KH, 817 Pacific Ave, Manhattan Beach, CA 90266, **k6kh@ aol.com**

Separating Aluminum Tubing

I have found a method for separating telescoping lengths of aluminum tubing that may be stuck firmly together due to minor corrosion. First, I use a screwdriver to lift, very slightly, the segments of the slotted ends. I do this one segment at a time, to prevent permanent bending. Then I cup my hand closely under the joint and dribble a little rubbing alcohol on it (see **Figure 30**).

I repeat this 3 or 4 times over the course of a minute, then wait for a minute.

After allowing the alcohol to soak in, I use two pliers and/or locking grips to separate the two pieces. I find that the alcohol has penetrated to the very end of the joint, softening the corrosion products, allowing the tubes to be separated, at which point I can wipe off the soft residue before it dries.
— 73, Bob Wilkinson, W7VN, 19048 Woodton Ln, Brookings, OR 97415, **w7vn@arrl.net**

Pool Pole Antenna

The telescoping pole handle for my swimming pool brush broke and would no longer stay extended. The two aluminum tubes nest together nicely so I thought they could make a vertical antenna. I live in a neighborhood with CC&R so the antenna would need to be stealthy.

Building the Antenna

The two poles are both 8 feet long, making them about 15 feet tall when nested. However, the tubes do not nest snugly. The bottom tube has an inside diameter of 1.187 inches while the top tube has an outside diameter of 1.125 inches. The plastic locking device that fits in the space between the poles was broken and couldn't be used. I needed a way to lock the sections together.

I modified the bottom tube by sawing a single vertical, 7 inch slot down the tube wall at one end. The tubes have an anodized coating inside and out, which acts as an insulator. For conductivity between the two tubes, I used a small wire wheel (or you can use sandpaper) to remove the coating for 7 inches along the mating surfaces of the two tubes. Once the conductive surfaces were exposed, I slid the top tube inside the bottom tube. I secured this joint with a stainless steel hose clamp (see **Figure 31**).

Near one end of the bottom tube, I drilled a hole for a #10 screw. Again, I removed the coating around the hole on the inside of this tube. I slid a stainless steel internal star washer on to a 1 inch #10-32 stainless steel

screw, then I inserted this screw through the hole *from the inside* of the bottom tube. I dropped a flat washer on to the screw projecting out from the tube. Then I threaded on a stainless steel nut but did not tighten it. This screw would be used as the RF connection to the mast and would be tightened after the tube is inserted into the PVC base support.

I added a whip to the top. RadioShack sold 102 inch whips for a competitive price (catalog # 21-903). I obtained the bracket pieces for the whip from the CB radio section at a nearby truck stop (see **Figure 32**). I sanded away the anodized coating from the top 2 inches of the tube. I reinforced the inside of the top tube with a 1 inch dowel before clamping the whip to the top. The walls of these tubes are quite thin, so too much clamping force will easily distort them.

The Ground Mount

Next, I needed a mount for the antenna. Fully assembled, the vertical stands about 24 feet tall. I decided on a ground mount. I bought a 5 foot length of 1¼ inch Schedule 80 PVC pipe. This pipe has an outer diameter of 1.66 inches and an inner diameter of 1.278 inches.

Schedule 80 PVC pipe is generally not available at local home improvement stores. I had to go to a plumbing supply store. (Some plumbing supply stores will sell the pipe by the foot while others will require you to buy a 10 or 20 foot piece.)

This PVC pipe made a tight, sliding fit with the bottom aluminum tube. I cut a 7-inch slot in the PVC pipe. At the bottom of the slot I drilled a hole for the #10 screw that projects from the tube to be the RF connection point.

Digging Detour

I chose to make a hole 40 inches deep next to the trunk of an olive tree to conceal the antenna within the tree's branches.

The Schedule 80 PVC pipe dropped right into the hole made by a hydraulic ram I constructed. The dirt that was forced out of

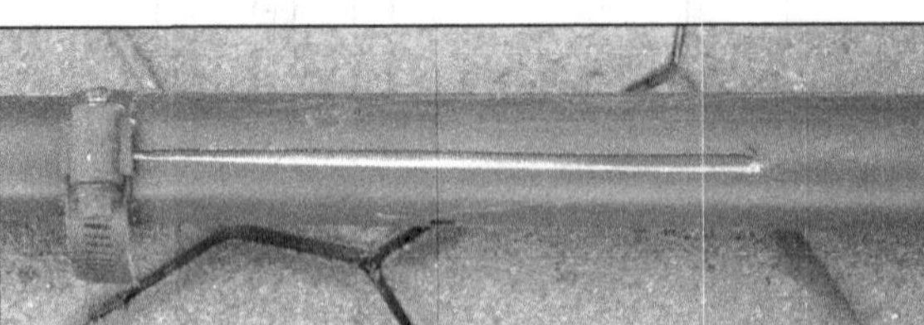

Figure 31 — A slot is cut in the wall of the lower tube and a screw clamp is used to compress the lower tube on the upper one holding them together. Green paint was applied to help conceal the antenna. [John Marshall, WA7BSR, photo]

Figure 30 — A little rubbing alcohol applied to the junction of two stuck antenna elements can soften whatever impurities have become lodged between the sections, allowing them to be more easily separated. [Lorraine Wilkinson, W7RFC, photo]

Figure 32 — This bracket attaches the 102 inch whip to the top of the upper pole. [John Marshall, WA7BSR, photo]

Figure 33 — A square aluminum plate is drilled and tapped for 40 radials. The black wire is the ground connection for the MFJ tuner. [John Marshall, WA7BSR, photo]

the hole was piled up around the pipe and watered back into the hole. I didn't use concrete since the soil where I live is similar to concrete. You may need to use a bag of ready-mix concrete to stabilize the Schedule 80 pipe in your soil. used a bubble level to ensure that the PVC pipe was vertical.

Ground Radials

I bought a 1 foot square, ¼ inch aluminum plate for use as a ground radial plate. I drilled and tapped 40 holes for #10-32 screws along the edges of the square and cut a large hole in the center for the antenna base to pass through. I used #10-32 socket head stainless steel screws and internal star washers for the radials.

I made 20 radial wires, each 15 feet long, from 18 gauge stranded, insulated wire. Each radial wire has a ring lug for attachment to the aluminum plate (see **Figure 33**).

Final Assembly

To reduce the antenna's visibility, I painted the aluminum tubes with a flat olive paint. On reflection, a dark gray-brown paint may have been a better choice for blending with tree trunks and branches. I painted the whip the color of the sky at the horizon. When I look up, the sky is a deep blue, but looking at the horizon the blue is more pastel. I used a flat non-reflective paint. The idea is that no one will notice the camouflaged antenna.

I slipped a 2 inch stainless steel pipe strap over the Schedule 80 pipe. I fed the whip end of the antenna up through the branches of the tree then fed the other end into the Schedule 80 pipe. I tightened the screw that projects outward from the PVC pipe, then I tightened the pipe strap at the top of the Schedule 80 pipe and the antenna was ready to go.

Testing

I attached an MFJ-993BRT automatic antenna tuner to the #10-32 screw at the bottom of the antenna and to the ground radial plate. For testing, I used my Icom IC-706 transceiver. On the first day of testing I made contacts to Santa Barbara, California; Boston, Massachusetts; Ottawa, Quebec; Vancouver, British Columbia and Japan.

EZNEC *Modeling*

Usually I model an antenna in *EZNEC* before building it but, in this case, I built the antenna first and then I modeled it. On the 160-12 meter bands the launch angle varies from 18-28°. It rises to 49° on 10 meters.

Some Further Observations

- This antenna has the lowest angle of radiation on 12 meters where the antenna is approximately ⅝ wavelength long.
- The primary lobe launch angle is high on 10 meters, but there is a minor lobe at 10 degrees elevation that is 10 dB down.
- This antenna may not match with an automatic tuner on 17 meters since it is close to ½ wavelength. I have not tested the antenna on 17 meters.
- Efficiency will decrease and Q will increase progressively from 60 meters down. Do not expect great results.
- The antenna performs well on 40, 20, and 15 meters, the bands that I use most.
- The antenna should perform well on 40-12 meters.

Conclusions

This antenna was inexpensive, easy to build, and performs well on the bands I am most active on. It has been in operation for more than 6 months and has stood up to the strong winds, summer monsoons, and haboobs that really churn up the desert. Finally, my wife and I hosted a neighborhood party and none of our guests noticed the antenna, so it meets its covert requirement. — *73, John Marshall, WA7BSR, 7828 W Caribbean Ln, Peoria, AZ 85381-3441,* **wa7bsr@arrl.net**

SteppIR Storm Safety

With the unusually high amount of rough weather experienced in North America this past year, SteppIR antenna owners may want to start taking an extra step of caution to minimize potential storm damage. When the antenna is not in use, retracting the tapes to the "Home" position could help minimize the probability of tape damage from excessive flexing of the fiberglass tubes during high wind conditions. This also minimizes the antenna's attractiveness to lighting strikes.

As a reminder of antenna status, prior to applying RF power, I use homebrewed "memory aides" fabricated by a call sign badge engraver typically found at hamfests (see **Figure 34**). Each of the two placards consists of two engraved call sign badges glued back to back. Simply flip the placard over when

changing the antenna "Home" status to reflect the current tape position. The second plaque is used as a reminder for any other antennas that may have also been disconnected. Assigning different colors to the signs can help identify antenna status at a mere glance and from a greater distance. — *73, Don Kasten, N4DK, 2056 Six Branches Dr, Roswell, GA 30076-3029,* **n4dk@arrl.net**.

Mast Bracket for Rotator Replacement

Like many ham installations, my rotator sits inside my tower with a thrust bearing at the tower's top. When my rotator needed replacement, I needed some way of keeping the mast from turning and shifting sideways when the rotator was removed. The solution I used was to make brackets using Unistrut, a steel channel used for supporting conduit, which is available from electrical supply stores. The Unistrut pipe clamps slide into the channel and adjust for length. They are available for all common pipe sizes. I used three brackets, one from each tower leg to the mast. The brackets work well and can be used for other tower sizes and mast diameters by changing the pipe clamp sizes. — *73, Dave Palmgren, N8DP, 6132 County 420 - 21st Rd, Gladstone, Michigan 49837,* **n8dp@arrl.net**

Fishing with a Magnet

After building an octagon shaped magnetic loop antenna using ¾ inch copper pipe I realized that remote tuning was essential. I built a motorized unit for the air variable capacitor and decided to run the control cable on the inside of the pipe as suggested in my *ARRL® Antenna Book*. I tried repeatedly to push the control wire through the pipe but found that negotiating the 45° elbows was a real challenge.

Instead of using a vacuum cleaner or other method for pulling a string through the pipe, I had the idea to use a small, round head screw with the string tied to it. Using a strong magnet, I was able to run the magnet on the outside of the pipe to easily pull the screw with string all the way from top to bottom. It was then a simple matter to attach my control cable to the string and pull it through the pipe. — *73, John Merritt, K4KQZ, 2430 Hidden Lake Cir, Columbia, TN 38401-5832,* **k4kqz@arrl.net**

Small Antenna Mount

My friend, Bill, N9CHN, moved to a new condo in Arizona where rules do not permit

Figure 34 — Signs like these can help you keep track of the status of your SteppIR and other antennas. [Don Kasten, N4DK, photo]

antennas to be visible from the street. The building has a flat, rubberized-membrane roof so we needed a mounting solution that didn't require making holes in the building.

After some brainstorming, we constructed the light duty antenna mount shown in **Figure 35**. The materials required are:

1 10 foot length of 2 inch PVC pipe
3 PVC Ts
5 end caps
4 45° elbows
4 #6 × ¾ inch stainless steel wood screws
25 lbs pea gravel

Also required are PVC cleaner and glue. Painting is also required to protect the PVC pipe from UV radiation.

We assembled the mount on the ground, putting all the components together with the exception of two adjacent end caps. We then moved it to the roof, turned it on end, filled the tubes with pea gravel and placed the last two end caps. The pea gravel added about 20 pounds to the structure and made it quite stable. We used screws, not glue, to hold the end caps in place, so the gravel could be removed if necessary. Before installing the end caps, drill a ³⁄₁₆ inch hole in each to allow for water drainage.

Any number of variations are possible. We chose the X configuration as it is stronger than an H and would also allow for the legs to be angled downwards in order to conform to a peaked roof. The legs could be fitted with threaded couplers for easy disassembly for portable use. The end caps could be glued in place and the unit filled with water instead of gravel for added weight, if freezing temperatures are not expected. It is possible that the vertical section could be made longer for some applications such as supporting the center of a dipole. [This design should only be used with small, low wind load antennas. Depending upon your situation, some form

Figure 35 — This PVC mount is simple and inexpensive to build. It is adequate for small, low wind-load antennas. [Ed Toal, N9MW, photo]

of guying may be prudent. — *Ed.*] — *73, Ed Toal, N9MW, W8471 State Road 39, Blanchardville, WI 53516-9663,* **n9mw@ tds.net**

Mobile Pass-Through Panel

One challenge that every ham faces is how to get a signal from the antenna to the rig inside the house. A whole range of homebrewed and commercial solutions exists. The mobile ham faces the same issue. Drilling holes is the usual solution but that is always inconvenient and often not aesthetically pleasing. The popularity of VHF for repeaters, satellites and APRS has increased the use of multiple rooftop antennas creating the need to bring multiple signals into the vehicle.

The easiest way to do this is to leave a rear window open a crack and run the cable through. This option upsets the heating and air conditioning systems, admits rain and snow to the interior and runs the risk of the cables flopping around or being pinched by the window.

A better idea is to put a thin strip of wood at the top of the window with slots for the cables. **Figure 36** shows one such installation. A piece of wood or other material the thickness of the glass (usually ⅛ inch) is cut about 2 inches wide and shaped to fit the particular window's profile. A slot about ¼ inch long is cut on the bottom for each cable entering the vehicle. — *73, Alex Burr, K5XY, 695 Stone Canyon Dr, Las Cruces, NM 88011,* **k5xy@arrl.net.**

Improvised Radial Tie Point

I was installing a vertical antenna and needed a way to connect the radials. Looking for inspiration at my local home improvement center, I came upon equipment grounding bars in the electrical section. An equipment grounding bar is an aluminum bar with wire terminals and screws used in home power distribution panels. These grounding bars can be effective tie points for antenna radials. I chose the nine terminal version, which, when configured in a square, will hold up to 36 radials (see **Figure 37**). — *73, Richard Steck, W9RS,* **w9rs@arrl.net**

A One Trim Dipole

Simple wire dipoles are among the easiest and cheapest antennas to make, but getting the length and resonant frequency exactly right is often left to trial and error. It is well known that, due to various factors, the formula for the length of a dipole, Length = 468 / *Frequency*, is often inaccurate. However, there is a way, using

Figure 36 — A small piece of wood or rigid insulation is easily shaped and notched to make a feed line pass-through for your mobile installation. [Alex Burr, K5XY, photo]

Figure 37 — A set of four equipment grounding bars can be bolted together to form a square around the base of your antenna and they will provide ample connection points for radials. [Richard Steck, W9RS, photo]

an antenna analyzer, that you can resonate the dipole at the desired frequency while trimming the length only once.

First, use the standard formula. For example, for a center frequency of 14.2 MHz, the length comes out to 32.96 feet or about 16.5 feet per side of the half wave dipole. Then, add about 6 inches to that length on each side and raise the antenna to its permanent location.

From your ham shack, attach the antenna analyzer to the end of the coax and find the resonant frequency (lowest SWR). Now, solve for the *real* constant, by multiplying the total length you initially used (in our case 34 feet), times the resonant frequency you found on the antenna analyzer (Constant = *Frequency* × Length). For example, if the antenna analyzer showed resonance at 14.4 MHz, the real constant of your antenna would be 34 × 14.4 or 489.6. Now, replacing the 468 in the formula with the real antenna's constant, we have Length = 489.6/14.2, which equals 34.48 feet or about 17¼ feet on each side (17 feet 3 inches). Since you used 17 feet 6 inches to start with, just trim 3 inches off each end and *voila!* The antenna analyzer should now show a resonant frequency of 14.2 MHz. — *73, Paul Voorhees, W7PV, 10090 Misery Point Rd NW, Seabeck, WA 98380-9784,* **w7pv@arrl.net**

Installing Coax Boots

I like to put waterproof boots on all my coax connectors but the boots have tiny openings at the coax end, making them a real chore to install. Last time I installed a boot I thought there must be an easier way. I remembered that I had used a cone shaped tool to install O-rings without nicking them and thought that there must be a similar way to slide the boot over the coax.

I looked on my workbench and saw one of the tips that come with tubes of silicone sealer — the long, pointy type that you use to apply the sealant. The tip has a nice taper with a large threaded knob at one end to screw onto the tube. I cut the threaded portion off with some tin snips (most anything sharp would work) and filed the cut to remove any burrs.

The tip fit perfectly over the coax. I applied some of the silicone grease that came with the boot onto the coax and the boot easily slid on (see **Figure 38**). A job that could take several minutes was over in seconds. It has the added advantage of keeping the silicone grease used to install the boot off the end of the coax, so as not to interfere with a clean solder joint. You just have to remember to put the boot on before you install the connector. — *73, Michael Buck, K6BUK, 203 Brockman Ct, Lincoln, CA 95648,* **mrbuck@ pacbell.net**

Chalk Reel Antenna

I do a lot of portable operating from hotels or while backpacking. I have found that a reasonably effective antenna for 40 meters and above is a simple long-wire antenna of approximately 50 feet in length. It tunes

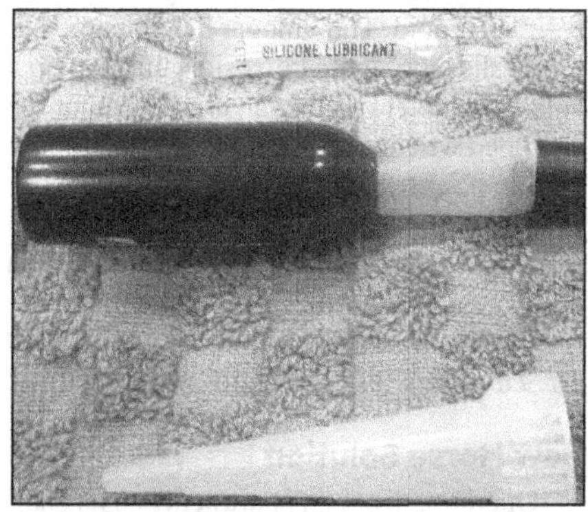

Figure 38 — The conical applicator tip from a tube of glue or sealant can be converted to a handy tool that makes fitting weather seal boots onto coax a simple job. [Michael Buck, K6BUK, photo]

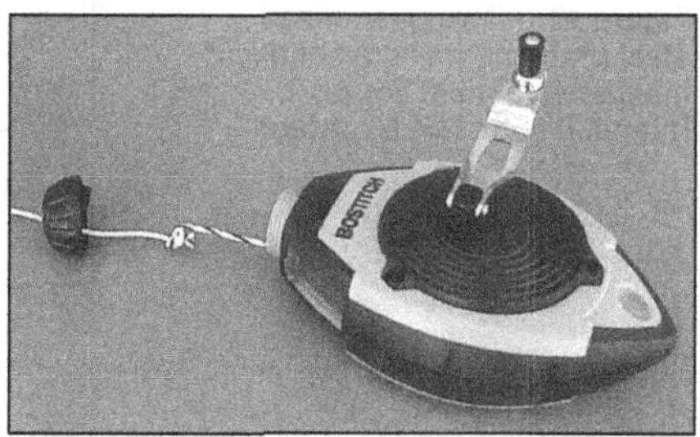

Figure 39 — An inexpensive chalk reel is a convenient way to carry an end-fed long wire. [Alan Amos Jr, KN1O, photo]

well with a simple **L** network or with the internal tuner in my K1. When operating from a hotel, I typically try to get a second or third floor room and use a tree or my car to support the antenna. Generally, I use the air conditioner to provide a ground reference. When backpacking I feed the antenna from the low end and get the other end as high up a tree as I can and simply lay another wire on the ground (where no one will trip over it) as the ground reference.

The biggest problem has always been keeping the wire from becoming a tangled mess. I have tried a variety of methods, some of which were more successful than others. While walking through the tool section of our local hardware store, an item caught my attention; a chalk line reel. The intended use is to hold 100 feet of string along with powdered chalk and the string is used to mark a straight line. I had just acquired a reel of Teflon-coated fine wire (#26 I believe) for portable antennas and the diameter isn't much greater than the string in the chalk reel. The reel is quite compact and is light enough to be practical even for backpacking.

I opened the reel and removed all but about a foot of the string and also removed the end cap that the string feeds through. I simply tied the remaining piece of string to the wire as shown in **Figure 39**. I fed the wire through the internal felt ring, which provides a little bit of drag on the wire to help it wind evenly on the spool. To connect the transmitter, I simply stripped off a short section of the insulation a few inches from the spool end of the wire to provide a place where I could use a small alligator clip to attach the transmitter. This solution has truly made me a "happy

camper" when operating portable; no more tangled wires! — *73, Alan Amos Jr, KN1O, 30 Bromfield St, Newburyport, MA 01950,* **radiokn1o@gmail.com**

Inner Tube Weatherproofing

I've successfully used this method for a few years during the course of antenna installations to protect coax connections from the elements. Rather than using electrical tape, coax seal or some of the more traditional methods of sealing coax connectors and splices, I cut up a bicycle inner tube and use it as a sleeve over the connection. A generic 26 inch bicycle tube can be purchased for about $3 at the local bike shop or discount store and will yield enough material for about a dozen connections.

I cut the sleeve to extend about 2 inches beyond each side of the joint then slip it over the connection, completely covering the joint. Then I secure it tightly to the coax on each end with a tie-wrap (see **Figure 40**). I always treat the sleeve with a protective coating such as Armor All or a similar product to prevent UV damage. The best part is that the connection remains flexible and watertight and the sleeve is easily removed and may be reused multiple times. — *73, Stephen Burns, N5VTU, 10208 S Summerlin, Conroe, TX 77302,* **n5vtu@arrl.net**

A Piano Plays Radio

I am one of those hams who lives in a townhouse community that does not allow antennas of any kind outdoors. I have found that piano wire in gauges from #0 to #5 works well for a low-visibility antenna. [Steel and piano wire gauge sizes 0 and 5 are similar in diameter to copper AWG sizes 31 and 27, respectively. — *Ed.*] It's available from many suppliers and an Internet search will provide a number of leads.

I went out on a foggy Sunday morning (a stealth antenna has to be raised in a stealthy way) and slung a weight with some 50 lb test line about 35 feet into a tree behind my home. I then ran the piano wire from my shack to the tree for a length of about 150 feet. It has proved successful in that you cannot see the very fine but *very* strong piano wire.

One thing to note, when buying piano wire, be sure to order it on a spool — *not* in a

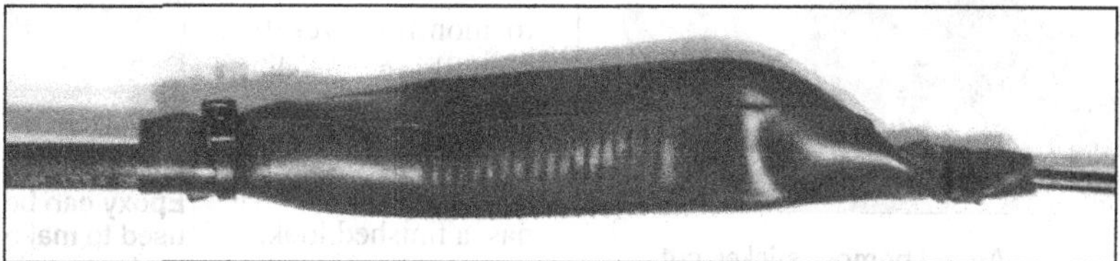

Figure 40 — A short piece of bicycle inner tube makes an inexpensive weatherproof seal for coax connections. [Stephen Burns, N5VTU, photo]

coil. I first bought the coiled wire and learned the hard way that it has a mind of its own. It took me some 2½ hours to get it untangled!

I run 100 W with an AH-4 autotuner and have made many contacts around the world. So, if you also live in a restricted community, give this idea a try. I think you'll be happy with the results. — *73, James Waters, W3BIF, 311 Cherry Ln, Kennett Square, PA 19348,* **captainjimwaters@comcast.net**

Mag Mount Sealer

After repairing my 2 meter magnetic mount antenna, I needed to apply a weather resistant seal to the bottom surface. Rows of electrical tape didn't appear to be the best solution. I had some leftover bumper stickers just waiting to be stuck onto something. I cleaned the base of the mount to remove any dirt that might penetrate the covering. Then I cut one of the bumper stickers in half, stuck it to the bottom of my newly repaired antenna, and trimmed the excess with my pocket knife (see **Figure 41**). Now the bottom of the mag mount was extra clean and smooth, and ready for a scratch-free roof mounting. — *73, Phil Grant, N1YPS, 119 Hoe Shop Rd, Bernardston, MA 01337,* **phill112643@verizon.net**

Framing Bracket Mast Support

My latest VHF antenna project, consisting of a VHF-UHF discone on 15 feet of 1.5 inch PVC pipe, had been mounted to the side of the house with the all-purpose house bracket material — also known as the tie wrap. This worked for a quick test, but when the test lasted a few months, my wife suggested that I design something a bit safer and more secure.

I started looking for an antenna house bracket similar to the ones I'd used before. What I found was either cheaply made, or ridiculously priced, and would need modifications anyway. I started searching the local home improvement store and, while in the lumber and decking section, I came across

Figure 41 — An old bumper sticker, cut and trimmed to fit, protects your car's roof against scratches from your mag-mount antenna. [Phil Grant, N1YPS, photo]

Figure 42 — Two galvanized steel brackets make a solid and inexpensive mast support. [Lionel Booth, N5LB, photo]

reinforcing brackets used in framing for wood construction. I selected a couple of galvanized steel brackets that looked promising (see **Figure 42**) and for about $6 I had something to work with. Many other shapes and sizes were available, all at reasonable prices.

The brackets turned out to be an easy solution to my specific situation. They are very strong and galvanized to reduce corrosion. After drilling, I bolted them together and then bolted the assembly to the overhang on the side of the house at 10 feet.

Now my quick and dirty test setup can continue with the approval of my wife, who is no longer afraid of getting hit by a falling discone. The home improvement stores stock many items that, with a little ham ingenuity, can be fashioned into useful supports for that all-important reasonable cost.

Always be careful when erecting antennas and keep in mind that the results of any solution may vary, but in the best ham tradition we experiment and test, with utmost caution and care for those around us affected by our efforts. — *73, Lionel Booth, N5LB, 3226 Wellspring Lake Dr, Fulshear, TX 77441-4486,* **n5lb@arrl.net**

Whip Tip

While trimming some Larsen whips for 10 meter and 2 meter mobile antennas, I realized that the left over material was sufficient to make a 70 centimeter antenna and a 1080 MHz quarter wave antenna (I use 1080 MHz to monitor aircraft.). Not liking the sharp edge of the cut whips, I added a safety tip, which gave the antennas a finished look.

I had been doing some fiberglass work, so I dipped the end of the whips into the ep-

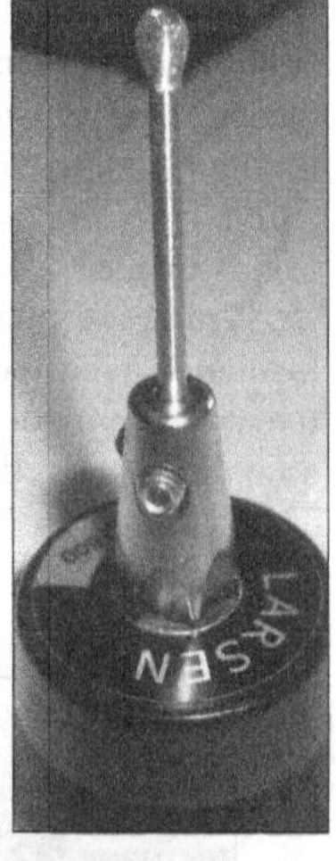

Figure 43 — Epoxy can be used to make a professional-looking safety tip for a whip antenna. [Dale Harris, N5QO, photo]

oxy resin. While holding the tip downward, gravity and surface tension shaped the epoxy into a teardrop shape. I dipped the end several times to build it up to a good size. After the epoxy set, it made a wonderful safety tip for the whip (see **Figure 43**).

Five-minute epoxy works very well. You can repeat the process after the resin has set if you want a larger tip. It works best if you allow the epoxy to begin setting up before dipping again. Just don't wait too long or the tip won't be able to flow into a teardrop shape. The next time you need to put a safety tip on a whip, give epoxy a try. — *73, Dale Harris, N5QO, 200 Chadwood St, Hot Springs, AR 71901,* **n5qo@yahoo.com**

Removing Wire Corrosion

I was faced with the task of resurrecting a Garmin handheld GPS-12XL, the alkaline batteries of which had burst and corroded. To make matters worse, corrosion had wicked into one of the power conductors. The damage could not be cleaned with heat and rosin, and the wires were fragile enough that scraping would only do more damage.

I recalled a similar situation at work. I consulted our materials and processes engineer, who suggested using an acetic acid solution. Acetic acid was not available at my local pharmacy, so I tried the next best thing — plain white cider vinegar, which is 5% acetic acid. The vinegar did an excellent job of dissolving the corrosion. Just set up a small container of vinegar, dip the end of the corroded wire in, and let it soak. It took about half an hour to eat away the corrosion. After the corrosion is dissolved, rinse the wires in running water or a solution of baking soda and water to neutralize any remaining acid. — *73, Bob Chadwick, W1MTX, 1785 Teak Rd SE, Palm Bay, FL 32909,* rchadwic@att.net

Saw Horse Solution

Operating from an apartment has its challenges — especially when it comes to antennas. I found that the MFJ-1622 Apartment antenna (**www.mfjenterprises.com**) was a good solution for my situation. It consists of a whip, coil, and counterpoise wire that covers 40 – 2 meters. I could work the 2 meter repeat-

Figure 45 — A tilt-over connector and a homebrew mast make this mobile vertical ready for NVIS operation at a moment's notice. [Joseph Park, WB6AGR, photo]

Figure 44 — A sawhorse makes a convenient base for a portable antenna. [Mark Starr, KK4RVL, photo]

ers from indoors but maximum performance could only be obtained by moving it outside.

My outdoor solution is shown in **Figure 44**. I enclosed the coil inside a weatherproof box and mounted it to a painted saw horse. I ordered a custom "DANGER" sign, as there is no ground for the system. All of the coax and the counterpoise wire are marked with construction flagging, for the sake of safety, as the area where I placed it has pedestrian traffic. The tape measure is used to determine the proper lengths for the whip and counterpoise as I change bands. The structure is light enough for me to keep it indoors and move it outside when the bands open up. It only cost about $22 to construct. — *73, Mark Starr, KK4RVL, 2023 Blue Rock Dr #104, Tampa, FL 33612,* **kk4rvl@qsl.net**

DX to NVIS

Near Vertical Incidence Skywave (NVIS) antennas have become an important part of disaster communications. When setting up a portable communications post, mounting a dipole very close to the ground can greatly enhance local, short-distance communications. When operating from a mobile, ori-

enting the vertical antenna horizontally can achieve the same effect.

I operate from my truck with a vertical antenna that uses a screwdriver tuning coil. Orienting the antenna horizontally was a problem because the coil rises and falls several inches as it tunes. I needed a mount that would allow the antenna and tuner to move together while tuning and still allow it to be used in an NVIS configuration.

After giving it some thought, I realized that a tilt-over connector would be a simple solution (see **Figure 45**). I mounted the tilt-over connector to the top of the screwdriver tuning assembly, with the connector oriented so the antenna tilts into a position parallel to my truck's body. As the screwdriver rises and falls, the tilt-over joint flexes, allowing the antenna to be tuned while in the horizontal configuration.

The only other problem was a support for the "top" end of the antenna. I used two

Figure 46 — A bungee cord anchors the end of the antenna in place when driving. [Joseph Park, WB6AGR, photo]

3 foot pieces of PVC pipe to make a short telescoping mast. I cut a notch in the top section to hold the antenna and a "pin hole" through both sections to support the upper section when raised. The lower section is mounted in a stake hole and fits over the top section so I can lower the antenna for driving and raise it somewhat for stationary operating. Finally, I added a bungee cord from the tip of the antenna down to the stake hole to keep it from bouncing off the mast while on the road (see **Figure 46**).

With this arrangement, I can use the antenna for NVIS work as needed and if I want to work DX, I can raise it into a full vertical position (if your vertical is tall, it may require guying) and lock the tilt-over connector in place. — *73, Joseph Park, WB6AGR, 2713 Alysheba Ave, Modesto, CA 95355-8435,* **wb6agr@arrl.net**

The Snagger

If you use wire antennas that are held up by ropes, and if, like me, you aren't meticulous about maintenance, you have probably experienced a sinking feeling as you look at your antenna lying on the ground and then up at the frayed and broken support rope dangling from your tree. When this happened to me recently, I noticed that the loose rope was low enough for me to touch it with a 12 foot pole from the ground. What I needed was a way to grab the rope with the end of the pole.

I took an empty plastic jug from our recycling bin, cut off the handle, and cut an opening and a slit in the side of the jug, thereby creating the "Snagger." I mounted the Snagger

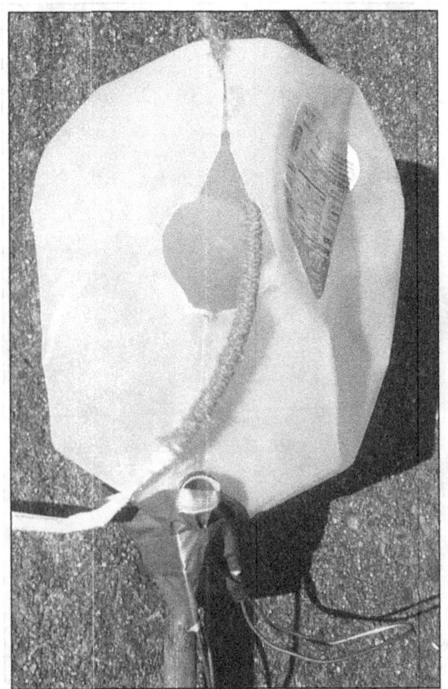

Figure 47 — Fixing a broken antenna or raising one at Field Day often involves a rope or wire hanging out of reach in a tree. A slot cut into a gallon jug can be used to grab the end of that recalcitrant rope. [Bob Dickerman, WA1QKT, photo]

on the end of the pole, reached up to position it against the dangling rope's end, jerked down, and snagged the rope's end on the first try (see **Figure 47**). I was able to pull the old rope down, attach a replacement, and pull the new rope back up through the tree. — *73, Bob Dickerman, WA1QKT, 32 Alexander Hill Rd, Northfield, MA 01360,* **rld@dickermanelectronics.com**

Thrust Bearing Umbrella

Many off-the-shelf bearings, commonly used as thrust bearings for antenna rotator

Figure 48 — The reflector "umbrella" is attached to the mast above the bearing using a rubber pipe cap and is positioned so its bottom edge rests lightly on the bearing support bolts. [Roy Hook, W8REH, photo]

applications, are not equipped with any means of preventing water from entering the bearing and causing problems. I repurposed an aluminum reflector from a Lithonia Lighting Luminaire light fixture to create an umbrella over my thrust bearing. The top of the reflector is attached to the 2 inch mast pipe using a modified 2 inch flexible rubber pipe cap. I used a 1⅞ inch hole saw to cut a hole in the center of the cap, which provides a tight seal at the mast. The band clamp included with the flexible pipe cap was not needed because the rubber cap fit very snug to the top of the reflector.

The bottom rim of the reflector is kept in position by the thrust bearing's mounting bolts, which protrude upward (see **Figure 48**). The reflector easily turns with the mast and the contact between the metal reflector and the thrust bearing bolts should help to prevent any static charge buildup. The small gap between the bearing and reflector permits any moisture to escape and provides access to the bearing's grease fitting without moving any of the components, a plus when working at 60 feet in the air. — *73, Roy Hook, W8REH, 6611 Steitz Rd, Powell, OH 43065,* **w8reh@arrl.org**

Antenna Pulley from Discarded CDs

My design for a horizontal delta antenna allowed the antenna to "float," suspended by large-diameter pulleys at the vertices (see **Figure 49**). By floating the delta I would equalize the tension in the three legs and the large-diameter pulleys would reduce bending stress on the antenna wire.

I decided to make my own pulleys. I realized that high-tech "coasters" (discarded CDs)

Figure 49 — Large diameter pulley reduces bending stress on the antenna wire. [Barry Shackleford, W6YE, photo]

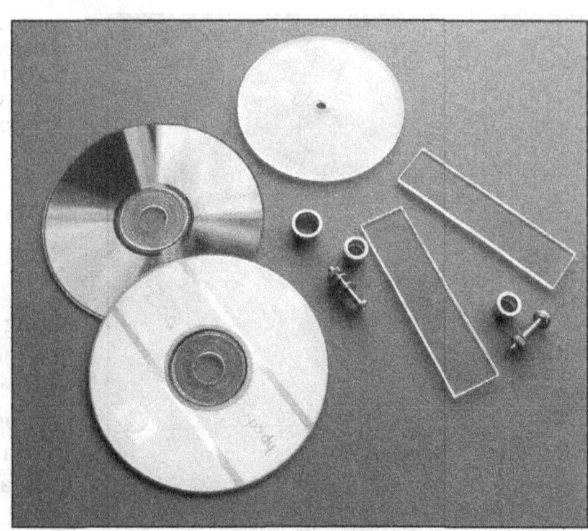

Figure 50 — The pulley is formed from two CDs and a plastic disc. The bearing and spacer are made from aluminum tubing and the frame is two pieces of ⅛-inch acrylic. [Barry Shackleford, W6YE, photo]

seemed to be an appropriate diameter (see **Figure 50**). Two CDs form the outer portion of the pulley. For the inner portion, I used a ⅛-inch thick, 4-inch diameter plastic disc. You should size the disc thick enough to allow the wire or rope you will run through the pulley to fit loosely between the CDs. I used an inexpensive hole saw to cut the disc; a jig saw could also be used.

The bearing for the pulley is made from some scrap telescoping aluminum tubing. The frame is made from two 4 × 1 inch pieces of ⅛-inch thick acrylic plastic held together with 8-32 stainless steel screws, washers, and nuts.

The sleeve portion of the bearing (see **Figure 51**) is made by using a pipe cutter to cut a length of ⅝-inch aluminum tubing equal to the combined thickness of the inner disk and both the CDs. The shaft portion of the bearing is ½-inch OD aluminum cut about ⅟₁₆ inch longer than the sleeve. The frame spacer is cut to the same length as the shaft. Sand all ends square after cutting.

The pulley is comprised of the two CDs cemented to either side of the 4-inch plastic disc. I used a gel-type polyurethane adhesive. Before gluing, I roughed up all contact surfaces with fine sandpaper. Take care to center the inner disc on the CD and

Figure 51 — The pulley bearing assembly (left) and pulley frame spacer (right). The bearing sleeve is ⅝-inch OD aluminum tubing and the spacer and shaft are ½-inch OD tubing. [Barry Shackleford, W6YE, photo]

then align the second CD with the first in a "sandwich." After gluing, clamp the assembly between blocks. Spring clamps are preferable to **C**-clamps, which might apply too much pressure.

The "donut hole" will need to be enlarged to accommodate the bearing. I have found that the scraping action of an ordinary woodworking spade bit is ideal for drilling large-diameter holes in plastic. After drilling, press the bearing sleeve into the hole and secure it with several drops of an instant-bond glue.

Final assembly of the pulley consists of first inserting the bearing shaft into the bearing sleeve and lubricating it with a generous dab of lithium grease. Then attach the two frame pieces to the pulley with 8-32 stainless steel screws passed through the bearing shaft and frame spacer. I used washers under both the nuts and screw heads. Lastly, cut the two screws flush with the nuts and file off any rough edges. Give the pulley a twirl; it should spin freely. — *73, Barry Shackleford, W6YE,* **w6ye@ arrl.net**

Bow and Arrow Tips

After a number of years of using a bow and arrow to provide supports for antennas, I have found a few refinements that make

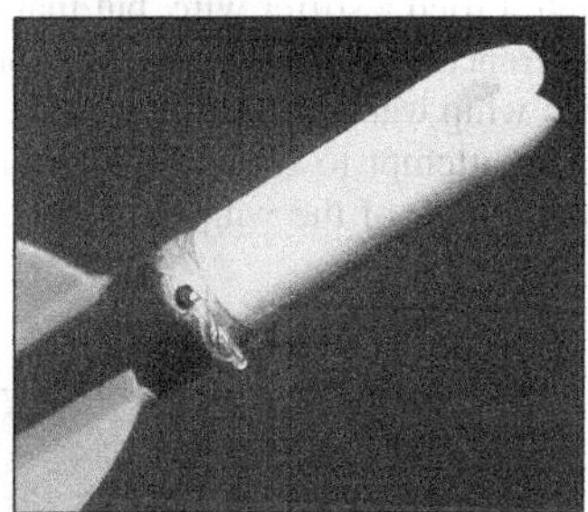

Figure 52 — Drill a hole in the base of the arrow shaft for the monofilament line. [John Lehman, WA8MHO, photo]

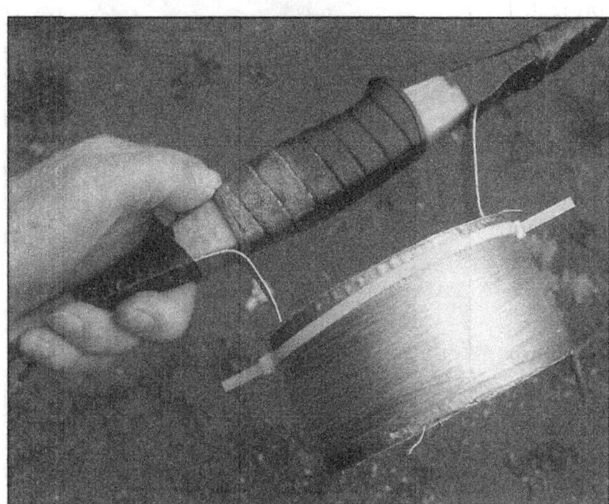

Figure 53 — I cut a piece of PVC pipe to use as a reel for the monofilament line and attached a pair of metal strips to the inside of the reel to attach it to the bow just in front of the bow's handle. [John Lehman, WA8MHO, photo]

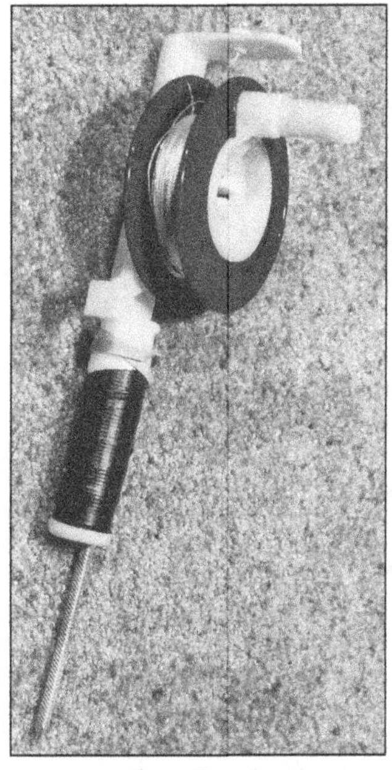

Figure 54 — A kite reel makes an excellent dispenser for the nylon cord used to lift the halyard line. [John Lehman, WA8MHO, photo]

the process easier and less frustrating. First consider the arrow: select a lightweight fiberglass arrow without a point. Drill a small hole for the monofilament line at the rear of the shaft between the feathers and the notch (see **Figure 52**), then wrap solder around the tip end to add weight. This will help to pull the arrow and line down and through the tree canopy. Some experimentation may be needed; adjust the weight so it is heavy enough to fall through the branches but not so heavy that your bow can't get it up through the tree. Once you have the weight right, cover the solder with tape.

I use a fiberglass recurve bow of at least 40 pounds; compound bows will give you less control when shooting into lower trees and send the arrow too far. A line reel or loop is attached perpendicular to the bow (see **Figure 53**). The arrow is fed through the center for shooting. I use the metal lip of one of the reel supports to keep the line from unwinding until shot.

In selecting a trajectory, I stand close to the trunk and shoot up through the canopy and out into the open area on the antenna side, this way there is less chance for the arrow to get caught in the tree.

After the line is shot to a suitable location, a heavy nylon cord will have to be pulled through, to be used to pull the halyard line. For this I use a kite string reel (see **Figure 54**) that can be found at toy stores or sporting goods stores, or fabricated using a short length of PVC pipe. I taped a spike to the kite string reel's handle so it can be stuck in the ground, and a rubber band holds the ratchet lock off when dispensing the line.

Tie the monofilament and the nylon cord together using a simple overhand bend (see "The Knots of Ham Radio"[1]). This knot is used when joining two different kinds of line. Keep a knife handy, because when pulled tight it will be difficult to untie. After pulling through the nylon cord, it is then used to pull the halyard line through the tree. Again the overhand bend knot is used, if the knot gets stuck in the tree, then some tape and a tie-around can be used to make a smoother knot. — *73, John Lehman, WA8MHO, 2553 Bell Rd, Lexington, OH 44904,* **wa8mho@ arrl.net**

Burying Radials

I have found that it's not hard to put radials in a lawn if you follow a few simple steps. The tools you'll need are a half-moon shaped edging spade, a short dandelion digger with the inside V filed smooth, a set of 6 inch landscape staples (about one for every 2 feet of radial length) and a rubber hammer for the staples. A pair of knee pads might also come in handy.

First, be sure the ground is damp before starting. Then, push the spade into the ground until the lip at the top is flush with the soil. This makes a slot about 5 inches deep. Continue cutting into the lawn for 5 – 10 feet.

Using the dandelion digger, push the radial into the slot in the ground. Push a staple in at the start of the run to secure the end. Tap the staple down as far as it will go using the rubber hammer. Stretch the wire as you go to keep it straight. Then, put a staple every 2 or 3 feet or as needed. Once you reach the end of the first slot, cut another 5 – 10 feet and repeat.

Also, I recommend using aluminum clothesline to make your own landscape staples. One foot can be bent into a 6 inch staple and will not rust in the ground like a steel staple will. — *73, Robert Dixon, W8HGH, 1560 Hawthorne Cir, Harrisonburg, VA 22802,* **w8hgh@arrl.net**

An Inexpensive Wire Antenna Winder

Our club uses wire antennas for many portable operations such as Field Day, State QSL Parties, and emergency operations. We have been looking for an inexpensive device that will allow easy winding of the wire (and

[1]Collins, WX3A, "The Knots of Ham Radio," *QST*, Jun 2006, pp 57 – 58.
[2]Przedpelski, KØABP, "RFI by Remote," *QST*, Mar 2014, p 60.

Figure 55 — A piece of board, a few dowels, and a couple of pieces of pipe can become this simple wire winding device to keep those Field Day antennas under control. [Tony Padavich, N9YPN, photo]

no sharp edges that would cause kinking), easy unrolling of the antenna, and a small storage profile. Not being able to find such a device — I made one (see **Figure 55**).

The PVC handles allow for fast and easy rolling and unrolling of the wire without causing kinks or sharp bends. It is inexpensive to make and has a small storage profile. Construction is easy. You can easily adapt the design to your needs and the only tools required are a table saw to slit the PVC pipe and a drill to make the holes for the dowels.

The wood portion is a 4 × 10½ inch piece of ¼-inch oriented strand board (OSB). The dowels are ⅜ inch diameter × 2½ inches long; four are required. The handles are ¾ inch diameter PVC, that is 7½ inches in length; two are required.

To assemble the winder, first, using a table saw, cut a slot in one wall of the PVC that is 4⅛ inches long. Push the OSB into the slot of the PVC and apply some wood glue to both sides. I used Titebond III for this project. The dowels are applied 1 inch from each side next to the PVC. Align the dowels on either side of the board and then glue them down.

Note that these dimensions are the ones I used to accommodate my club's wire antenna. Your antenna's size may dictate different dimensions. If your club uses a full wave 160 meter antenna, then the board and dowels will have to be longer to accommodate the larger volume of wire.

We have been very happy with the use of these winders. Now it is very easy during event teardown to wind the wire antennas and have them ready for the next event. No more spending time days after the event untangling and unkinking the antennas. — *73, Tony Padavich, N9YPN, 4019 28th Ave, Rock Island, IL 61201,* **n9ypn@revealed.net**

Radial Plate in a Flash

My previous vertical antennas generally had four or five radials, which resulted in

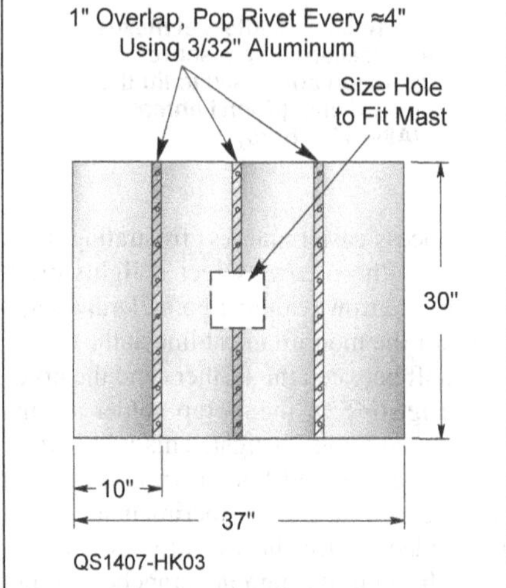

Figure 56 — Drawing of radial plate.

lackluster performance. But even with five radials, making the connection at the antenna was a problem.

For my latest vertical antenna project, I decided that I had enough space to give the antenna a large collection of radials, which I hoped would improve the performance. My old vertical didn't seem to warrant investing the money in an undoubtedly high-quality commercial radial plate. Plus, I wanted to test the installation before making a significant investment, so another solution was needed.

As I often do for antenna projects, I made a trip to the local home improvement store, where I came across 10-foot rolls of aluminum roof flashing, gauge unknown, for a very reasonable $10. I was certain I could use the flashing to make a radial plate. Tin flashing was also available, but I thought it would rust faster than the aluminum would corrode.

I cut the 10-foot length of flashing into four 30 × 10 inch strips (see **Figure 56**). These I then overlapped 1 inch and fastened the

strips together using pop rivets to form a single 37 × 30 inch plate. To obtain a good electrical connection you must use the smallest rivets available, and plenty of them, as the flashing is rather thin. Testing of the plate at multiple points and edges showed good conductivity.

To finish the plate, holes were drilled around the edges for #8 hardware. I used brass machine screws, washers, and nuts to secure to the plate and wing nuts to connect the radial ring lugs. Finally, I cut a hole in the center large enough to accommodate the antenna's mounting mast.

How does it work? I have 30 radials connected (10 × 30 feet and 20 × 18 feet) and, after 6 months, this vertical is working like an antenna rather than a "Cantenna." I am planning on installing more radials for 80 meters; the plate has plenty of space around the edges. The total cost of the flashing and screws was less than $20. Given its performance, I think this radial plate may be permanent. — *73, Lionel Booth, N5LB, 3226 Wellspring Lake Dr, Fulshear, TX 77441,* **n5lb@arrl.net**

Whistling Whip

I installed a new 220 MHz mobile whip and was immediately bothered by a very high pitched whistle when I drove over 40 mph. I tried a stiffer wire, but that made no difference. Something about the roughly 13 inch whip wanted to make noise.

In an attempt to change the resonance (audible type!) of the whip, I added a cork

Figure 57 — Adding a cork to the tip of a mobile whip will eliminate any annoying sonic resonance. [David Kaplan, WA1OUI, photo]

to the top (see **Figure 57**), and my problem immediately disappeared. This has proven to be a cheap way to fix the problem and it was fun getting the cork, too. — *73, David Kaplan, WA1OUI, 36 Drumlin Rd, West Simsbury, CT 06092,* **wa1oui@arrl.net**

Recycled Wire Spreaders

With Christmas over, I was left with a variety of packaging materials. While looking at these leftovers, I noticed some plastic strips that had holes punched in them, which had been used to secure toys in their packages.

I had the idea that the plastic strips could be reused as parallel line spreaders, to build a linear loaded or folded dipole, or even as central support or end insulators for a lightweight dipole.

I decided to use them to construct a linear loaded dipole for the 60 meter band. I used cable ties to keep the wires in place, although a shot of hot-melt glue would serve as well. — *73, Angel Pelaez, EA4DUT, PO Box 49002, 28080 Madrid, Spain,* **ea4dut@gmail.com**

Tensioner Weight Control

Whether they are used for emergency communications or just a fun day of hamming, our wire antennas need to stay up. Trees will sway in the wind and no wire will stop your green supports from swinging in opposite directions. So, we use pulleys; running the support rope or cable through a pulley and down to a weight has been the solution for many years. However, this solu-

Figure 58 — A section of plastic pipe can be cut and glued to form an antenna counterweight, which is then attached to a section of mast to keep it under control. [Patrick Hamel, W5THT, photo]

tion has its own problem. I once saw two cement blocks that were being used to tension an antenna. The blocks were hanging at the end of a rope and doing a good imitation of a wrecking ball as they swung against the tree. It seemed to me a better solution was needed.

After considering the issue, I came up with a simple solution — plastic pipe. I took a piece of pipe about twice the diameter of the antenna support mast. I cut two 1-inch collars off one end and sealed the other end.

Next, I cut the collars in half. I mounted two of the half-moon pieces to the tube using pipe cement and screws, and then I positioned the support mast in the half-moon and roped it in place. I glued the other half-moon pieces to those mounted on the pipe, waited for the glue to harden, and removed the rope. Finally, I added enough weight to the inside of the pipe to provide the correct amount of tension to the antenna wire.

Figure 58 shows the tensioner at one end of my 630 meter inverted **L** antenna. The tree at the other end will swing a few feet in a storm, but (so far) not so much that it pulls the pipe up to the guy ring at 10 feet. The wind drag is minimal.

If you have a house, fence, or something else on the end of the antenna you want to protect, this method will keep those weights under control. — *73, Patrick Hamel, W5THT and WD2XSH/6, 1157 East Old Pass Rd, Long Beach, MS 39560,* **w5tht@arrl.net**.

2 Meter Vehicle Vertical Beam

Two easily made and quickly attached magnetic whips instantly turn a 19-inch rooftop vehicle whip into a high-gain three-element beam. Use it as a simple foxhunting antenna, or to reach a distant repeater in an emergency situation. Keep these add-on whips in your vehicle, and when needed, place the longer (20-inch) whip behind your 19-inch whip as a reflector and the shorter (18-inch) whip in front as a director, spaced 16 inches element-to-element. Voila, a low-profile vertical Yagi for your mobile.

Small, round $\frac{3}{4} \times \frac{1}{8}$ inch neodymium super magnets, with countersunk holes for flat-head screws, are the secret to easy construction. Use $\frac{1}{2}$ inch 10-24 stainless or

Figure 59 — The whip bases are made from neodymium magnets, 10-24 screws and nuts, a crimp lug, and the whip element itself. [Ernie Sloan, W6ND, photo]

brass screws and nuts together with a #10 crimp-on screw-eye lug to attach the whip to the magnet (see **Figure 59**). I obtained the magnets from eBay, but there are other Internet sources. The whips do not need to be physically attached to the vehicle's body; capacitive coupling makes the connection.

I used $\frac{1}{16}$ inch stainless steel whips. With flux and adequate heat, stainless can be soldered. Steel piano wire is also satisfactory and solders more easily. Solid soft copper wire will bend in the wind.

The whip dimensions are from the tip of the whip to the surface of the vehicle. The recommended spacing of 16 inches yields a $0.2\,\lambda$ spacing, which is a near-optimum Yagi design. I modeled it in *EZNEC* for a front-to-back ratio of 19 dBi for foxhunting. Forward gain is 11 dBi. Horizontal beamwidth is 98°. Adding these new elements will introduce some mismatch, but at this spacing the loss is small and the SWR adequate for modern transceivers. As a final touch, I added a single layer of electrical tape to the bottom of each magnet to preserve the vehicle's finish.

With the vehicle parked and pointed at a distant repeater, these simple accessory whips could someday be a lifesaver. For a little extra fun, I can add an attenuator and join a local 2 meter foxhunt. — *73, Ernie Sloan, W6ND, 5032 Michael St, Santa Maria, CA 93455,* **w6nd@arrl.net**

Mobile Antenna Carrying Case

On a trip I like to bring along several hamstick-type mobile antennas. During the day the 20 or 15 meter bands are great when they are open, and a short mobile antenna can be quite efficient on these frequencies. At night, 40 meters is my choice.

A disassembled hamstick-type antenna is a little over 4 feet long. There are two parts: the helically loaded base and a stainless steel whip. I made a tubular carrying case that will hold several such antennas. I can stow the tube with antennas underneath other luggage in my hatchback car and access an antenna easily without removing everything else.

Inexpensive mailing tubes with a 3-inch diameter are available in 3-foot lengths at any stationery store. To make my carrier, I bought two of them and cut a 16-inch section off the end of one to make an extension for the other. I butted the short piece and the full-length tube together, and wrapped a piece of cardboard about 6 inches wide around the joint to make a union, overlapping each mailing tube piece by about 3 inches. I used duct tape to hold the union together and attach it to the end of the longer tube. I then pulled out the short tube and cut two notches about an inch long into the open end so I could squeeze the short tube's end slightly, to make it easier to slide it in and out of the union attached to the long tube. I used more duct tape to permanently fasten

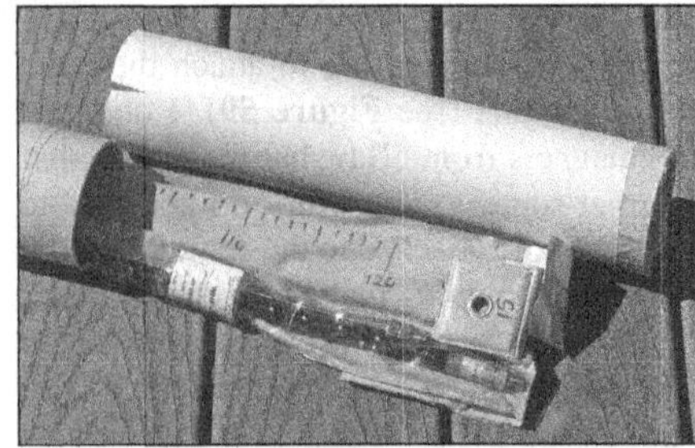

Figure 60 — A couple of inexpensive mailing tubes, appropriately modified, makes a convenient way to store and transport mobile/portable antenna elements. [Al Woodhull, N1AW, photo]

Figure 61 — The coax connector plate made from aluminum plates welded together at a 60° angle. [Dennis Klipa, N8ERF, photo]

Figure 62 — Mounted just below the top of the tower with U-bolts, the coax connector plate provides a convenient transition place to change from stiff to flexible types of coax. [Dennis Klipa, N8ERF, photo]

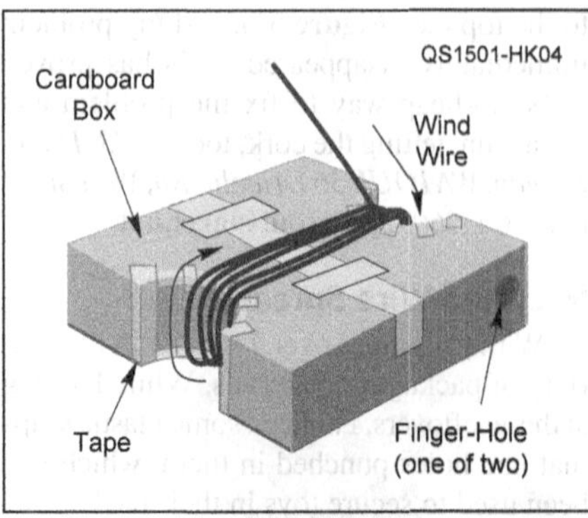

Figure 63 — A cardboard box can be easily converted into a wire antenna holder that will prevent tangles when storing or transporting the antenna to its mounting location. [Arthur McAlister, KD6SF, photo]

the plastic plugs in the ends of both pieces.

This makes a good 4-foot, 4-inch storage and transportation container that can hold as many as five antennas in their plastic sleeves. The sleeves keep the bases and the whips for each band together. I label each sleeve with a marker to identify its band. Referring to **Figure 60**, you can see that I have also used a marker to make a 120 centimeter ruler on the backs of some of the plastic sleeves. This is to measure the extension of the whip so I can quickly readjust the frequency from the CW to the SSB end of a band.

My antenna case is easy to handle. This is not something rugged enough for checked luggage on a plane, but it is a handy way to carry a set of antennas in a car. In my old Civic wagon I was able to use bungee cords to fasten the antenna tube to the front seat overhead grab handle and the rear seat coat hanger hook. In my current car, that doesn't work but I can put the antenna tube under my luggage and access it from the rear hatch. — *73, Al Woodhull, N1AW, 199 Eden Tr, Leyden, MA 01337,* **n1aw@arrl.net**

Coax Connector Plate

While assembling a tower and antenna system for our new radio club, John, W8QN, suggested that we should install a plate at the top of the tower where we could terminate our LMR-600 and LMR-400 coax runs. This would allow us to run more flexible coax to the antennas. The loss in short runs of Flex LMR-400 would be acceptable at UHF and VHF and would prevent damage to the stiffer LMR-600 from rotator motion. This setup would also permit us to ground the shield of the coax to the top of the tower.

We planned five runs of coax to the top of the tower, two LMR-600 and three LMR-400. With that in mind, I designed the plate shown in **Figure 61**. I made the plate from ⅛ inch aluminum and MIG welded them together with an internal angle of 60°. The plate is mounted to the tower with ¼ inch stainless steel **U**-bolts that are attached to the horizontal portion of the **Z**-brace on the tower (see **Figure 62**).

The ½ inch diameter, N-type, chassis mounted barrel connectors have two flat sides so that they will not twist when you are tightening your coax connector. The holes for the connectors were drilled slightly undersized and then filed to enlarge them, leaving corresponding flat sides in the holes. Finally I added a couple of stainless steel eye bolts to which I could attach pulleys in case we wanted to add some wire antennas to the tower at a later date. — *73, Dennis Klipa, N8ERF, 644 E Whitethorn Dr, Midland, MI 48640,* **n8erf@arrl.net**

Antenna Wire Winder

I had an idea to help simplify the task of winding wire antennas without being troubled by kinky antenna wire. It's been a help to me over the years to keep the antenna from wrapping itself around my legs while I'm climbing a ladder to hoist them in place. One would think that a wire antenna would be coiled up on spools, but homebrew antennas typically have traps or coils that don't take well to being bent.

This winder is made from a cardboard carton that measures about 1 foot square and about 4 inches deep. To make the winder, simply use a sharp knife or a large pair of scissors and cut two **V**-shaped notches, one on each side of the carton; such that there are two notches on opposite sides (see **Figure 63**). The width of the **V** should be about 4 or 5 inches.

Using this tool you can coil up your antenna, leaving the traps and coils just hanging on the sides of the carton as you climb the ladder. When starting the winding process, start at the end-insulator and work yourself toward the center of the antenna (for a dipole configuration). In use, the antenna wire tends to stay pretty much grouped together. The end-insulators can be stuffed into an open end of the carton. I find it helpful to use some duct tape to temporarily hold the wire in place until you're prepared to unwind it.

To assist winding and unwinding, punch some finger holes in the ends of the carton. Using those holes, you can wind or unwind the wire just as if it is on a windlass.

Once in place, you and your helper can unwind the wire. Start at its center and, using the finger holes, rotate the carton and the antenna wire will fall off your "spool" as you (and your helper) move toward the mounting masts. It takes longer to describe the use of the carton than it takes to prep your "spool." — *73, Arthur McAlister, KD6SF, 7570 Dartmouth Ave, Rancho Cucamonga, CA 91730-1534.*

Antenna Stretcher

Need some extra inches on that wire dipole or long wire antenna? Rather than hauling it down and putting it back up several times, build one or two of these handy, inexpensive, "one-time" antenna stretchers (see

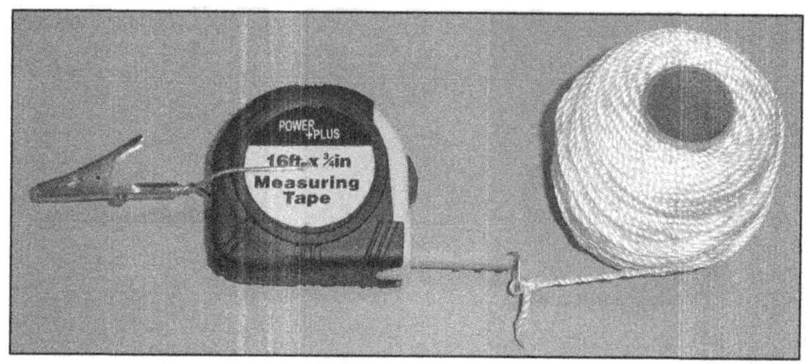

Figure 64 — A steel measuring tape and an alligator clip make a handy device to help resonate that wire antenna. [Jim Bailey, W6OEK, photo]

Figure 64). Just lower the antenna, attach the alligator clip to the end and raise the antenna back to its normal height. Note: the string and tape measure do not support the antenna; they just hang on the ends. Next, use the strings to pull out tape until you achieve the tuning you want. The length of wire that you will need to add to the antenna will be the same as the tape measure reading plus the fixture length. Read the tape measure with a pair of binoculars. [The self-locking type of tape measure would be a good choice also. — *Ed.*]

Use the ubiquitous dollar-store ½-inch bargain tape measures to make the stretchers. Smaller-width, lightweight tape measures are best. It is not necessary to take them apart. The inner end of the internal winding spring is secured by a slot in the hollow center pin. Simply remove the belt clip and drill through the mounting hole until the bit comes out the label side. Then insert a 9-inch length of 14 AWG solid wire through the hole and bend it back along the sides of the case. Twist the ends together and solder on an alligator clip. Verify that the wire makes solid contact with the metal spring and tape by a continuity check between the alligator clip and the pull-out end of the tape. Finally, attach a length of lightweight pull cord to the tape's end cleat. — *73, Jim Bailey, W6OEK, 1836 Sequoia Dr, Santa Maria, CA 93454,* **w6oek@arrl.net**

Converting Camouflage Spreaders

While at a local hamfest, I happened upon a set of fiberglass camouflage poles in a green duffel bag. The parts were originally designed to support various kinds of camouflage netting. The price was too good to pass up, so I bought them, thinking they would make an excellent portable antenna mast.

At home, I opened the bag and inspected the contents. Along with the poles came a strange object that looked like a four-armed paddle when it was opened up. It actually consisted of four separate arms, connected together at the center by a coupling nut. "I know what to do with the poles," I thought, "but what are the paddles for?" It turns out they are for making antenna wire spreaders and guy rings.

The round end of the paddles is the same size as the $6 guy rings sold for use with the fiberglass masts. All you need to do is cut the round part off the end of the paddle, mark the center, and drill out the holes to get a guy ring. I copied a previously purchased ring so it was easy to duplicate (see **Figure 65**).

Next, drill a ⅜-inch hole at the end of one of the leftover paddle arms. Then drill two ¼-inch holes through the arm at a point close to the large mounting hole (see **Figure 66**). Use this paddle arm as a pattern so you can make the holes in the remaining arms identical. Once you have drilled the holes, place the arms on the coupling nut and configure them in a **T** or cross. Drop a cotter pin or ¼-inch bolt through the ¼-inch holes to keep the arms from pivoting, and tighten the coupling nut.

You can string the wire though the spreaders and make an end-fed antenna or a center-fed dipole. Spreading the wire into an elongated loop in this way improves the antenna's bandwidth (see **Figure 67**).

The wire I have on this 75 meter antenna is about 190 feet long, but I am only occupying 96 feet of space. My tuner has no problem finding a match. — *73, Stephen Bellner, W8TER, 6853 Deer Ridge Rd, Apt 6, Maumee, OH 43537,* **w8ter@arrl.net**

Powerpole Rotator Upgrade

I purchased a used Hy-Gain rotator, which was old enough to have terminal strip connectors on the back of the controller and the rotator housing. I wasn't too pleased with the idea of loosening eight screws each time I needed to disconnect the controller for an impending thunderstorm, then re-cabling the terminal strip after it passed.

I looked around for a better way and

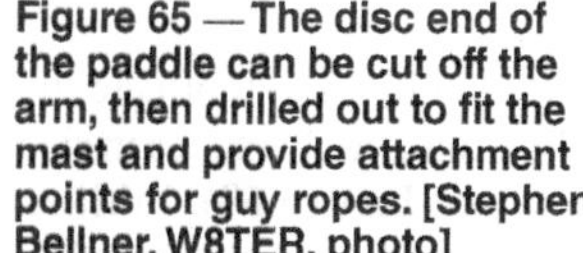

Figure 65 — The disc end of the paddle can be cut off the arm, then drilled out to fit the mast and provide attachment points for guy ropes. [Stephen Bellner, W8TER, photo]

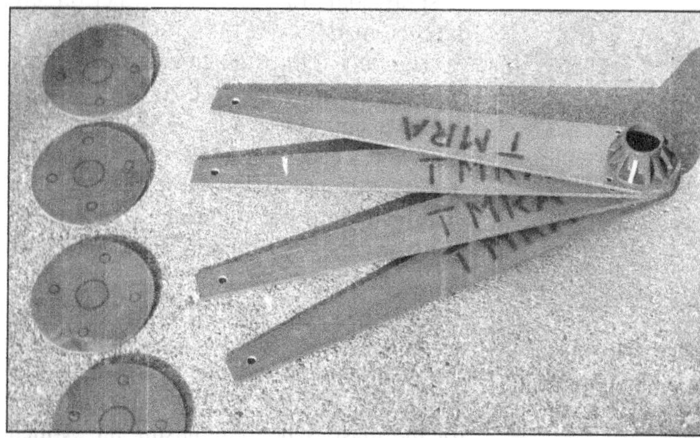

Figure 66 — After the disc is cut off of the arm, drill a large hole in the arm's disc end to support the antenna wire, and two smaller holes in the coupling nut end to lock the arm configuration. [Stephen Bellner, W8TER, photo]

Figure 67 — Three of the arms arranged in a T formation make great spreaders for dipole or loop antennas. [Stephen Bellner, W8TER, photo]

Figure 68 — The final result of the conversion yields a quick-disconnect arrangement with the added advantage of color-coding of the various lines, which will be helpful in any future troubleshooting. [James Philopena, KB1NXE, photo]

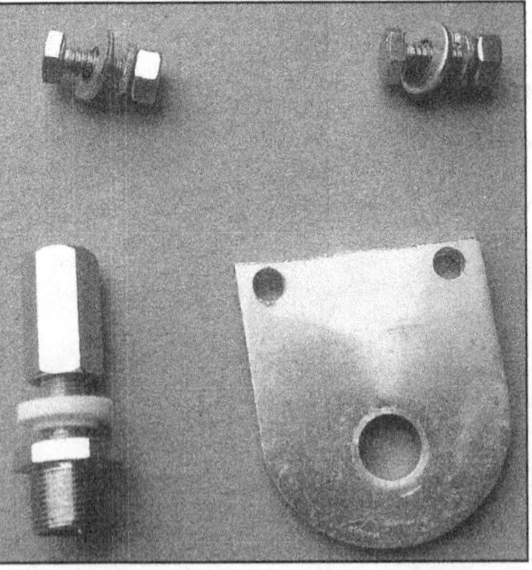

Figure 69 — This home-made bracket and a couple of stainless steel screws make a sturdy base support for a mobile antenna. [John H. Lehman, WA8MHO, photo]

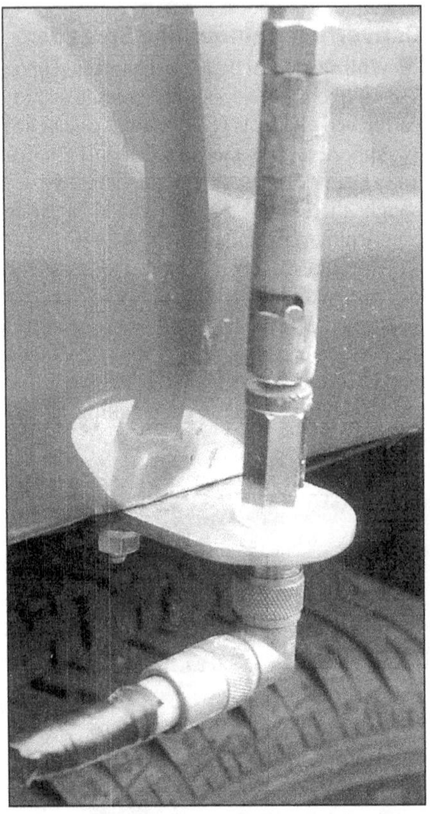

Figure 70 — Attach the bracket to the wheel-well lip using two stainless steel screws. When mounting the antenna connector, don't forget to install the step washer to insulate the antenna from the car's ground. [John H. Lehman, WA8MHO, photo]

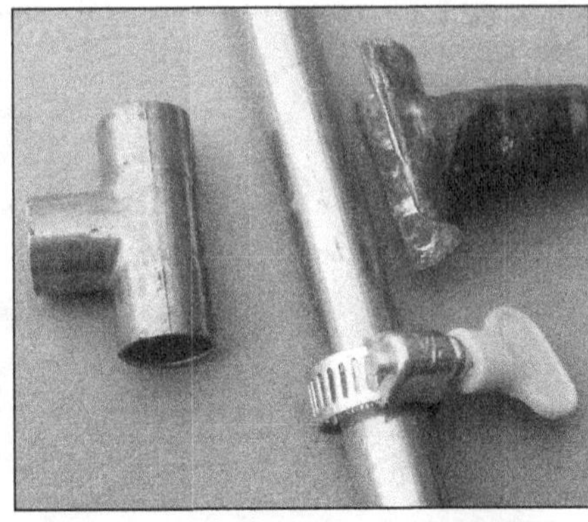

Figure 71 — A plumbing T cut lengthwise and secured to the antenna with a hose clamp provides an attachment point for a horizontal support. [John H. Lehman, WA8MHO, photo]

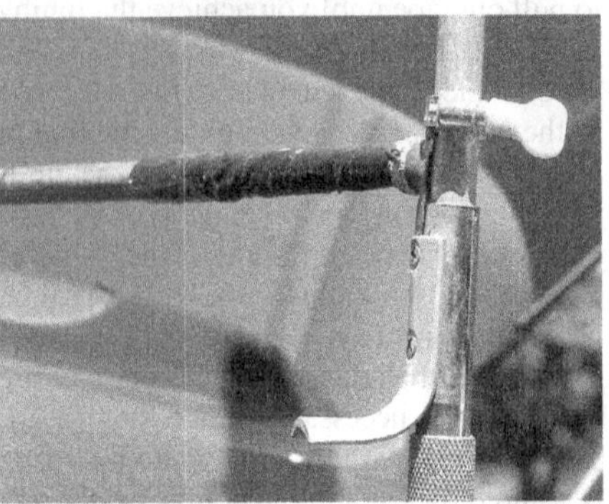

Figure 72 — The upper support is made from a cut-off plumbing T. A fiberglass shaft is attached to the other side of the T with a screw and tape to support the upper part of the antenna. [John H. Lehman, WA8MHO, photo]

found a solution in Anderson Powerpoles. Several vendors have multiple color connector shells, as well as latching panel receptacles for ganged Powerpole connectors. I used an eight-pole panel receptacle with the corresponding latching plug housing. I also ordered multiple colors of the Powerpole connector shells to correspond to the standard wire colors in readily available eight-conductor rotator cable.

The panel receptacle didn't fit the hole left after the removal of the controller's terminal strip. I used a nibbler tool to modify the hole and then snapped the receptacle in place. I soldered (crimping is fine as well) the eight Powerpole contacts to the cables and then inserted them into the correct color Powerpole shell without changing the pinout. The end of the cable from the rotator was terminated in the same manner and placed into the latching housing (see **Figure 68**).

Now I have a perfectly modern, easy-to-disconnect rotator termination for my older Hy-Gain controller. If you want to take it to the next level, you can always create a Powerpole connector to ground the connections just as easily. Disconnect the housing from the controller and then attach it to a grounded eight-Powerpole receptacle. In less than 5 seconds, your rotator cable is grounded. — *73, James Philopena, KB1NXE, 265 Frost Hill Rd, Marlborough, NH 03455,* **kb1nxe@arrl.net**

Mobile Antenna Lip Mount

Need to mount a mobile antenna? If the body panels around the rear wheel wells of your vehicle are metal and not plastic, you can use this type of mount for a very stable installation. The body panel is usually bent over at a right angle around the wheel well

to form a lip. This lip is a good location to mount an antenna.

I used aircraft aluminum to fabricate the bracket, which is about $2 \times 2\frac{1}{2}$ inches in size and $\frac{1}{4}$-inch thick (see **Figure 69**). I drilled two holes 1 inch apart in the bracket, another two in the lip to match, and rounded off the exposed corners for safety. [After drilling the holes in the wheel well lip, you should paint or spray some rust-proofing primer on any exposed metal — *Ed.*]

The only critical dimension is the large hole for the antenna base. It is $\frac{1}{2}$ inch in diameter and must be checked for fit with the nylon washer provided with the mount. Note that the nylon stepped washer has a lip that fits into the hole to insulate the antenna from the car body's ground while an electrical ground is provided by the coax shield. The bayonet male part goes on top and the coax fitting on the underside (see **Figure 70**).

Most hamfests will be a source for the antenna base mount, the nut, and the bayonet for the mast. When purchasing the $\frac{1}{4}$-inch hardware, be sure to buy stainless steel parts.

The type of heavy mobile antenna this mount is designed for will also need an upper support. The upper support is fabricated from

Figure 73 — When using larger coils, it is helpful to add a guy line from the base of the coil to the roof rail or other attachment point. [John H. Lehman, WA8MHO, photo]

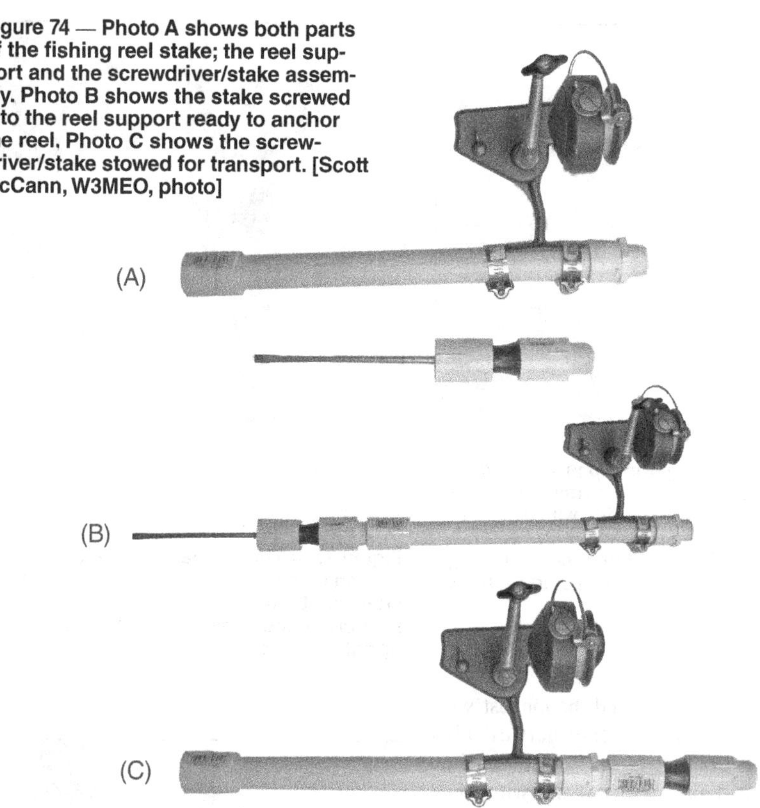

Figure 74 — Photo A shows both parts of the fishing reel stake; the reel support and the screwdriver/stake assembly. Photo B shows the stake screwed into the reel support ready to anchor the reel. Photo C shows the screwdriver/stake stowed for transport. [Scott McCann, W3MEO, photo]

a ⅜-inch copper plumbing **T** cut lengthwise to fit against and partly around the antenna mast (see **Figure 71**). It is secured by a small turn-key type hose clamp, which eliminates the need for a screwdriver. The bottom of the **T** is drilled through and a screw secures it to a small fiberglass rod, which is then anchored to the crossbar of the roof rail (see **Figure 72**).

If one of the large resonators is used, a nylon cord from the mast just below the resonator to the front of the roof rail is a good idea (see **Figure 73**). In my installation, the elbow joint in the mast is properly positioned for the top part of the antenna to lie down on the roof when putting the car in the garage. When installing this mount on your vehicle, some planning regarding any fold-down features would be prudent. — *73, John H. Lehman, WA8MHO, 2553 Bell Rd, Lexington, OH 44904,* **wa8mho@arrl.net**

Fishing Reel Support Stake

I concocted this device as an aid to raising wire antennas. Whenever I went to shoot fishing line into a tree, there was always a problem with supporting the fishing reel. I finally came up with this simple solution.

Referring to **Figure 74**, photo A shows a 10-inch length of ½-inch PVC pipe with a male adapter cemented to one end and a female adapter cemented to the other. A spinning reel is mounted by two hose clamps. Just below this is the ground stake section, which consists of a screwdriver to which male and female adapters have been epoxied to both sides of the handle.

The screwdriver is used to tighten the hose clamps when mounting the reel. With the reel mounted, the screwdriver is screwed into the female adapter of the pipe (see photo B) to make a ground stake for holding the reel. This makes it easy to throw or slingshot a sinker weight over a tree limb at Field Day or for any portable operation. I use ordinary fishing monofilament line on the reel; this is strong enough to pull up a heavier line or can be used as a halyard for a light portable antenna. The monofilament is plastic, which means no insulator is needed for the antenna and it is stealthy.

When the antenna is in the air, packing up is easy. The screwdriver screws into the male end of the pipe (see photo C) to stow the blade for transport. — *73, Scott McCann, W3MEO, 160 Shields Ln, Queenstown, MD 21658,* **w3meo@arrl.net**

Sweet Saltwater Ground

I came up with an unusual maritime mobile setup. The problem with any maritime mobile station is that you need to somehow connect your equipment to the boat's ground via a low impedance copper foil or braid. This is needed for a counterpoise or ground plane for your antenna. In effect, you are creating an untuned vertical antenna with a very large (sea water) ground plane. Locating this ground and installing the braid is not a trivial task, especially when all you want to do is operate from the dock or anchorage.

Figure 75 — The antenna side of the coax feed line is split into shield and center conductor just after the balun. The center conductor is connected to the antenna wire at the plastic insulator. [Greg Charvat, N8ZRY, photo]

Figure 76— An alligator clip is soldered to the end of the shield. To operate, the clip is attached to the swim ladder, which, when lowered, provides an excellent saltwater ground. Note the use of heat-shrink tubing on the coax to seal the end. [Greg Charvat, N8ZRY, photo]

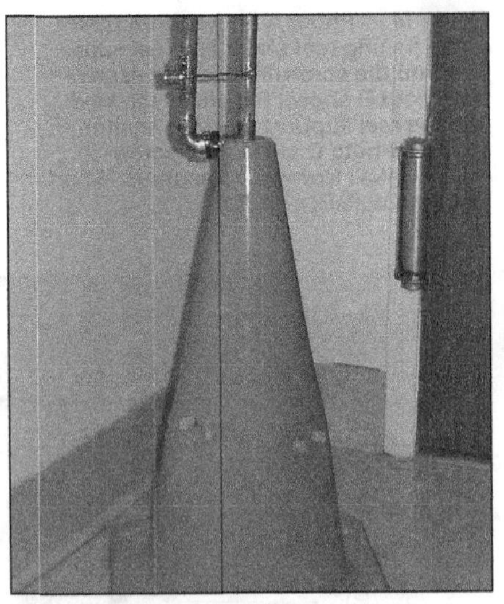

Figure 77 — The two pairs of wooden dowels can be seen poking out of the cone's side about halfway up. These form a support inside the cone to keep the antenna upright. [Joseph Tavares, KC1BGF, photo]

Figure 79 — When working with bundles of coax, this little gimmick can help identify which end is which. [Lynn Burlingame, N7CFO, photo]

For the antenna, I used the longest wire I could raise with my flag halyard also employing insulators on both ends. For the feed line, I made a shunt balun with some type 43 ferrite cores. The antenna was fed by a 7 – 8 foot run of RG-58, which is just long enough to get into the cabin. An old MFJ antenna tuner rounds out the system.

For the ground, I had to improvise. I soldered an alligator clip to the RG-58 shield coming off the balun. I then lowered the stainless steel swim ladder and connected the alligator clip to the ladder to form a counterpoise with the sea water (see **Figures 75** and **76**).

This setup worked very well. I was getting into Denmark and Poland on 20 meters with a 59 signal while operating from Long Island Sound. These great reports were obtained without having to dig through the boat and find a place to attach a ground or install an external antenna tuner.

This would make for an excellent Field Day/emergency operating location. This could be a solution for other sailors out there who do not want to go to the trouble and expense of permanently installing HF gear. — *73, Greg Charvat, N8ZRY, 630 Old Clinton Rd, Westbrook, CT 06498,* **charvatg@gmail.com**

Quick J-pole Mount

I needed to find a way to mount my J-pole antenna. I live in an apartment so I didn't have a lot of room. I made a trip to a local hardware store and noticed some orange plastic safety cones, which gave me an idea. I bought a cone, a dowel, and cable ties. I cut the dowel into four pieces and mounted them in the cone as two parallel dowels at right angles to the second pair to hold the bottom of the antenna (see **Figure 77**). I cut a couple of slits in the top of the

Figure 78 — A flap is cut out of the top of the cone and the "J" of the antenna is attached to the flap with cable ties. [Joseph Tavares, KC1BGF, photo]

cone to form a flap and bent the flap down to a right angle. I used the cable ties to secure the J part of the antenna to the flap on the cone (see **Figure 78**). It works very well and has a small footprint. The mount is all non-metallic, so it doesn't affect the tuning of the antenna. — *73, Joseph Tavares, KC1BGF, 850 Pleasant St, Apt 323, New Bedford. MA 02740,* **unijat2@verizon.net**

Lost in the Bundle

Have you ever installed a bundle of coaxial cable and lost track of which cable was which? Here is an easy way to identify a cable. Short out an SO-239 panel mount by connecting the center connector to the ground side as shown in **Figure 79**. Install the SO-239 one end of the cable in question and then check the continuity on the other end of the cable with a VOM or tone tester. — *73, Lynn Burlingame, N7CFO, 15621 SE 26th St, Bellevue, WA 98008,* **n7cfo@n7cfo.com**.

Potato Gun Line Controller

After building a pneumatic tennis ball antenna launcher, I quickly discovered that feeding the initial line at high velocity without tangling is a problem. Retrieving and rewinding the line can also be troublesome. This simple line feed adapter provides an inexpensive solution to both problems.

The basis is a heavy plastic 5-inch flowerpot found in your local garden store. It's about ¼-inch thick, has smooth, undecorated sides, and no drip tray. The important dimension is the base diameter, which should be at least an inch larger than the launcher's tube diameter. I

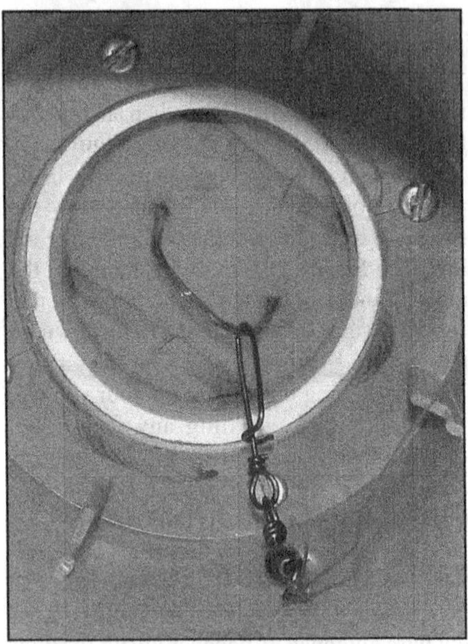

Figure 80 — This is an inside view of the flowerpot. The flowerpot is screwed to the wooden base. A tennis ball is seen in the barrel. The fishing line passes from the ball to the paper clip, which keeps the line from unraveling. [Mike Pittaro, KJ6VCP, photo]

Figure 81 — The remainder of the fishing line is wound around the base of the flower pot. When fired, the line feeds smoothly off the pot. [Mike Pittaro, KJ6VCP, photo]

Figure 82 — A chain wrapped around a recalcitrant ground rod and anchored with a vice grip will provide a solid hold on the rod to jack it out of the ground. [Shanon Herron, KA8SPW, photo]

used 2-inch PVC for my launcher, so a 5-inch pot was perfect.

I made the base out of a piece of ¾-inch plywood that was about 6 inches square. I drilled a hole in the center of the plywood with an adjustable hole saw, just large enough to provide a snug fit on the outside of the launcher tube. I also rounded the corners of the plywood base.

On the bottom of the flowerpot, mark the center and concentric points for four or five mounting screws. Use the hole saw to drill out the flowerpot center to the same size as the tube or slightly larger. Then drill clearance holes for the mounting screws (see **Figure 80**). I used a scrap piece of 2 × 4 for support inside the pot while drilling from the bottom.

Center the pot on the plywood and attach it with some ½- or ¾-inch sheet metal screws. A bead of hot-melt glue between the pot and the plywood prevents the line from snagging in the small gap. A folded paper clip epoxyed inside the pot provides a convenient anchor to keep the line from unwinding before launch. Tie the end of the launching line to the base of the pot, wind up some line, and you're ready to launch (see **Figure 81**).

After the tennis ball has been launched, the adapter is easily removed and provides a rigid handheld winder to help pull the line back in. — *73, Mike Pittaro, KJ6VCP, PO Box 507, La Honda, CA 94020,* **kj6vcp@ arrl.net**

Ground Rod Removal Tip

Field Day is over and it's time to break down. It all goes smoothly until you get to the ground rod. It was hard to drive into the ground, so now how do you get it out?

If you have a short length of chain, a pair of locking pliers, and a jack, there is a neat trick to pull that rod out of the ground. Use the locking pliers to secure the first link to the ground rod (see **Figure 82**). Wrap the chain link around the rod, arranging each link so it lies flat against the rod. After wrapping it around three or four times, attach the end to the jack. When the jack

is ratcheted up, the links will tighten against the rod and the friction will enable the rod to be quickly raised with minimal effort. Note that the chain will cause the rod to rotate slightly until the vice grip is stopped by the jack, so position them next to each other. — *73, Shanon Herron, KA8SPW, 5736 Lathers, Garden City, MI 48135,* **ka8spw@arrl.net**

Vertical Antenna Truck Mount

Mounting a vertical antenna on a truck can be a real challenge, because until now there just wasn't any good location to mount it. Mounting it to the bed wasn't a good option because most of today's truck beds are insulated from the frame. If you mounted the antenna vertically, you stood a good chance of breaking it off on a low-hanging tree. Alternately, you could take it down and remount it every time you wanted to use it, but this is inconvenient. Mounting it to the bumper would interfere with the tailgate or loading/ unloading activity. After some consideration, I came up with this simple and relatively inexpensive solution (see **Figure 83**).

Using this method, a vertical antenna can be mounted safely and cheaply with nothing more than a wrench and a quick turn of a nut to put it up using three simple and readily available antenna brackets.

Most trucks have what are called stake holes in the side of the pickup bed. I inserted a piece of square pipe into the stake hole. I anchored the pipe with screws through the truck stake holder from inside the bed, and then cut the pipe to a height just below the top of the truck.

To mount the antenna to the pipe, I went to the local CB shop and bought three antenna holders (see **Figure 84**). Two of the three were designed to mount on the horizontal piece of a truck body and one was a 90°

Figure 83 — A quick, inexpensive drop-down mount arrangement that allows you to mount a vertical antenna on your truck. [Mayor Nelson, KD5LBE, photo]

Figure 84 — This arrangement of three CB-style mounting brackets makes for a simple drop-down mount for your vertical antenna. [Mayor Nelson, KD5LBE, photo]

holder. One of the horizontal types was mounted to the vertical pipe in the stake hole. Because the truck body is insulated from the truck frame, I added a piece of ribbon cable and a piece of heavy ground wire, connecting both to the frame at a clean ground location. I wrapped the ribbon cable with electrical tape so it would not rub on the truck bed or cab.

The second horizontal-type holder is mounted to the first through the antenna mounting holes. The 90° holder is connected to the other holder with a single $\frac{7}{16}$-inch bolt that comes with the holders. Loosen the bolt and pull the antenna straight up or at any angle and tighten the bolt to hold it firmly in place. Loosen the bolt again to lay the antenna down to any angle, and then tighten the bolt to secure it. When in the lowered position, I made a little loop of baling twine to secure the end and prevent it from bouncing around. One wrench is all that's needed to raise or lower the antenna, quick and easy. — *73, Mayor Nelson, KD5LBE, 8 Deerwood Dr, Morrilton, AR 72110-4416,* **kd5lbe@arrl.net**

Clothespin Spreaders

I designed a fan dipole for 40 and 20 meters. I needed a way to support the 20 meter elements underneath the 40 meter wires. For a little more than $3, I found a pack of 36 plastic clothespins in the housewares section of a local department store.

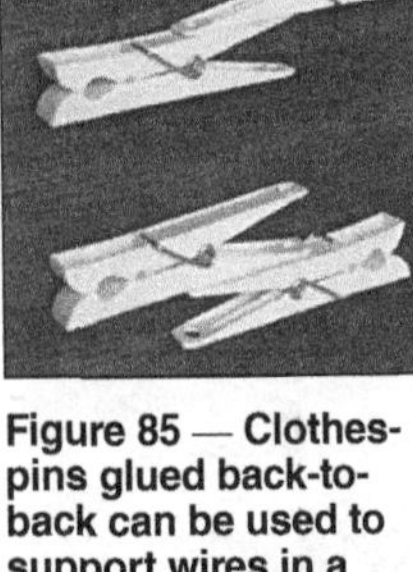

Figure 85 — Clothespins glued back-to-back can be used to support wires in a fan dipole or as clip-on anchors for other applications. [Bob Haynes, WB4AKA, photo]

I needed three spreaders for each 20 meter element and there were just enough white clothespins in the package to do the trick. Silicone glue was used to fasten six pairs of clothespins, back-to-back (see **Figure 85**). I used the white clothespins hoping to minimize complaints from neighbors because white is not a very noticeable color. The clothespins should be inspected yearly for signs of deterioration. — *73, Bob Haynes, WB4AKA, 4316 Lithia Pinecrest Rd, Valrico, FL 33596,* **rhaynes5@ tampabay.rr.com**

Rotator Wiring

In order for a rotator to work, a cable of four, six, eight, or even 10 wires is needed to control the positioning of the rotator's head. When wiring up the rotator, using the resistor color code (see **Table 1**) to match wires to connection points simplifies the installation. For example, use brown, red, orange, yellow, green, blue, and violet, and gray for connections 1 through 8 on an eight-wire rotator. This eliminates having to install wire numbers on each end of the cable or needing a "cheat sheet" of the rotator connections, which might become lost, necessitating a trip up the tower. Note that heavy-duty rotators usually require a pair of larger gauge wires to supply brake current. These brake wires are usually colored black and white. Their larger size makes them easy to distinguish in the cable, so they don't need a color-code designation. — *73, Terry Webb, NØTW, 103 Fulford Rd, Monticello, FL 32344,* **n0tw@ arrl.net**

Thrust Bearing Template

I purchased a Yaesu GS-065 thrust bearing to replace the one on my Hy-Gain HG-54HD tower. The existing thrust bearing was triangular with three mounting bolts. The Yaesu bearing has four mounting holes, which required the drilling of four new holes into the top plate of the tower. I found that Yaesu doesn't provide a template for the bearing. I searched everywhere and checked several ham blogs. No one has seen any templates for it. I tried to make a template myself without success.

Then I had a thought; I placed the bearing on my printer/scanner, and printed a picture of the base. I was amazed at how good the printed copy was. I used a ruler to verify that the spacing on the picture was the same as the spacing on the bearing. It matched up perfectly.

I decided to go a step further. I got a thermal laminating pouch used to laminate papers and used one of the two laminating pages. The outside is smooth plastic and the other side had a matte finish. I cut it to $8\frac{1}{2} \times 11$ inches and placed it in the printer.

Figure 86 — The photo-template after the drilling was completed. [Joe Seibel, WA3SRU, photo]

Figure 87 — The thrust bearing mounted to the tower with the mast centered within it. [Joe Seibel, WA3SRU, photo]

I rescanned the bearing base and set the printer to darken the image. The result was a very legible printed copy, which I let stand until the ink was thoroughly dry.

A friend helped me align the clear template over the hole in the top of the tower. Once we had it aligned, we taped it in place. We used a center punch to make a start point for the drill. My friend started with his heavy-duty ½-inch drill, and we kept increasing the drill bit size until the bolt fit. We carefully did the same for the other three holes (see **Figure 86**. We did a test fit and it was perfect.

I repainted the tower top with cold galvanized paint. When that was dry, we mounted the bearing, inserted the mast, placed the mast in the rotator, and centered the mast in the bearing (see **Figure 87**). Finally, we tightened down the rotator bolts and adjusted the rotator to keep the mast centered. I ran the rotator through its 450° of rotation, and it turned smoothly with no binding. Using this simple method, I produced a homemade template that simplified the job and resulted in a successful project. — *73, Joe Seibel, WA3SRU, 233 Lea Ln, Warminster, PA 18974-5223,* **wa3sru@arrl.net**

Table 1
Resistor Color Code —
Significant Figures

Color	Value
Black	0
Brown	1
Red	2
Orange	3
Yellow	4
Green	5
Blue	6
Violet	7
Gray	8
White	9

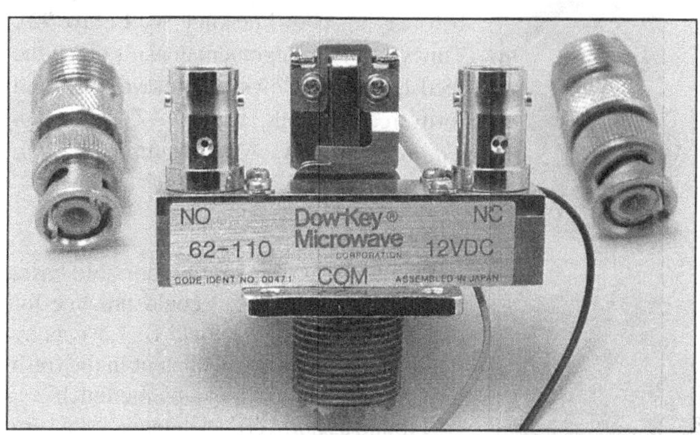

Figure 88 — The Dow-Key Microwave 62 series coaxial switch is a good choice for switching V/UHF antennas. [Don Dorward, VA3DDN, photo]

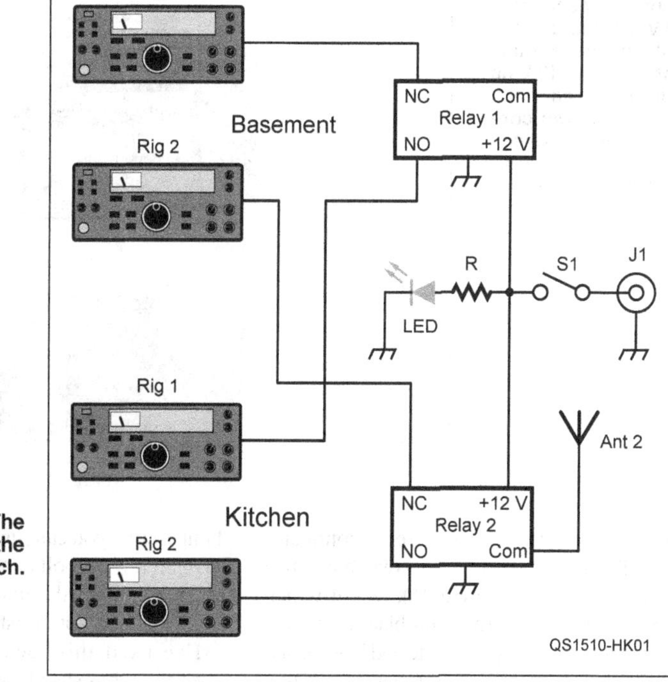

Figure 89 — The connections for the dual-relay switch.

A Wireless Remote Antenna Switch

Like most other amateurs, I have one main operating position, proudly referred to as "the shack," which is located in a corner of my basement. However, I spend a lot of time in the kitchen on the main floor, and found it convenient to set up a secondary operating location there for my VHF and UHF transceivers.

At first, I used small antennas set up near the kitchen until it occurred to me that it would be better to "split" the feed lines for my existing outside antennas, for use either in the main shack or in my "kitchen shack."

My first attempt used two Daiwa CS-201 manual two-port coaxial switches, mounted in the basement shack location, one to switch the 220 antenna and the other to switch a dual-band 144/440 antenna, both of which are located a good distance up my tower.

I wired these manual switches so that one position connects the outside antennas to the basement rigs, and the other position diverts the antennas to the kitchen. This approach was simple, but in order to use the kitchen position, I had to run downstairs and throw the Daiwa switches, and then remember to switch them back! What I needed was a means of doing this by remote control.

I searched for another solution and came across a surprising variety of single-pole, double-throw (SPDT) coaxial relays on eBay for reasonable prices. Eventually, I found a coaxial relay that can be switched electrically, can operate at VHF and UHF frequencies, and can handle the 50 W output of my rigs.

I ordered two of the Dow-Key Microwave 62-110 relays, which will operate from low HF frequencies to 1 GHz at up to 150 W. The particular switches I ordered had an SO-239 connector as the common connector and BNC females for the normally open (NO) and normally closed (NC) contacts

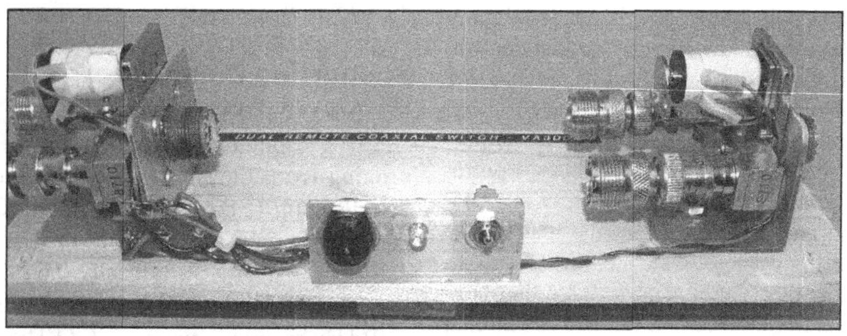

Figure 90 — This is the completed dual-relay unit. Relays 1 and 2 are on either end of the board. The control panel contains, from the left, power connector J1, the LED, and manual switch S1. [Don Dorward, VA3DDN, photo]

(see **Figure 88**). This was not a problem as I simply added two RF adapters to convert the BNC females to SO-239. These coaxial relays operate on 12 V dc and consume 170 mA each (a total of 340 mA) when activated.

Figure 89 is a simplified block diagram of my coaxial antenna switch assembly, with two of the Dow-Key coaxial relays, one for each antenna. The common (COM) connector of one relay is connected to the 220 MHz antenna and the common of the other to the dual-band antenna. The NC output of each relay is connected to the basement rigs and the NO outputs are connected to the kitchen rigs. The LED in Figure 89 can be any you have available; size resistor R to match the particular LED you use. **Figure 90** shows the two relays mounted on a 3½ × 10 inch piece of wood. I connected the relay coils in parallel and fed them with 12 V dc from an ac to dc adapter (wall wart) that had

a 0.5 A capacity. The panel in the center has the dc power jack (J1), "relay on" LED, and an **ON/OFF** switch for manual control.

The remote control aspect of the design uses a wireless remote switch to control the relays. These wireless switches come in ac and dc versions. AC switch examples are the Woods 32555, Globe 78901, or SVAT WRC101. The wireless dc switch I used was the **clickandlights.com** RS101. There are many other versions of both available.

The ac version is the simplest to install. You just have to plug the wireless switch into an ac outlet and the wall wart or other power supply into the switch. Using the wireless remote control, you can turn the relays on or off as necessary.

The dc option is a little more involved, requiring you to solder the wireless switch's input connections to the 12 V power supply's output and the switch's output to a dc

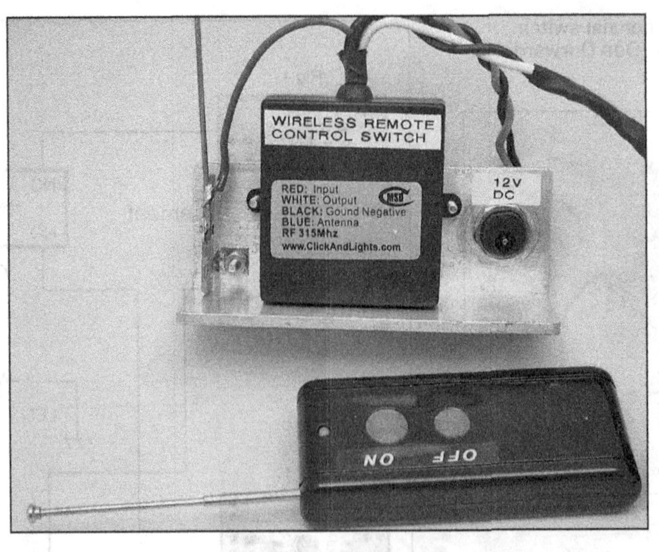

power plug (P1) to match relay connector (J1). The RS101 wireless dc switch has five wire pigtails (see **Figure 91**), two of which are the 12 V input, (red and black), two are the switched dc output (white and black), and one for connection to a short antenna (blue). Other switches may have other arrangements.

Another point to consider is that some wireless switches are toggle types with only one button on the wireless remote control, that is, when the switch is off, pressing the button will turn it on; if the switch is on, pressing the button will turn it off. Others have separate **ON** and **OFF** buttons. I prefer this type because you can press the **OFF** button anytime you forget the state of the switches.

Note that if you want to forgo the wireless control option, you can add an optional second switch (S2) to provide hardwired remote operation. When using this option, S1 and S2 need to be SPDT switches wired in a three-way configuration.

I have found the remote operation to be quite reliable over several months of use. I have also found the control range to be more than adequate at my location with about 30 feet of horizontal separation, and main floor to basement with mainly wood construction.
— *73, Don Dorward, VA3DDN, 1363 Brands Crt, ON L1V2T2, Canada,* **va3ddn@arrl.net**

Bungee That Mast

If you are about to mount an antenna mast on your home's outside wall, you're going to wish you had three hands because there's going to be a lot of busywork happening up there.

You'll probably be mounting a saddle clamp on the wall to hold the mast in place and if you install eyebolts on each side of the clamp frame, you can run a small bungee cord between the eyebolts and across the face of the clamp. Then the mast can be held erect and in place as you finish the rest of the task.

I've used this idea myself numerous times and it has saved me at least one injured knuckle. — *73, Mike McAlister, KD6SF, 7570 Dartmouth Ave, Rancho Cucamonga, CA 91730-1534*

Supporting Balanced Window Line

I have a long run of 450 Ω balanced line running from the house to my extended double Zepp (EDZ) antennas. [The EDZ is a dipole-type antenna whose legs are each longer than ½ λ, typically 0.64 λ per leg. The EDZ will radiate with a gain of 3 dBd but requires a tuner. — *Ed.*] The feed line needed some support to stay overhead as it crossed the backyard and out of the way of traffic. At first, I just tied a bowline knot on the end of a length of paracord to support the feed line, but I found that the knot would pinch the window sections on the feed line together — not an ideal arrangement. I came up with this simple idea to avoid that situation.

I took apart an old ballpoint pen and cut the barrel section into three equal lengths. Then I ran the paracord through those three pieces before tying a bowline knot (see **Figure 92**). This support arrangement doesn't pinch the feed line, and allows it to move back and forth freely with the wind. — *73, Cameron Conover, AJ4TW, 4234 Brentonshire Ln, High Point, NC 27265,* **aj4tw@arrl.org**

A Radial Idea

The thought of operating portable has a certain appeal for me. I could practice for possible emergency or Field Day events by making some contacts while out in the fresh air with a good view and a picnic lunch.

I found a short vertical antenna for a very good price. I knew radials were required for verticals, but not much else, so I started to do some research.

One of the interesting facts I discovered was that radials for short vertical antennas can be about the same length as the antenna itself. My antenna is a bit less than 8 feet tall, which means that the radials can be the same length (see **Figure 93**). Another important factor is the number of radials that one should use. Most of my research indicated that somewhere around 32 radials would provide the best bang for the buck. A smaller number leads to lower performance and a larger number, while giving somewhat better performance, may not be worth the cost. [The number and length of radials is a complex subject. Interested readers can find a more detailed discussion in Chapter 3 of *The ARRL Antenna Book. — Ed.*]

It occurred to me that I could use a steel tape measure as a radial. Using a 25-foot tape, I could find the center and have two 12½-foot radials, which would be more than long enough for my 8-foot antenna. Best of all, I can usually coil and uncoil these things without too much trouble and a minimum number of knots.

To create your own radial system, first decide on the number of radials needed and then divide that number by two, as you get two radials for each tape measure. Obtain

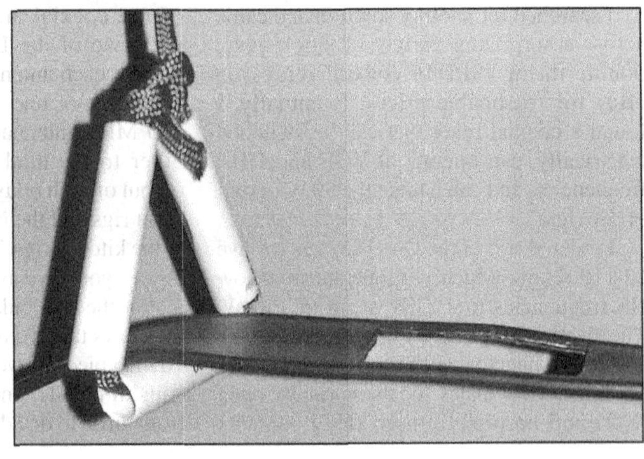
Figure 92 — A plastic pen case and some paracord will make a balanced-line support that won't detune the line and will allow it to flex in the wind. [Cameron Conover, AJ4TW, photo]

Figure 93 — Steel tape measures can make an effective radial system for a quick-deploy vertical antenna. [Mike Brown, WA7JLL, photo]

Figure 94 — A piece of aluminum sheet forms the base plate. A ¼-inch bolt in the center is used to anchor the radials. [Mike Brown, WA7JLL, photo]

Figure 95 — The assembled antenna. The bracket sits on the radials, holding them together while also supporting the transmission line connector, which connects the shield to the radials and the center conductor to the antenna. [Mike Brown, WA7JLL, photo]

Figure 96 — This measuring stick makes it easy to place the 40 radials. [Mike Brown, WA-7JLL, photo]

the required number of tape measures and modify them by removing the belt clip, lanyard, and the rubber covering. This makes the devices smaller and lighter.

Next, find the middle of the tape (12 feet, 6 inches in this case) and make a hole at that point. I used a hole punch (the type used for paper), which does a very quick and neat job, and the hole is just big enough to fit over a ¼-20 bolt.

Then you need to remove the paint from the area around the hole, so all the radials will make good electrical contact. I used a sander with a relatively rough grade of sandpaper. The paint came off easily, but it also ground down the edges of the metal so they became very sharp. I recommend wearing work gloves to keep from being cut. A chemical paint remover may be a better option, but do follow the directions, as some of those products can have nasty side effects.

Once you have prepared all your tape measures, you need to build some kind of a device to connect them all together. I found a piece of flat metal, roughly 7 × 7 inches, to use as a base plate. I attached a ¼-20 × 1½-inch bolt in the middle of it and drilled some holes in the corners of the base plate so I could place some metal tent stakes through the holes and secure the base plate to the ground, if necessary (see **Figure 94**).

Lay the tapes out, making sure the bolt from the base plate goes through the hole in the tapes. To make the connection to the antenna, I used a corner brace left over from another project and adapted it for this project by drilling a hole on one side for the base plate bolt and, on the other side, I drilled a hole that accepts a bulkhead type BNC connector. Of course, you could use other coax sizes. A short length of coax connects to the antenna itself, while the coax from the transceiver connects to the other. I used a wing nut here for convenience (see **Figure 95**).

In order to lay out the radials equally, I had to determine the circumference of the radial circle and divide by the number of radials. As a reminder, the circumference can be found by multiplying the diameter (in this case 25 feet) by 3.14. My system used 40 radials, with each radial tape 1 inch wide. I added 1 then divided the circumference by 41, and converted the result to inches. Remember that this figure is the middle of the tape, which is, in my case, 1 inch wide. So the distance from the side of one tape to the side of the next tape is 1 inch less.

The idea is to quickly lay out the radials with an even distance between them, so I cut a couple of lengths of wood to the proper length and painted the ends red. When out in the field, I lay out the first radial, and then place the stick to show me where the next one should go (see **Figure 96**). Because I am actually placing two radials at the same time, a stick for each end can be helpful.

Once all the radials are in place, mount the antenna, connect the coax to the rig, and you're on the air. At the end of the day, tearing down the station is a breeze. I have yet to produce any indescribable knots or tangles!

This has been a fun project. Do be prepared for passersby giving you a strange look; they tend to notice all that yellow lying on the ground. So far no one has ever tripped over it, but precautions should be taken. — *73, Mike Brown, WA7JLL, 539 Teco St, Grand Junction, CO 81504,* **wa7jll@arrl.net**

Rock Your Tuner

I have a 43-foot vertical monopole antenna. Originally, I matched it to the transceiver with a substantial matchbox/antenna tuner located in the shack. The conventional wisdom is that it is more effective and efficient to match these big sticks with a remotely located tuner at the feed point. To this end, I purchased and installed the MFJ-998RT Automatic Remote Antenna Tuner.

The "weathertight" steel and plastic cabinet appeared insufficient to protect it from my harsh Michigan weather. It lacked a proper rubber or silicone seal or gasket and was more of a top cover than a weather-tight cabinet.

I resolved to house the device within a more weatherproof enclosure and, upon the suggestion of my wife, settled on a plastic rock that looks like a large fieldstone. The fake rock is an Emsco Medium Plastic Architectural Rock, model number 1794171, which cost about $30.

I hand-cut **L**-shaped mounting brackets from 1 × 1-inch aluminum angle iron and mounted (suspended) the entire tuner unit inside the rock (see **Figure 97**). I used various ¼" × 20 nuts and bolts with 1-inch fender washers to attach the frame to the rock (see **Figure 98**). I was worried the rock would dull the tip of the drill bit, but it was easy to drill four ⅝-inch mounting holes in the rock. I made all fittings from a single 3-foot-long piece of angled aluminum stock, which cost about $3. Note: to ensure a weatherproof seal, the screw hole must be smaller than the screw head. A poorly made hole that is deformed and/or larger than the screw head will likely allow water to enter the unit.

Perhaps not the most elegant solution, but it does have "wife approval," and it provides additional protection from bad weather. The fake rock disguises the installation (see **Figure 99**), minimizes the visual impact of the remote tuner, and hides it from the curious. — *73, James Richards, K8JHR, 5787 Charolais Dr SW, Wyoming, MI 49418,* **k8jhr@arrl.net**

Installing Antenna Ropes with a Fiberglass Mast

A telescoping mast can be an effective way of installing antenna support ropes in trees at a height of 30 feet with good precision and reasonable safety. Instead of flinging a projectile through the air, a weight is hoisted over the desired tree limb and dropped. An old brass water hose nozzle is heavy enough to pull down 80 feet of fluorescent or white nylon twine, when jiggled to reduce the static friction. Alternately, I've used old pill bottles filled with 2 – 5 ounces of water.

I use fluorescent duct tape to attach a wire guide loop to the mast (see **Figure 100**) and to attach the nylon leader to a Dacron antenna support rope. I also use fluorescent-colored

Figure 98 — A close-up of the bracket arrangement that attaches the tuner frame to the curved exterior of the rock shell. [James Richards, K8JHR, photo]

Figure 99 — The completed project provides a spouse-approved, weathertight shell for a remote tuner. [James Richards, K8JHR, photo]

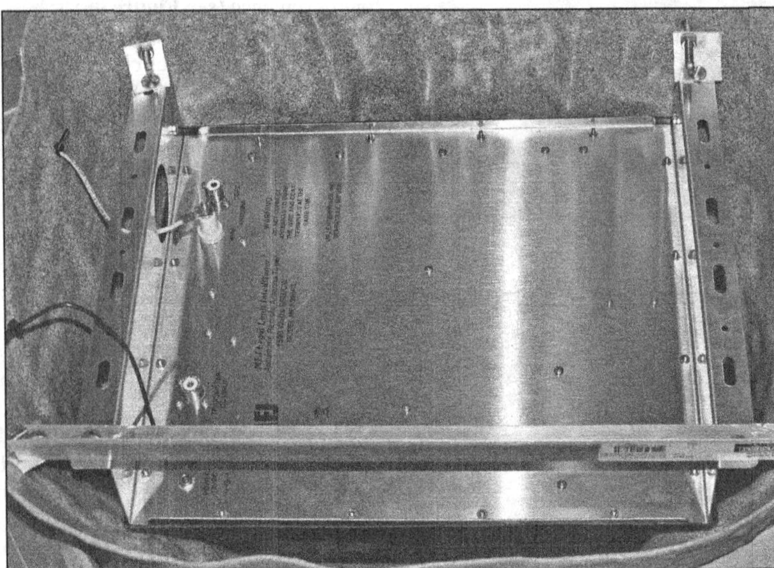

Figure 97 — The support framework made of angle brackets holds the tuner safely within the rock shell, protected from the weather. [James Richards, K8JHR, photo]

Figure 100 — A loop of #14 electrical wire is taped to the top of fiberglass mast to act as a line guide. [Zack Lau, W1VT, photo]

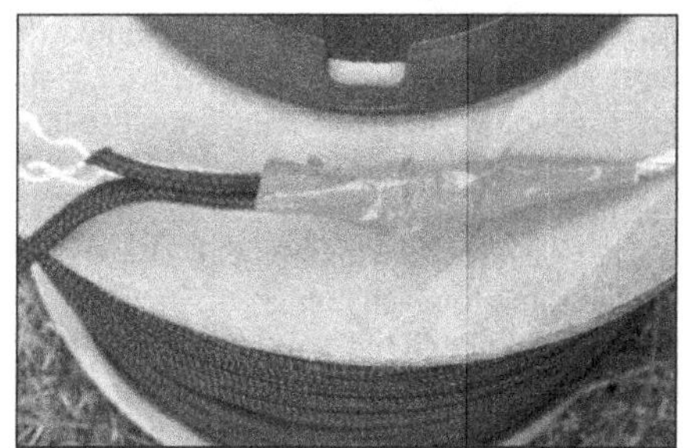

Figure 101 — Orange duct tape prevents the knot from untying, reduces the chance of snagging in the tree, and is clearly visible amid the foliage. [Zack Lau, W1VT, photo]

nylon twine for improved visibility to pull up the black Dacron antenna rope. It helps if the sky is overcast; bright sun or windy conditions make the job more difficult.

When rigging the loop, avoid creating a gap between the loop and the mast that would allow the twine to snag. Duct tape will hold the loop firmly, but will come apart eventually. This can be useful; if the pole gets stuck in the tree, put tension on the taped connection and wait for the adhesive to let go.

Taping the nylon twine to the Dacron rope (see **Figure 101**) not only provides a secure connection, but allows you to make a streamlined bump that is much less likely to get snagged in the tree.

The accuracy of this method allows you to select the best branches for your antenna. A thick branch close to the center of the tree will minimize any movement that would stress the antenna. It should also be sufficiently clear of other branches and foliage to get the rope over the branch and down through the lower branches. If you have the time, it may be possible to repeat the exercise, first to get it over the highest branch, and then again to pull the rope over lower branches. It can be useful to have a second pole to guide the weight around lower tree branches that would snag the rope.

Another advantage of this method is that you can poke the pole though foliage, pulling the weight through the leaves, which can allow you to access otherwise inaccessible branches.

Keep in mind that the pole acts as a lever, so the weight at your end of the pole feels much heavier when the pole is at an angle. The 33-foot fiberglass pole I use has a textured finish that is easy to grip. If your pole has a smooth finish, it may help to wear rubber garden gloves or tape the end to improve the hold.

After some practice, I find I can use a telescoping pole to lift the weight nearly straight up, extending the mast into the tree while giving the weight more twine. When doing this, some tilt is desirable to prevent the weight from wrapping itself around the pole. Such a near-vertical lift maximizes the

height that can be obtained by minimizing the bending of the pole.

Once at the desired height, position the weight over the desired limb, and then let the weight fall, jiggling the line to reduce line to limb friction. Once the line is positioned, tie off the weight end, lower the pole, and run the line completely through the loop, instead of removing the loop.

I find it best to avoid pulling the weight back over tree limbs if you discover you have dropped it over the wrong branch. It is usually best to let the weight drop, untie the weight, pull the twine back through the tree, and start over again. This is somewhat time consuming, but better than having a weight wrap around some branches and get stuck high up in the tree. If the twine does get stuck, the safest way to cut nylon twine that has been looped around a branch is to use the pole to thread another piece of twine through the loop the stuck piece makes with the branch and pull the second piece back and forth. It will eventually cut through the stuck piece.

I've found this procedure practical with a 30-foot pole, but a 40-foot pole can be hard to control. The 30-foot pole will position your antenna at a height that is a half wavelength on 20 meters, and also high enough to be useful on 30 and 40 meters. — *73, Zack Lau, W1VT, Senior Lab Engineer, ARRL, 225 Main St, Newington, CT 06111*, **zlau@ arrl.org**

Anchoring Tape Measure Elements

I see many hams building PVC tape measure beam antennas for foxhunts and other portable applications. The usual method is to attach the tape measure segments to the PVC boom with a hose clamp. The hose-clamp method straddles the outside of the PVC and deforms the tape while looking somewhat crude. I wanted a cleaner-looking method and came up with the idea of using short pieces of PVC pipe to hold the tape inside the cross fittings (see **Figure 102**). The tape goes through the inside of the fittings, the short PVC pipe

pieces are inserted to sandwich the tape to the inside wall of the fitting.

The 1-inch-wide tape is sandwiched between the ¾-inch cross fittings and some short (approximately 1 inch) sections of ¾-inch pipe (the fitting is sized to fit the pipe, after all). The same holds true for whatever size pipe you use for the boom; just get the appropriately sized fitting. With the tape in place, it is a tight fit that will not move nor deform the tape (see **Figure 103**). The driven element cross fitting will have two holes drilled in it where the coax is soldered on. — *73, Robert Evans, WBØSVS, 117 Boardwalk Gardens Dr, O'Fallon, MO 63368*, **beach_bob@msn.com**

Two Meter Mag-Mount Ground Plane

This 2 meter mag-mount antenna project is low-cost and perfect for portable operations. When complete, you will have a small, portable 2 meter antenna that will go anywhere, be easy to set up, and provide communications for everything from a fun day out to a disaster deployment.

The parts needed are listed in **Table 2** and the cost is less than $7. I found all the parts at Home Depot, except the tape measure, which you can get for free from Harbor Freight — just look for the coupon. The tools you will need are a screwdriver, adjustable wrench,

Figure 104 — The cover plate with the mast mounting hole drilled into the edge. [Rich Russo, KB3VZL, photo]

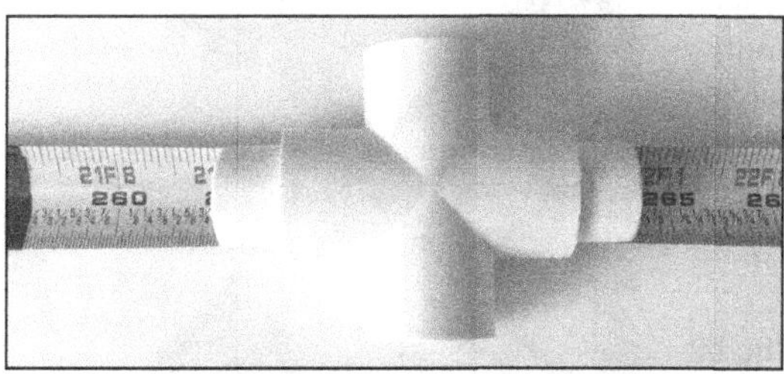

Figure 102 — Top view of the T-connector with the measuring tape element passing through it. The smaller diameter tubes coming out either side hold the tape in place. [Robert Evans, WBØSVS, photo]

Figure 103 — This side view shows how the small diameter tube fits into the T-connector to hold the measuring tape element in place. [Robert Evans, WBØSVS, photo]

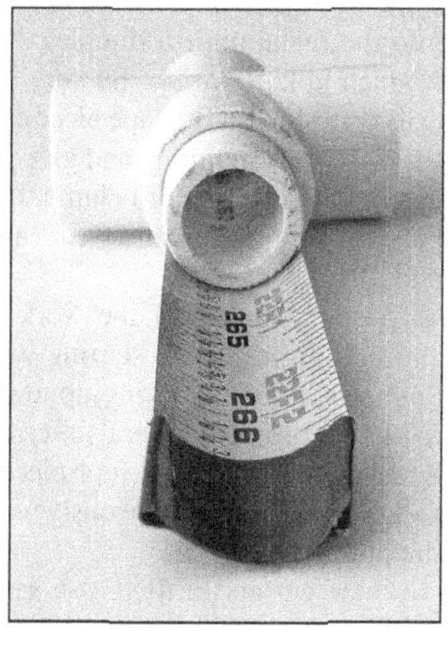

Quantity	Description	Part Number	Cost
1	4" square electrical box cover, galvanized steel	Home Depot 050169007525	$0.60
4	10 × 24 × ¾" machine screws	Home Depot 33041	$ 1.18
4	10 × 24 hex nuts	Home Depot 18521	$ 1.18
4	10 × 24 wing nuts	Home Depot 802351	$ 1.18
8	#10 flat washers	Home Depot 887480024715	$ 1.18
1	¼ × 20 × 1" machine screw	Home Depot 803481	$ 1.18
1	¼ × 20 connector bolt	Home Depot 822251	$ 1.20
1	1" wide tape measure	Harbor Freight 69030	$ 0.00

Figure 105 — An electrical box cover and a metal tape measure can be used to turn your mag-mount antenna into a portable base unit improving the range and effectiveness of your communications. [Rich Russo, KB3VZL, photo]

(second) tape measure, shears, an awl, and ³⁄₁₆- and ¼-inch drill bits.

The square blank box cover requires one ¼-inch hole to be drilled ¼ inch in from the edge and 1 inch in from a corner (see **Figure 104**). This is to mount the coupling nut that is used for the support. You also need to drill or awl four holes in the tape measure to mount it on the square cover plate.

Use the shears and cut four 24-inch pieces from the 1-inch-wide tape measure. These are the antenna radials. Measure ½ inch from one end of each piece and mark a spot to be drilled with the awl by pushing into the metal until it dimples. Now drill a ³⁄₁₆-inch hole. *Caution*: Be very careful and drill very slowly. The tape piece can catch on the drill, start spinning, and give you a nasty cut. It is a good idea to clamp the tape to a piece of scrap wood and wear safety gloves and goggles.

Also, snagging of the work by the bit can be minimized by starting with a small diameter bit and working up to the desired diameter through several intermediate diameters. Clamping the four pieces together and drilling all simultaneously will speed up the process.

Once cut and drilled, use sandpaper or steel wool to remove the paint from the tape for about an inch from the holed end. I put a piece of electrical tape over the cut to protect my fingers from the sharp edge, but wearing gloves would be even better.

The box cover has a slot cut into each of its four corners. The radials are mounted to these slots. Place one of the 10 × 24 × 1 inch machine screws in a slot. Secure it with a hex nut and tighten it. Next, place a washer on the nut, then the tape measure radial, and a second washer. Secure the assembly with a wing nut. Do the same for each corner.

Calculate the length of the ground plane radials for the frequency you want to use. Mine came out to be 18¹¹⁄₁₆ inches long. Measure from the center of the steel plate out and mark the radial at whatever value will resonate your frequency. Check the SWR, and then use the shears to slowly trim the 24-inch radials to resonate at your frequency. Once trimmed to the correct length, place a piece of electrical tape over the cut ends, which are very sharp.

Next, place a ¼ × 20 bolt through the hole you drilled into the plate. Screw on the coupling nut and tighten it (see **Figure 105**). This allows the unit to be mounted atop an old camera tripod that I use for this antenna. A variety of other mounting arrangements are possible.

To use the ground plane, attach the plate to the tripod or other support using the ¼ × 20 mounting stud. The antenna's mounting magnet attaches to the center of the steel cover. If necessary, adjust the antenna SWR to its minimal value. If you want to get fancy you could paint it, but it's not necessary. — *73, Rich Russo, KB3VZL, 105 Colonial Ave, Norristown, PA 19403,* **kb3vzl@arrl.net**

Radial Reel

While the radiating element is important to a portable antenna system, a good counterpoise or ground radial system is also a necessity. The reel device described here allows you to deploy a radial wire quickly with a flick of the wrist, which is particularly helpful over wet or uneven terrain.

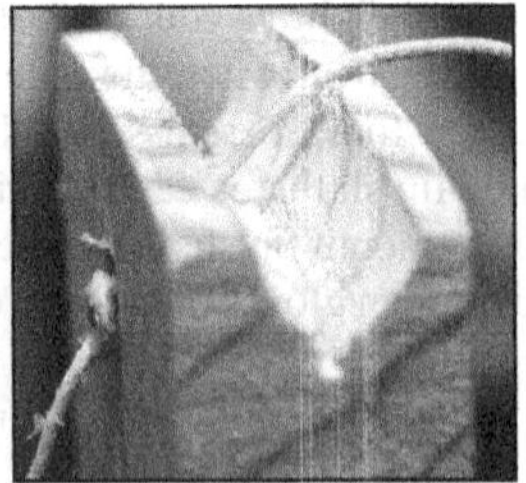

Figure 106 — This wooden cross not only makes a convenient radial storage device, but is very helpful for deploying and retrieving the lines. [H. Scott McCann, W3MEO, photo]

Figure 107 — A close-up view of the anchor end of one of the legs, showing the V groove and the radial anchor point. [H. Scott McCann, W3MEO, photo]

To build the reel, cut a cross out of a piece of ¾-inch shelf lumber. The legs are approximately 1 foot long, end to end, and 2 inches wide (see **Figure 106**). The exact dimensions are not critical but the cross should be symmetrical to roll out well. Next, cut a **V** groove at each end of the legs (a backsaw or dovetail saw would be the best choices, but nearly any saw will work). Round out each groove with a rat-tail file (see **Figure 107**). Drill a small hole large enough to fit your size wire in the side of one of the legs. Push the end of the wire through and secure it with a knot. I wound about 40 feet of number 22 AWG stranded flexible wire, which is enough for one ground wire.

To deploy the wire, peel off enough wire to get to your tuner or rig and anchor it to something solid (not your rig!), then pitch the reel sidearm like a Frisbee. Make sure you have the reel correct side up so the wire will unreel. When the event is over, put a small screwdriver through the center hole to allow the wire to be reeled in. — *73, H. Scott McCann, W3MEO, 160 Shields Ln, Queenstown, MD 21658-1278,* **w3meo@ arrl.net**

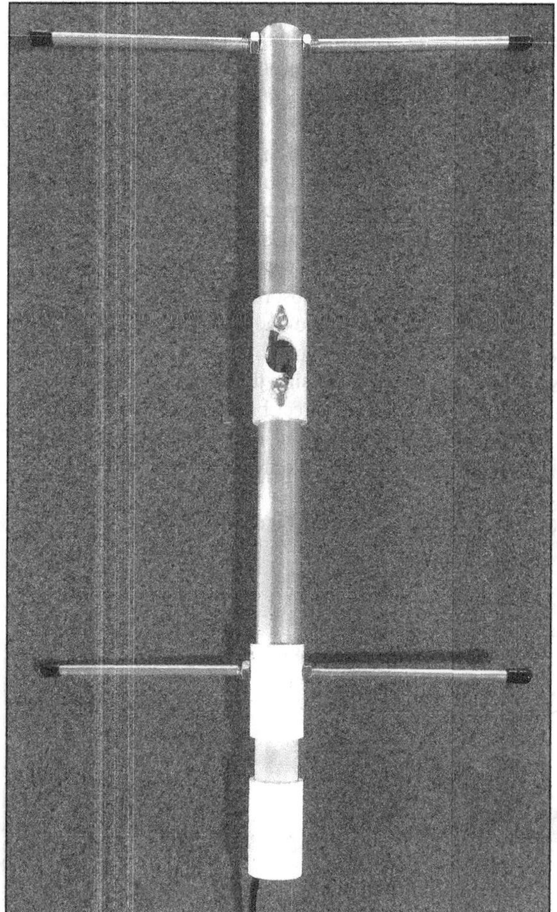

Figure 108 — Schedule 40 aluminum *pipe* will fit into PVC fitting to make insulating sections useful in antenna design. Aluminum *tubing*, however, doesn't fit PVC fittings. [John Portune, W6NBC, photo]

Matching Aluminum Tubing to PVC Fittings

Many hams have been frustrated during homebrew projects while trying to make aluminum tubing fit neatly into PVC water pipe fittings. The antenna in **Figure 108** illustrates how easy it is, if one selects the correct type of aluminum.

Metal dealers carry two types: aluminum tubing and aluminum pipe. Tubing is measured in outside diameter (OD), pipe in inside diameter (ID). Tubing won't fit snugly into PVC pipe fittings; Schedule 40 aluminum pipe will.

Knowing this has been a boon to me for several recent homebrew V/UHF antennas. The PVC fitting to pipe equivalency holds true for all normal sizes of PVC and aluminum. For example, buy ¾-inch PVC fittings and ¾-inch Schedule 40 aluminum pipe will fit nicely into it.

The antenna shown is the prototype of a 220 MHz antenna. The design uses PVC sections to isolate the upper and lower elements electrically while providing strong mechanical support. — *73, John Portune, W6NBC, 519 W Taylor St, Spc 111, Santa Maria, CA 93458,* **jportune@aol.com**

Copper Wire Tie-Down

If you use a messenger cable to route coax, rotator, and control wiring to your tower, you can use #14 AWG solid copper wire to attach these other wires to your messenger cable. Unlike cable ties, #14 is unaffected by sunlight and can be removed and reattached many times. If you need to add a cable, simply untwist each tie-wire, route the new cable through, and then reattach the tie-wire. I have also found that #14 wire is very useful for keeping coax cable rolled into a neat diameter when coiling it up. — *73, Terry Webb, NØTW, 1103 Fulford Rd, Monticello, FL 32344-4352,* **n0twterry@ yahoo.com**

Fitting Ferrite Beads

Have you ever tried to fit #73 ferrite beads onto RG-58/U coax to make your own 1:1 current balun? The #73 beads almost fit, but you really need to force them on. We're only talking about a few thousandths of an inch difference. [The exact diameters of ferrite beads and coax cables vary. Try applying a silicone lubricant before using this procedure. — *Ed.*] I purchased a balun kit that comes with 13 inches of RG303/U, which wasn't long enough for my application when fitted with PL-259 connectors and adapters on both ends. Another issue is the need to reduce the weight of the balun's PVC tube and additional hardware at the center of a dipole. The following procedure will allow you to do that using RG-58/U feed line.

The solution is either to make the hole larger or the peg smaller. I wasn't going to attempt to enlarge the holes of the ferrite beads, but I was able to change the diameter of the RG-58/U coax. I put one end of some RG-58/U in a vice, and used 100 grit sandpaper to sand down the jacket of the coax over an area long enough to fit the necessary number of beads. Just fold the sandpaper around the cable and work it back and forth evenly for the desired length (see **Figure 109**).

When you're done, you'll need to cut off the flattened end of cable that was in the vice, so remember to account for that part when you figure the length. Once it's sanded down, you can slide the beads on easily. After the beads are placed on the cable, put some shrink-tubing on it, or wrap it with Scotch 33+ electrical tape. This little trick has given me more options and lot less frustration.

Note that RG-303/U will handle the legal limit of power. Some brands of

Figure 109 — A little careful work with some sandpaper will make the job of fitting ferrites onto RG-58 much easier. [Gary Drasch, K9DJT, photo]

RG-58/U will, but some will not. Do not attempt this procedure with RG-8X. — *73, Gary Drasch, K9DJT, 307 W Barry Ave, Port Washington, WI 53074,* **k9djt@sbcglobal.net**

On the Level

When I installed the Universal Towers base for my tower, I needed to make sure it was level. In browsing through my local home improvement center, I found the Swanson Post Level (Model PL001M), which proved to be ideal for attaching to the tower for leveling. — *73, Lloyd Korb, K8DIO, 8960 Gettysburg Dr, Twinsburg, OH 44087,* **k8dio@arrl.net**

Mag Mount for a Mag Mount

A 2 meter magnet (mag) mount antenna can have many uses beyond just sitting on the roof of your vehicle. While I was setting up a 2 meter rig in a large barn-type structure for a radio club event, I noticed the wooden barn had an internal steel girder framework. This gave me an idea. When I got home, I mounted a 2½-inch diameter magnet onto a 1½ × 2½ -inch steel **L**-bracket (see **Figure 110**). [Make sure the **L**-bracket and hardware you use are metallic throughout. For the mag-mount antenna to work, you need an RF path to whatever the **L**-bracket magnet is stuck to. — *Ed.*]

When I returned to the barn, I placed the bracket against the girder about 10 feet up. I now had a horizontal surface to mount my mag-mount antenna. The building became the counterpoise. The rig worked perfectly, talking in the event attendees on our local 2 meter repeater. The magnetic bracket has

Figure 110 — A strong magnet attached to an angle bracket can be attached to many vertical metallic surfaces found in the field, making a quick mount for your mag-mount antenna. [Rich Russo, KB3VZL, photo]

come in handy on a few other occasions and is a great gadget to keep in your go-kit. — *73, Rich Russo, KB3VZL, 105 Colonial Ave, Norristown, PA 19403,* **kb3vzl@arrl.net**

Easy-Off Connector Seal

With my antenna installations, I always design them *not* to be permanent. I design my installations to be easily serviced or changed to accommodate experimentation. I use stainless-steel hardware exclusively with soldered crimp lugs for making connections. I used to use the coax-sealing tape available from multiple sources, but soon found that it complicated disassembly of the connections by getting into the threads of the hardware. I came up with this solution.

After the connection is tightened, apply one layer of blue painter's tape over the exposed metal parts. Then cover the area with coax-sealing tape, molded to the part's shape. Finally, apply black RTV silicon rubber over the coax-sealing tape. The completed connections are protected from the elements and with just one slice of a knife, you're able to peel off the RTV, coax-sealing tape, and painter's tape with no residue to complicate unscrewing the stainless steel hardware. Yes, it takes a few minutes longer to make the connections but, trust me, when it comes time to open things up, you will realize the benefits of your extra efforts. — *73, Evan Rolek, K9SQG, 1295 Oakleaf Dr, Beavercreek, OH 45434-8002,* **k9sqg@arrl.net**

Ladder Line Construction

Many methods of constructing homemade ladder line require drilling holes in spreaders and threading the wire conductors through the holes.[1] This can become tedious and time-consuming if you are making long transmission lines, which might require 100 or more spreaders. A relatively quick and easy construction method is to use 4 millimeter diameter black plastic irrigation tubing and 2 × 150 millimeter cable ties (see **Figure 111**). The cut tubing is placed between the two conductors. A cable tie is passed through the tube, around the conductor at the other side, and back through the tube. The cable tie is then pulled tight, holding the conductors against the ends of the tube.

The tubing is easily cut with a utility knife to the required line spacing. For 450 Ω line using PVC insulated 1 millimeter diameter wire, the spreaders should be approximately 22 millimeters (⅞ inch) long. For 600 Ω line, the spacing is approximately 75 millimeters (3 inches) and two cable ties will be needed, being passed through the tube in opposite directions and linked at their ends. — *73, Robert Hancock, VK5AFZ, 30 Tottenham Court Rd, Port Elliot, South Australia 5212,* **rfh48@tpg.com.au**

Buried Cable Water Wick

No cable likes being immersed in water, and trying to arrange a conduit to drain condensed water is often impractical. However, a conduit provides an easy way to work on, change, or replace cables, and often (such as when buried under a cement slab) is the only reasonable way to allow for cable access. Condensation is a problem, because the concrete or dirt surrounding the conduit is generally below ambient temperature, causing the conduit to attract moisture.

To solve this problem, one should start with a waterproof conduit, which in most cases means a PVC or ABS conduit with properly glued couplings. Then, when pulling the cable through, pull a piece of cotton rope with it. Make sure the cotton rope is open to the air at each end, and that each end is protected from exposure to moisture (such as rain). Any water condensing in the conduit will be absorbed by the rope, which will wick it to the ends, where it can evaporate. Don't use synthetic rope — it doesn't absorb moisture. Using this method, you can often avoid the direct burial of cable. I have used it for several decades in installations ranging from public address systems and electronic organs to power distribution. — *73, Wilton Helm, WT6C,* **wt6c@arrl.net**

[1]Shackleford, W6YE, "Custom Open-wire Line — It's a Snap," *QST*, Jul 2011, pp 33-36.

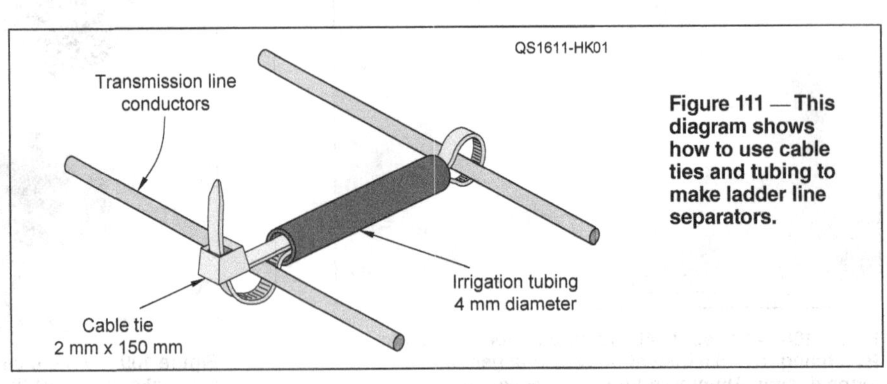

Figure 111 — This diagram shows how to use cable ties and tubing to make ladder line separators.

AN AC POWER FAILURE ALARM FOR REPEATERS

One of the local VHF Amateur Radio repeaters is located out in the country at a camp. There is no telephone line available and the power line snakes through some heavily wooded areas. For that reason the local electric power company service is not very reliable.

The repeater owner, Lee Lewis, N3NWL, has equipped the site with a generator and means to supply backup power to the radio system during extended outages. The repeater is also equipped with a heavy-duty 12 V battery that handles short-term power requirements. If, after several hours, the electric service is not restored or the generator isn't connected, the battery discharges and the repeater goes off the air. Lee has had this happen several times and the battery, damaged by being fully discharged, required replacement.[1] Since there are few full-time residents in the area it might be days before the electric company is notified of the power outage.

We discussed the situation and decided to

[1] For more information on batteries refer to *The ARRL Handbook for Radio Communications* available from your ARRL dealer, or from **www.arrl.org/shop**.

build an alarm that would add a tone to the audio of the repeater when the electric power was interrupted. It would be adjusted to be loud enough to be noticed by users while not interfering with normal voice traffic. Club members would be advised to notify N3NWL that the tone was heard so that he could visit the camp and connect the generator.

The next question was how to switch the tone in and out of the circuit. We decided to add a 12 V relay to the power circuit of the alarm that would pull when power was present. The 9 V battery for the tone board would be wired through the *normally closed* contacts of the relay, which would be *open* when 12 V was present. The completed circuit would be wired to the 12 V power supply that powers the repeater (see **Figure 1**).

My junk box had an old Potter & Brumfield 12 V relay that had an octal base pin arrangement. Just about any 12 V SPDT (single pole double throw) relay would work. I was concerned that the relay might heat up and fail from being energized continually. As a test, I connected the relay coil to a 12 V car battery in the shop and let it run for several hours. There was no apparent increase in temperature of the relay. I was satisfied that the relay would operate in continuous service inside the housing of the alarm unit without any temperature problems.

These homebrew devices may serve for years before needing service so I drew the

schematic for this project and placed a copy of the completed schematic in the housing.

The tone board is a standard 555 IC with minimal parts to produce a tone of about 850 Hz. The 330 Ω variable resistor at the output adjusts the volume of the tone and therefore the modulation of the repeater audio. Drain from the 9 V battery is about 15 mA when the tone is switched on so it will last quite a long time. I used hook and loop material to attach the relay and battery to the inside of the housing with two small tubular standoffs to mount the circuit board (see **Figure 2**).

There are two connections to the housing. One is the two-wire line to the output of the 12 V power supply that powers the repeater. It would also be possible to use a 120 V ac relay connected directly to an electrical wall outlet if the repeater has an internal power supply. The other line connects the tone output to the audio input of the repeater controller. It could be strapped across the terminals for the ID message or the audio from the repeater receiver. It also should be wired on the output side of the carrier operated relay (COR) so it doesn't hold the repeater in transmit mode when active. Since there are quite a variety of controllers used with repeaters some experimentation might be needed to marry the circuit to a particular controller application.

Set up and testing are pretty straightforward. Turn on the tone circuit on-off switch. Once the 12 V line from the circuit is con-

PETER CARR, WW3O

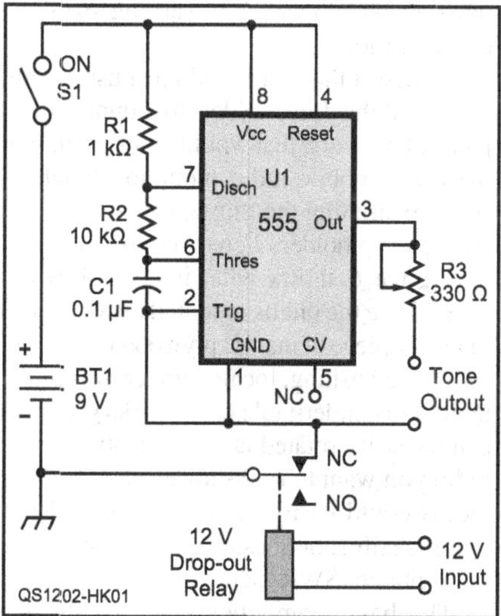

Figure 1 — The schematic of the power failure alarm.

Figure 2 — The power-failure alarm all boxed up and ready to be hooked into the repeater's circuitry.

nected to the power supply and the tone is wired to the controller, just unplug the 12 V power supply. The backup battery should power the repeater and the 12 V relay will drop out. This connects 9 V to the tone board, which sends a tone to the repeater controller. Then, ping the repeater from a handheld radio and listen to the squelch tail or repeater ID message. The tone from the alarm circuit should be heard on the signal. Adjust the variable resistor on the tone board for best modulation and replace the cover on the housing. If, over time, the tone weakens or quits completely it is time to replace the 9 V battery. [If your installation permits, connecting the tone generator to the 12 V battery will provide power for it as long as the repeater is active. — *Ed.*]

There are many repeaters located in remote locations that provide excellent coverage to their users. The downside of their location is the maintenance problems for the hams who service them. I would hope that this circuit will help them with the work they do so the rest of us can continue to enjoy this part of the hobby. — *73, Peter Carr, WW3O, 329 Little Ave, Ridgway, PA 15853-1220, ww3o@arrl.net*

Breath-Activated Key

What do an old relay, a balloon and a kazoo have in common? No, this isn't the start of a shaggy dog story. This is better; it's an old ham story. I have been getting older, as most of us seem to do and I get tremors in my hands making it very difficult to use a key of any kind. Fortunately, my mouth still works (my spouse will verify that) so I decided to make a key that uses air to close the contacts.

The parts needed are shown in **Figure 3**. First take a plastic kazoo and cut the top off of it. Smooth the hole with some fine sandpaper. Next, take a balloon with about a ¾ inch wide neck and cut the neck off at about 1¼ inches from the lip. Pull the neck of the balloon over the kazoo to cover the hole in the top. Plug the rear of the kazoo with whatever is available. I used the lens from a small panel light that was the right diameter.

Open up the old relay and take the contacts out. If you use a small relay and it is a double pole double throw type, you will get enough contacts to make eight or 10 keys. Pick two good contacts and one stiffener. One contact with a stiffener on the top should be soldered to the end of a wire. Make the wire long enough to reach your key jack, plus a little extra so you can move around. The other contact is soldered to the other wire. Add the proper plug for your transceiver.

The contacts are then laid on the balloon with the stiffener on top and glued in place (**Figure 4**). Space the contacts so that with a small amount of air the balloon will expand and push the bottom contact up into the top contact; the stiffener is to keep the top

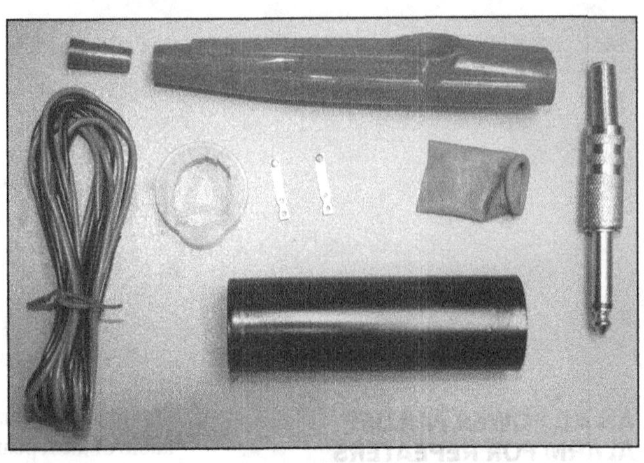

Figure 3 — The parts needed to create your own breath-activated key are, clockwise from left, wire, stopper, kazoo, transceiver plug, shell, shell cover, contacts and balloon.

Figure 4 — The contacts are glued to the body of the kazoo so that they extend onto the balloon. The top contact should include the stiffener.

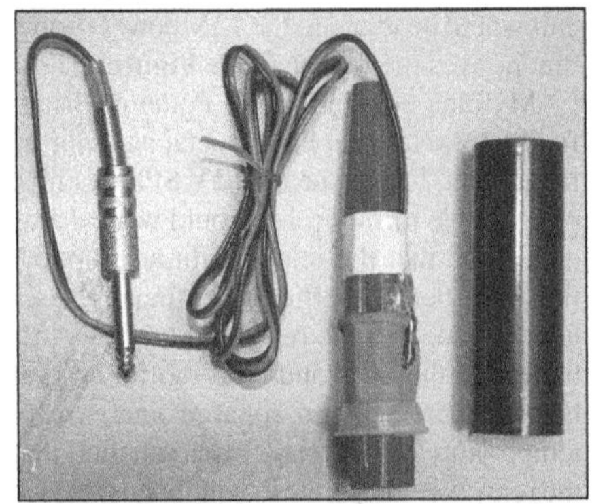

Figure 5 — The breath-activated key ready for you to plug into your transceiver and make CW music on the bands.

contact from moving a lot, making a better connection.

Basically the key is done at this point. I took a small tubular pill bottle, painted it black, cut a ⅝ inch hole in the cap, drilled another same-size hole in the end of the tube and pushed it on the kazoo. This protects the contacts and gives you something to hold on to (see **Figure 5**).

When the key is finished, plug it into the key jack, put it to your mouth like a trumpet, gently blow in the key and interrupt the airflow with your tongue. It'll take a little practice to make dots and dashes with your tongue, but you'll catch on pretty quickly. You will be surprised at how fast you can send. I have sent good code at over 20 wpm. People on the other end can't tell it's not a straight key. You can see my key in action on my YouTube video at **www.youtube.com/watch?v=UtYswqOEMZE**.

It could be used for mobile hands-free CW. Just imagine the looks on your fellow travelers' faces as you drive down the road, both hands on the wheel, making kazoo contacts. — *73, Lee Cheever, W5IG, 2727 Leta Mae Cir, Dallas, TX 75234-6234, w5ig@swbell.net*

Marine "Under-Table" Mount

Many Amateur Radio operators are also avid boaters. No matter if it's a rowboat or even a large yacht, the issue of bringing along or mounting an amateur transceiver usually comes up sooner or later.

In my case, I have a 21 foot pontoon boat and I was trying to decide how to mount two mobile transceivers in an out-of-the-way, yet somewhat accessible location. I was looking at the floor-mounted pedestal table and it occurred to me that it would be neat to mount the radios underneath the table surface.

The pedestal table in my boat is about 23 inches in diameter with four molded cup holders. This or something similar is used on powerboats and yachts in varying sizes and configurations.

At first, I thought I could just use the underside of the drink holders to mount a round piece of plywood that would, in turn, support the usual mobile radio mounting brackets. After examining the thin plastic used in the molded cup holders I changed my mind.

Figure 6 shows what is a much better idea. I have the radios mounted on an 18 inch diameter piece of marine plywood fastened to a suitable bushing, located under the regular table. This under-table table works great and can be easily rotated lazy Susan style so the radio you want to use is always handy. Four radios could easily be accommodated and there is still room beside each radio for the microphone, SWR meter, etc.

The basic components are an 18 inch diameter piece cut from some exterior or marine plywood, into which is drilled a 3 inch di-

Figure 6 — The under-table completed and all decked out with radios ready for a maritime adventure. Note the pass-through holes for the wiring and the Velcro tie on the pedestal below to keep the wiring neat. [Don Dorward, VA3DDN, photo]

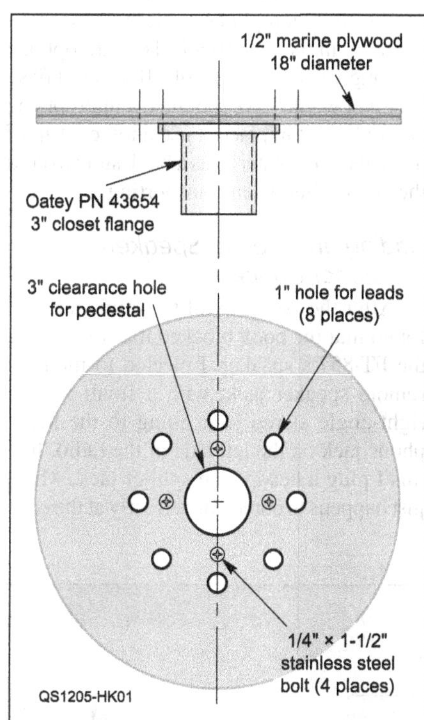

Figure 7 — This is a drawing showing the construction details of the table and flange.

ameter center hole to clear the pedestal and a 3 inch plastic closet flange made by Oatey (p/n 43654) that seemed made for the purpose. The lip of the flange is 3 inch OD but about 2½ inch ID. With a little shimming, it's a perfect fit for the 2¼ inch diameter pedestal. A simple screw type pipe clamp on the pedestal below the flange keeps it from slipping down and yet allows it to turn easily.

Figure 7 provides some dimensions and shows the addition of 1 inch clearance holes in the table for radio power and antenna leads. These cables are kept neat using a Velcro wrap on the pedestal. — *73, Don Dorward, VA3DDN, 1363 Brands Ct, Pickering, ON L1V 2T2, Canada,* **va3ddn@arrl.net**

Mount That Mic

For a number of years, I have placed my 2 meter microphone in a center console cup holder. Usually, I end up chasing it around the floor shortly after driving the car. My solution was to make a small removable flat surface with a RadioShack microphone clip (see **Figure 8**). The holder stays in place by means of a small brace. The surface is made from particleboard (4 × 2.5 × 0.25 inches). The brace is pine (3 × 1 × 0.75 inches). The brace length is the same as the diameter of the cup holder and its vertical edges are beveled for a snug fit in the round opening. Three wood screws hold the metal clip, flat surface and brace together (see **Figure 9**). — *73, Tom Hart, AD1B, 54 Hermaine Ave, Dedham, MA 02026,* **tom.hart@verizon.net**

Figure 8 — The microphone holder sitting in the center console. [Tom Hart, AD1B, photo]

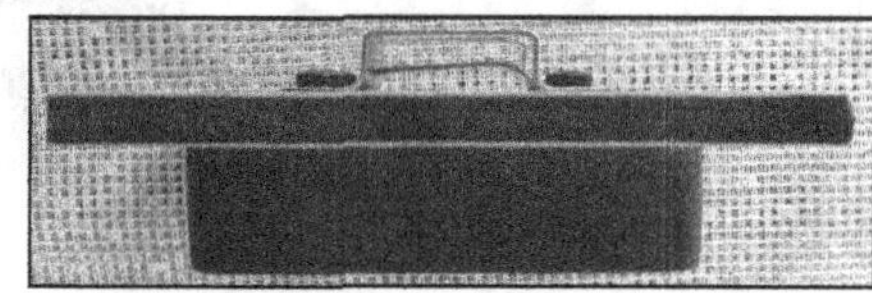

Figure 9 — This is a side view of the microphone holder showing the microphone clip on top, next the particleboard and finally the pine brace shaped to fit the cup holder. [Tom Hart, AD1B, photo]

Leg o' Ham

New car — new puzzle. How do I install my mobile rig, which fit fine in my 1997 Taurus, in a 2011 Camry? As hard as I tried, I couldn't find a good place to mount the remote head for my IC-208H transceiver without it interfering with a vehicle control or being unreadable because of the viewing angle. I even tried a cup holder mount but found that the head still ended up "floating" in the wrong place (and I lost a cup holder).

I concluded that the only place to safely operate the head, keep it out of the way of the airbag system and clearly read it was just above the dashboard vent. Yet, there was no easy way to mount it because of the dashboard slope, which aimed the head toward the floor. After toying around with a few ideas, toying around became the actual solution — I noticed that the versatility of LEGO building bricks was just what I needed.

Out of the hundreds of brick shapes I'd seen, one of them had just the right angle to level the remote head. With the bricks mounted to one of the smaller LEGO baseplates (just flexible enough to contour to the dashboard curve), I could give the whole thing a stable base and plenty of mounting surface area (see **Figure 10**). Luckily, both gray and black bricks were available and matched the car's interior close enough.

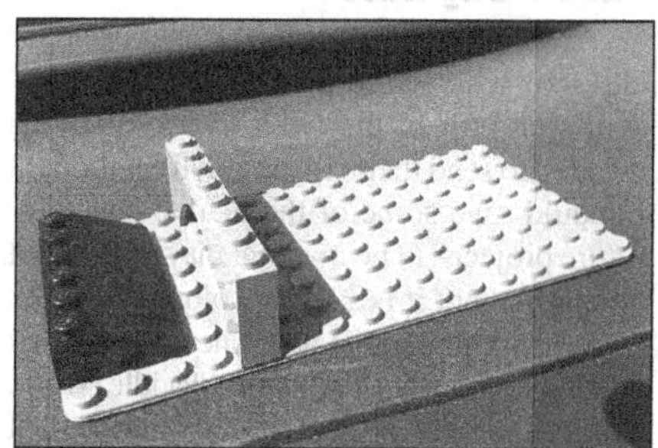

Figure 10 — Use these simple LEGO pieces to construct a mount for your transceiver's control head. [Mike Fesko, KB1LKH, photo]

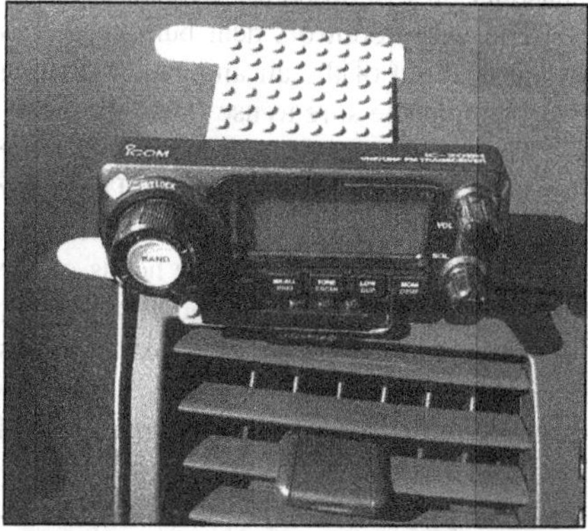

Figure 11 — Here the completed mounting holds the control head where it doesn't interfere with driving but is accessible for easy operating. [Mike Fesko, KB1LKH, photo]

Another consideration was not to damage the vinyl dashboard material while attaching the baseplate. For that, I used the leftover adhesive strips from a 3M coat hook. Hopefully, years from now all I will have to do is pull the tabs and the adhesive will come off clean. I secured the front end of the mounting baseplate to the plastic vent with a few small squares of hook and loop material, since I needed some rotational stability for button pushing and some height to fill the gap.

As you can see in **Figure 11**, the design is simple — a baseplate with a few angled bricks, plus a support arch to take the load if I push too hard on the buttons. If the support pops off, I'll glue it down, but for now it seems to be holding strong. None of the bricks are glued. There's plenty of room for my fingers to push buttons and tune with and I lucked out with the remote head cable — it managed to fall right in line with a dashboard/control panel seam. Some coax sealant, which also matched the interior color quite well, did a nice job of keeping it in place. Since taking the photos, I colored the white adhesive tab strips with a black marker to eliminate reflection in the windshield.

I'm not an electronics whiz so I'll never build much of a circuit, but with some imagination and playful experience, found I could whip a ham radio mounting problem. It was kid's play! — *73, Mike Fesko, KB1LKH*

Marble Bug Base

At some point, all of the trophies that our children won while participating in youth sports end up in the trash. When my sons decided to send theirs to the landfill, I thought I should save some of the marble bases for future projects. (The material may or may not be marble, but it is a white colored stone.)

When I decided to add a new base to my American Morse Express Porta-Paddle (**www. MorseX.com**) I chose a piece of marble that once graced a baseball trophy. By placing the marble in a plastic pan of water, I was able to use my drill press to make two holes for the screws that secure the paddle. [A carbide-tiped masonry bit will probably work best. — *Ed.*]

The water controls heat buildup, keeps stone dust out of the air and makes drilling very simple. Be sure to use safety glasses!

The screws go through the marble base into the bottom of the paddle. Then, for a final touch apply four stick-on felt feet. These allow clearance for the heads of the two screws that fasten the paddle. [Another option would be to countersink the holes and use flat-head screws. — *Ed.*] I am very pleased with the result and recommend recycled marble trophy bases for a variety of radio projects. — *73, Tom Hart, AD1B, 54 Hermaine Ave, Dedham, MA 02026,* **tom.hart@verizon.net**

A Mini-Station Go-box

Do you need to move your radio from house to car every time you want to go mobile? Ever wish you had a convenient "go-box" for your HF radio that can be connected in only a few seconds?

Searching for an Answer

I have one radio, a Yaesu FT-857 transceiver, which I use when at home or when going mobile. I had been looking for a "right sized" box in which to mount the radio and CW paddle. After many false starts (including an empty computer case my wife and I don't talk about any more), I happened upon the following idea.

I found a pair of fake decorative books at a local hobby store and decided to modify them for my HF radio. I decided to mount the smaller book with the CW paddle and microphone on top of the larger.

Building the Mini-station

I removed the bottom end board from the larger of the books and installed the FT-857, using materials found in my junk box (see **Figure 12**). I added a right-angle PL-259 to the rear of the radio to keep pressure off the coax. A carpenters pencil (filed to the right thickness) was screwed to the floor of the book aligned with a ridge on the bottom of the radio. Then I secured the radio to the bottom of the book using a strip of hook and loop fastener, but I used it as a strap with a bolt, washer and nut at each end.

On the right side looking at the book cavity, the dc power cable and antenna connector were affixed to the rightmost board. I elected to use a PL-259 to BNC adapter to make all connections easier and faster for moving the radio back and forth between my shack and vehicle.

I bolted the smaller book to the top flap of the larger with small hardware. To mount the CW paddle, I drilled holes for the paddle's four feet and then held the paddle down with another strip of hook and loop bolted as before on each end to both the bottom of the smaller book and the top of the larger (transceiver) book (see **Figure 13**). The paddle cannot move.

I cut an access slot in the bottom edge of the smaller book to access the paddles and provide an exit for the microphone cord when using SSB. I then drilled a larger hole between the two books to allow the CW paddle and microphone cables to connect to the radio in the other book. The microphone fits snugly behind the paddle. If it hadn't fit so snugly, I would have mounted a microphone clip to hold it in place. I use another strip of hook and loop material as a book strap to hold the books shut when transporting.

Adding an External Speaker/Headphone Jack

After the assembly of the mini-station, I found that the book blocked the sound from the FT-857's speaker. I elected to mount a remote speaker jack, with a small ⅛ inch right-angle stereo jack going to the headphone jack on the left side of the radio. Into this I plug a headphone splitter jack, which just happens to come out perfectly at the edge

Figure 12 — In this view the larger book is open to show the FT-857's mounting arrangement. Note the use of the hook and loop strap and right-angle connector. The wires going through the open cover lead to the paddle. [Sam Moore, NX5Z, photo]

Figure 13 — This interior view shows the paddle bolted and strapped in the smaller book. There is plenty of space left for the microphone. [Sam Moore, NX5Z, photo]

of the transceiver book. A screw angled into the book binding secures it firmly to the left side. When mobile, I connect the HF radio to the auxiliary jack of my car stereo. This arrangement allows me to change the volume without taking my eyes from the road. At home, I either plug in headphones or an external speaker.

Success!

All the connections were chosen to be on the front of the mini-station (bottom edge of the books), so it would be easy to connect and disconnect. When I am in my small truck, the books fit snugly in a hollow between the two seats, resting on the floor. (I plan to make a small seatbelt to keep it in place in case of a crash.) I can rest my arm on the truck armrest and the radio and paddle are right at my fingertips. It is probably the most comfortable station I've operated!

An added bonus of this setup is that, looking into your vehicle, if you tuck in or cover the wires, a would-be thief would see nothing but two books. You have your own clandestine radio mini-station. Another plus — you don't have to leave your expensive radio in the vehicle. You can take it inside the house or motel without fuss or problem, since it takes only seconds to connect or disconnect.

This mini-station will also serve as a great portable station, to "grab-and-go" when going to the lake, etc. The only other things you require are a battery and antenna. I wish I had done this a long time ago. — *73, Sam Moore, NX5Z, 22 Cundiff Dr, Sherman, TX 75092-6326*, **drsammoore@aol.com**

Door Stop Mount

I purchased a nightstand to use as an operating station. I am using a mobile UHF/VHF radio as my fixed station in my RV and I did not want to cause damage to the nightstand by bolting the mobile mounting bracket to the tabletop. I found a couple of wedge shaped rubber doorstops for less than $1 each and some hardware for less than $2, and attached the mounting bracket to the doorstops. The new table top mount is very stable, will not tip over and does not slide across the smooth table surface. Additionally, the angle of the mount is perfect to make the LCD display easier to read. — *73, Chris Seright, KE5ZRT, 1414 Sunrise Dr, Space 2b, Amarillo, TX 79104*, **ke5zrt@gmail.com**

Iambic Key Cover

I needed a cover for my MFJ-564B iambic paddle to prevent dust from accumulating on the contacts. I found a package of sticky-notes enclosed in a plastic box. The box had an open top and a slot in the front to make removing the notes easier. This was ideal, as the slot allowed access to the paddles (see **Figure 14**). The plastic box was almost exactly the size of the 4 × 4 inch iambic key base and had

Figure 14 — A sticky-note box can serve as an inexpensive paddle cover. [Randy Miller, AA5OZ, photo]

sufficient height to clear the key when placed over it. If you're concerned about dust getting on your paddle contacts, be on the lookout for a small plastic sticky-note box to help keep those contacts clean. — *73, Randy Miller, AA5OZ, 4122 Mary Ann St, Lake Charles, LA 70605-4102*, **randy500a@gmail.com**

Magnetic Panel Markers

To make band changing quicker I found magnetic refrigerator signs to be quite useful. Simply cut $^{1}/_{16}$–$^{1}/_{32}$ inch wide strips and position them on the (steel) panel. If you use white material you can color code these markers to specific bands or modes using a felt tipped marker. Repositioning is simple and does not deface the panel. — *73, Roy Lehner, WA2SON, 5464 Oakwood Dr, North Tonawanda, NY 14120*, **wa2son@toast.net**

Tame Two Bugs for Five Bucks

I was leafing through a catalog from an expensive hardware store and noticed a pair of brass staircase gauges shown in full size. These gauges are intended to be clamped to the edge of a rafter square to facilitate making the repetitive cuts required when building stairs. I thought, "*Wow*, these gauges could be *bug tamers*!"

What is a Bug Tamer?

A bug tamer is an extra weight added to the vibrating arm of a semiautomatic key (bug) to tame (decrease) the dit speed. Alternately, a bug tamer could be an extension to the vibrating arm permitting the original weight(s) to be moved farther from the pivot, also decreasing the dit speed. In this article, the bug tamer is a 27 g brass staircase gauge clamped to the existing vibrating arm along with the original weight(s).

Testing the Bunnell J-36 Bug

I recently acquired a J-36 bug manufactured by the J.H. Bunnell Co during World War II. The friend who sold it to me said it was a nice bug but a bit fast for him. It was a bit fast for me too so I went looking for staircase gauges

at a home improvement center. Johnson Level and Tool manufactures them and a pair cost me $5. I drove straight home to test my new toys.

Digging through my *Handbooks*, I found a formula relating CW speed to dit frequency:

$$\text{WPM} = 2.4 \times \text{dits per second} \qquad [\text{Eq 1}]$$

I connected a 9 V battery in series with my Bunnell J-36 and monitored the dit train with my oscilloscope. The time in milliseconds from the start of one dit to the start of the next dit is the *dit period*. The frequency in dits per second was equal to 1000 mS divided by the dit period in milliseconds.

For the Bunnell J-36, with its 22 g cubic weight placed at the maximum radius (minimum speed), the dit period was 70 mS, which per Equation 1 is:

$$2.4 \times (1000/70) = 34 \text{ WPM}$$

Placing the cubic weight at the minimum radius (maximum speed) yielded a 40 mS dit period, which per Equation 1 is:

$$2.4 \times (1000/40) = 60 \text{ WPM}$$

Ditching Dits

It is no wonder that I was uncomfortable with the J-36. Obviously, it would be hard work to get my dah speed up to match the dit speed! I clamped my 27 g brass staircase gauge, along with the standard 22 g cubic weight, to the arm of the J-36. The minimum dit speed dropped to 27 WPM (90 mS dit period) and the maximum dit speed dropped to 34 WPM (70 mS dit period).

In **Figure 15**, the cubic weight is at maximum radius and the bug tamer is at minimum radius giving an 80 mS dit period for a speed of 30 WPM on the J-36.

Using the Bug Tamer with Other Bugs

The bug tamer can be attached to a bug having either a rectangular arm (J-36 and

Figure 15 — To slow down a bug, attach a brass staircase gauge to the bug's vibrating arm. [T. M. Hamblin, VE3HIE, photo]

Table 1
Speed Ranges With and Without the Bug Tamer

Type of Bug Standard Weights	Speed Range Standard Weights	Speed Range With 27g Bug Tamer
Bunnell J-36 One 22 g cubic	34-60 WPM	27-34 WPM
Lionel J-36 One 25 g cubic	30-53 WPM	20-30 WPM
Vibroplex J-36 Lightning Bug One 25 g cubic	20-32 WPM	16-20 WPM
Vibroplex Original Two 14 g cylindrical	26-34 WPM	22-24 WPM
Vibroplex Original Deluxe Two 14 g cylindrical	30-44 WPM	27-30 WPM

Lightning Bug) or a circular-rod arm (Blue Racers,[1] Originals, Les Logan, Johnson Speed-X, etc) A comparison of the standard weights is given in **Table 1**.

If you decide to remove the standard weights, be sure to put them where you can find them. I suggest fastening them to your bug with a cable tie or similar device. If you eventually decide to sell your bug, the buyer will want the original weights (and your bug tamer).

If you have a fast bug that is making you uncomfortable, try using the brass staircase gauge bug tamer. It's not an expensive experiment! — *73, T. M. Hamblin, VE3HIE (sign HN), 9798 Trew Rd, RR1, Campbellcroft, ON, L0A 1B0, Canada,* **ve3hie@arrl.net**

Easy Rotator Correction Calculator

Many hams have antennas with rotators that don't point in the direction the indicator shows. It is a big job to climb up the tower and reposition the mast to the correct bearing, especially in bad weather. Large arrays tend to be shifted out of calibration frequently due to their high wind loads.

I devised a solution to this problem. My M[2] KT-34 (**www.m2inc.com**) was pointing south when the rotator indicated it was somewhat north. Rather than doing the tower chore again, I made up this error calculator so that I know where to point the beam from the shack.

All that is required is a common CD and CD case. Turn the CD over so that the *data*

side is facing out and fix it back in the case. Using labels or a fine felt marker, label it N, S, E and W, and add some tick marks for intermediate bearings. Do this about an inch from the CD's outer edge.

Then do the same on the outside of the clear CD case, at about the level of the CD's edge (see **Figure 16**).

To use the calculator, first aim the beam at any of the four main bearings according to the rotator's indicator. Then go outside and see where the beam is actually pointing. Open the CD case and rotate the CD to coincide with the direction the rotator says. The calculator is now ready for use.

Let's say the beam is facing south when the indicator shows north. You want to work VP6T on Pitcairn, which is more or less southwest from you. The rotator calculator now shows you must turn the beam to northeast on the indicator.

If you can't climb the tower to readjust your beam, just adjust the calculator to the current beam heading and you are good to go. — *73, Pat Bunsold, WA6MHZ, 1615 La Cresta Blvd, El Cajon, CA 92021-4072,* **wa6mhz@cox.net**

Hearing Aids as Headphones

If you're considering purchasing hearing aids (HAs), consider some of the new technology available. In addition to all the new tuning and intelligibility algorithms, many modern HAs have integrated Bluetooth wireless technology that allow them to interface with cell phones and television audio.

Apparently Bluetooth isn't yet available for Behind the Ear (BTE) units. In the case of my Oticon HA, there is a USB-charged receiver that you can slip in a pocket or wear with a lanyard antenna (see **Figure 17**). Its range to the HA is a short, but adequate. For the first use, it has to be linked to your HA by your audiologist, but afterward it can be paired by typical pushbutton methods to any Bluetooth device. Oticon offers a landline telephone interface and a line level audio box that plugs into a TV's audio outputs. I typically use the TV interface for enhanced clarity.

The user can link the pocket receiver with a dozen or so standard Bluetooth devices. Interestingly, it also has a 2.5 mm audio input jack. With the right adapter cable, you can connect your rig's headphone output to this unit and convert your HAs into custom

Figure 17 — **Modern Bluetooth enabled hearing aids permit the user to interface their devices to a variety of external audio sources, including your transceiver. [Don Pomplun, K2BIO, photo]**

Figure 16 — **If you're having problems keeping your beam aligned, this handy calculator will help keep your aim true even if your rotator indicator isn't. [Pat Bunsold, WA6MHZ, photo]**

[1] With the Blue Racers, the maximum and minimum speeds were both 24 WPM with the bug tamer because the arm is just long enough to hold the original weight(s) and the bug tamer. There is no room to adjust them so the maximum and minimum speeds have to be the same.

headphones. The Oticon receiver has an option for listening only to its audio source or to mix in the HA's microphones so you can hear other noise in the room (like a call to dinner).

Though still obscenely expensive, with limited medical coverage options, my audiologist included these accessories in the HA purchase price. If it's time for you to purchase or upgrade, these are features to look for. — *73, Don Pomplun, K2BIO, 521 Van Buren Pl, San Ramon, CA 94583,* **k2bio@arrl.net**

Inexpensive Microphone Boom

Why spend over $100 on a microphone boom? For $15-20 you can buy a boom style desk lamp at office supply stores. Simply remove the lamp assembly and cord from the boom and install the microphone using your existing floor or desk stand holder (see **Figure 18**). I used two washers to make up

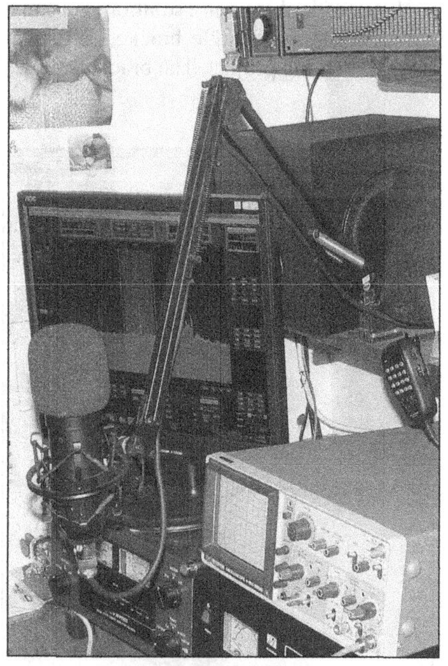

Figure 18 — For a few bucks and a little time an inexpensive flexible-arm desk lamp can become a boom for your microphone. [Joe Vlk, W8DCQ, photo]

Figure 19 — A couple of washers will compensate for the difference in size between the lamp assembly and the microphone holder. [Joe Vlk, W8DCQ, photo]

for the difference in thickness between the lamp head and the microphone holder (see **Figure 19**). Strap your microphone cable to the boom arms, leaving some slack at each elbow. Depending on the weight of your microphone, you may have to change the boom's counterbalance springs. — *73, Joe Vlk, W8DCQ, 3967 Shoshone Ct, Oxford, MI 48370-2933,* **w8dcq@arrl.net**

Give Your Rigs Some Room

As my ham radio interests extended up into the V/UHF area, I decided to obtain a set of three transverters for these bands. In using them, I found that they tend to heat up. I tried using a computer fan to improve cooling, but found that it wasn't sufficient to maintain a proper temperature. My mounting arrangement that had all three units stacked on top of each other only aggravated the situation.

Seeking a solution, I went to my local home improvement store to look for some inspiration. I found the perfect solution — a paint roller screen. They are less than $3 each, come in different sizes and can be easily bent to fit your needs. You can even paint them to match your gear.

After removing the plastic feet from my gear, I simply added some double-sided tape and attached them to the screen (see **Figure 20**). The coax and other wiring keeps the gear in place. — *73, Sebastian Acosta, W4AS, 19340 Franjo Rd, Cutler Bay, FL, 33157-8818,* **w4as@arrl.net**

Dollar Store Key Cover

Here is a great idea that I thought I would pass along to CW enthusiasts. If you are like me, being a 100% CW operator, your key, paddle or bug, is usually one of the most important items in your shack. Vibroplex and other dealers offer dust covers to protect and preserve these investments, but such commercial dust covers cost $50 or more. After spending $200-300 on a key, $50 doesn't

seem like much to spend to protect these high quality keys but I found this inexpensive and effective alternative.

I love to visit my local dollar stores. On one visit, in the office supplies section, I came across a transparent plastic office basket. It seemed lightweight and sturdy, and the perfect size to cover an entire key. So, for a few dollars, I purchased three of them.

I set the office basket next to my paddles and estimated where to cut out for the paddles and the cable. I performed the cuts in the plastic with a box cutter knife. The plastic cut easily.

This office basket was the perfect size and is very attractive (see **Figure 21**). They can be purchased in a number of different colors and sizes. They keep the dust off the key while adding a bit of color to your operating table. — *73, Bill Parker, KA3IXF, 3314 Old Capitol Trail, Apt K12, Wilmington, DE 19808,* **wwjp123@comcast.net**.

Figure 21 — This small trapezoidal shaped plastic bin is inexpensive and can be easily modified to protect your key or paddle from dust, dirt or spills. [Bill Parker, KA3IXF, photo]

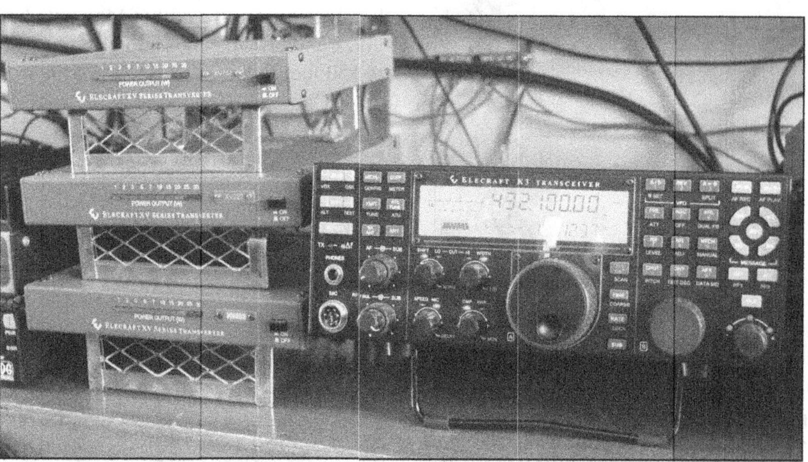

Figure 20 — Using modified paint roller screens to separate stacked equipment improves air flow and cooling while also eliminating a cluttered look in your shack. [Sebastian Acosta, W4AS, photo]

Handheld PTT

For those of us that don't use VOX, and would like a PTT switch remote from the rig, the following handheld PTT switch might be of interest. My idea was to come up with an inexpensive handheld device that would provide both momentary and on-off PTT capability in one unit.

From RadioShack I obtained a momentary switch with a black top (part number 275-0609), a SPST on-off switch with a red top (part number 275-0011) and a shielded 6 foot cable with ⅛ inch mono plug on one end and stripped wires on the other (part number 42-2434). [You may need a different connecter to mate with the accessory PTT jack on your rig. Consult your manual. — Ed.] Parts from the hardware store included two 1 inch PVC caps, PVC cement and 5 minute, two part, epoxy cement.

I drilled two ⁷⁄₁₆ inch holes in the dome of one cap to accommodate the switches and one ⅛ inch hole in the dome of the other to accommodate the cable. I then installed the switches into the holes using epoxy because there was no room to secure them with the supplied washer and nut. I then wired the two switches in parallel, ran the stripped end of the cable through the PVC cap and soldered the two leads to the switches. I tied a knot in the cable to provide strain relief where the cable exits the PVC cap.

I used PVC cement to join the two PCV caps together to complete the project (see **Figure 22**). [The PVC cement will permanently attach the caps. Before sealing the ends, inspect the solder joints and test both switches to make sure they are working properly. — Ed.] The mono plug end of the cable plugs into the breakout box that connects with the accessory jack on my rig. The interface to the PTT input on your rig may vary.

The switch unit fits comfortably in my hand. Depending on the nature of my contact, I use my thumb to activate either the momentary or on-off switch. — *73, Thomas Nolan, W3EX, 624 Store Rd, Harleysville, PA 19438-2718,* **w3ex@arrl.net**.

J-38 Backpacking Modification

I like to operate low power CW portable from remote locations and I prefer to use a J-38 key. The J-38 is too tall for my pack, plus the adjustment knobs get caught on straps and cables.

To solve these problems, I replaced the adjustors with 8 × 40 slotted screws cut just long enough for my preferred settings. The 8 × 40 screws are used for mounting rifle sights and should be available at any gunsmith or well-stocked hardware store.

I also found that reversing the key on the base reduced the overall length and helped prevent unexpected tip-up. The key is lower, more stable and only a screwdriver away from its original condition. — *73, Scott McCann, W3MEO, 160 Shields Ln, Queenstown, MD 21658,* **achess@juno.com**

Color Coded Cables

For a long time I've tried to find an efficient method of marking cables for easy identification. Paper tags never worked well. Engraved aluminum tubing sleeves lasted forever but were difficult to install and read. Commercial lettering tape was difficult to read at a distance. Then it came to me: resistors are color coded so why not cables?

Heat shrink tubing is now readily available, comes in all sizes and in a rainbow of colors. I use it for all types of cables although, in some applications, one needs to plan ahead and place the heat shrink tubing on the cable before installing the desired connector.

I use color codes to indicate the length and/or the location of a cable (see **Figure 23**). When getting ready to pack for the field its handy to glance at a coil of coax with brown, green and brown bands and instantly know it is 150 feet long.

I use a length of white heat shrink tubing as a base color so black and brown bands stand out. For coax in the shack, one of my methods is to identify all cables coming from the attic antennas with a green base color. Outdoor HF antenna cables have a yellow base color. Other colored bands on the base color tell me the specific antenna. The possibilities are limited only by your imagination. — *73, Walter Martin, KB5HOV, 10944 Grissom, Suite 715, Dallas, TX 75229,* **kb5hov@alumni.southwestern.edu**

Angle Bracket Panel Supports

I was having problems working on the control and metering panel for an amplifier. Every time I tried to do something to the panel, it was either in the wrong position or it fell over. I tried supporting it with wood blocks and with books, but those solutions either took up too much real estate on my bench or simply weren't stable enough.

I started looking for a solution and came across some right angle brackets left over from a shelving project. The brackets' holes

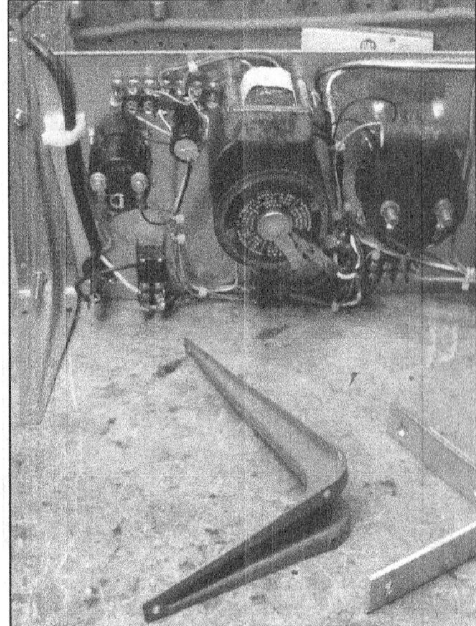

Figure 24 — Angle brackets make a solid and steady support when working on equipment panels. [Steve Gilbert, K1SG, photo]

Figure 22 — The completed PTT switch fits nicely in the palm of your hand. The black button functions as a regular PTT while the red button is a push-on, push-off type for those long ragchews. [Thomas Nolan, W3EX, photo]

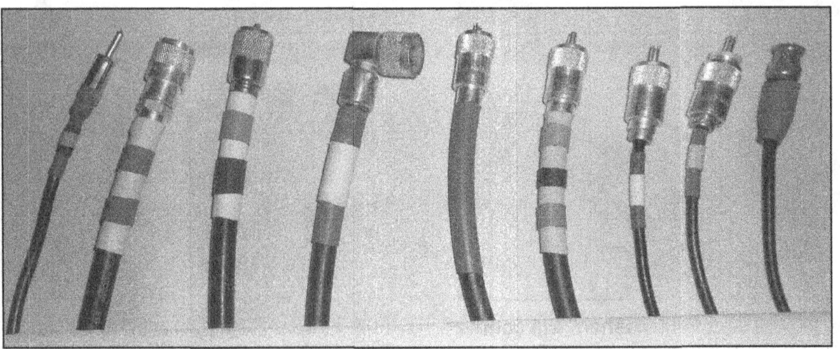

Figure 23 — Colored heat shrink applied to the ends of cables can be used to indicate length or application at a glance. [Walter Martin, KB5HOV, photo]

were in the wrong places to fit the notches in the panel but a drill fixed that. I used 10-32 hardware, the same type I use to attach the panel to the rack, and now my rack panel stands upright (see **Figure 24**).

Note that metal bookends can be used for similar purposes, but they may not be as strong. In this case, I was mounting transformers and small Variacs to the panel and needed solid support. — *73, Steve Gilbert, K1SG, 75 W Elm St, Hopkinton, MA 01748-2102, **k1sg@arrl.net***

Shack in the Rack

I am always on the lookout for ways to transport my public service radios. Any method has to meet two criteria as a way to organize the radios, wires and associated support gear. The first is equipment protection and the second is gear organization.

The first criterion is self-explanatory. Public service jump kits need to be ready to go in a hurry. Grab and go kits are subject to bumps and bruises sustained in disaster environments. They need to be kept safe from the bent knobs, smashed displays and stressed connectors that are the hazards of taking the rigs out of their safe shack environment. The second criterion, organizing all pieces of equipment, means that once deployed, the operating environment takes up minimal space while providing the greatest communications flexibility. This is especially tricky given how many critical cables, connectors and wires are associated with these devices.

Stand and Deliver

A few years ago I was organizing my work station in my home office and needed a way to keep the monitor off my docked laptop. I discovered the Allsop Metal Art Monitor Stand (**www.allsop.com**). It comes in two sizes (15 × 11 and 19 × 12 inches) and have simple U-shaped tubular steel legs held together with spot welded, perforated sheet steel on the top colored an anodized black or gray. I thought it would be easy to hang a radio off the bottom and use the top for the laptop platform (see **Figure 25**).

Mounting the 144/440 MHz radio's bracket, speaker and wattmeter wasn't a problem, but the power supplies (for radio and laptop) were more of a challenge. Zip tie wraps were the perfect solution. With so many holes on the top, mounting is easy for almost any shape.

The results were fantastic. I was able to put all the gear I needed on the front side and tuck the power supply away in the back. I secured all the cables as neatly as possible under the stand with only the necessary pigtails emerging from the sides and back (see **Figure 26**). These included the RF output, SignaLink USB cable, connector for digital

Figure 25 — An Allsop monitor stand makes a great base for a digital go-kit station. [Jon Rudy, K3QF, photo]

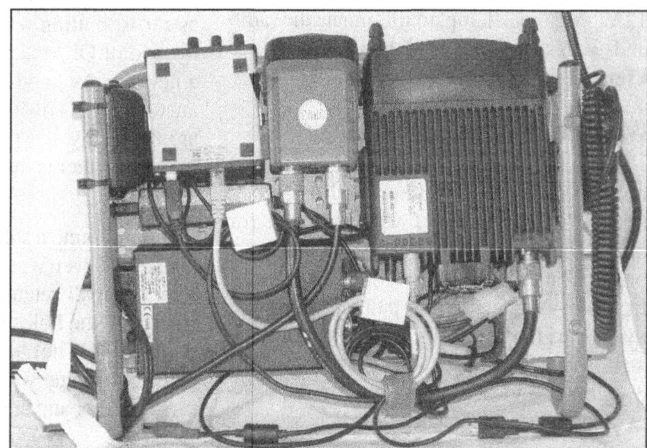

Figure 26 — A hook and loop strap turns a tangle of wires into a tidy bundle secured to the bottom of the stand. [Jon Rudy, K3QF, photo]

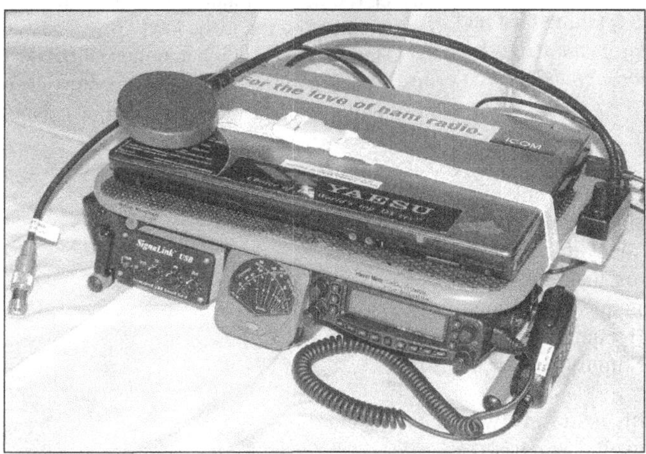

▲Figure 27 — Folded up and ready to travel, the digital station even has room for a gooseneck lamp. [Jon Rudy, K3QF, photo]

▶ Figure 28 — A wrap rack provides a three tier frame to support a power supply, transceiver and tuner, with space on top for a wattmeter and room on the side for a SignaLink and paddle. [Jon Rudy, K3QF, photo]

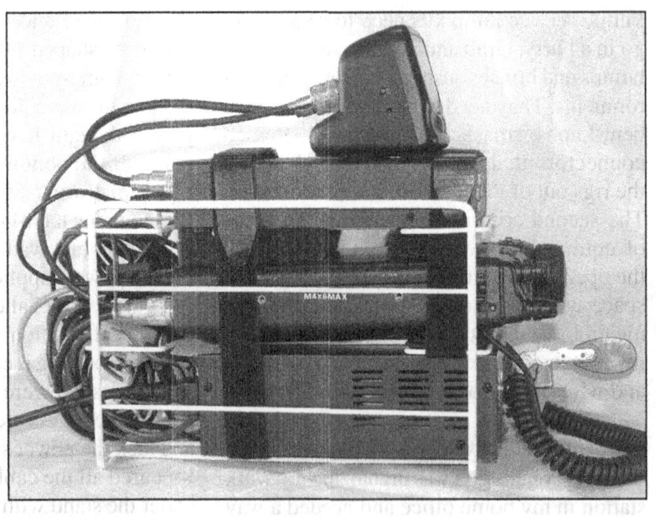

Figure 29 — Hook and loop straps keep everything secured to the solid steel wire frame. [Jon Rudy, K3QF, photo]

Figure 30 — A few pieces of PVC, some junk box parts and some sandpaper comes together to make a mobile mic holder. [Ted Pirolli, KD8HYT, photo]

operation, ac power cord and laptop power. I was even able to side mount a surplus 12 V gooseneck lamp, to illuminate the radio and workspace in low light situations (see **Figure 27**).

Wrap Rack

The second rack came from a most unlikely place — the kitchen. I happened upon a "wrap organizer," which is a rack for organizing boxes of foil, cling wrap and wax paper. The Schulte Wrap Organizer is made from heavy gauge white painted steel wire that forms three shelves, the top two having an I shape. I took a gamble ordering it for my jump kit idea, but for less than the price of a burger meal, I thought it was worth a try. After all, my spouse could use it in the kitchen if it didn't fit.

I was delighted to find that my power supply, FT-857 transceiver and tuner all fit neatly in the spaces provided (see **Figure 28**). The open nature of the setup allowed me to neatly wrap and secure cables to the back side. I used stretchy hook and loop straps to hold it all together for a grab and go shack in the rack setup (see **Figure 29**).

Using some trusty plastic tie wraps, I secured the SignaLink USB modem (**www.tiger tronics.com**) to the side and mounted a microphone clip to the radio using one of the mounting holes. The wattmeter is held on the top with sticky hook and loop fasteners so it is removable for transport. The keyer paddle is tie-wrapped to the side. I'm not entirely happy with the paddle and am considering other arrangements. As with any homebrew design, it is still a work in progress. — *73, Jon Rudy, K3QF, 608 W High St, Manheim, PA 17545,* **jonk3qf@gmail.com**

Powerpole Clip

Anderson Powerpole connectors (**www. andersonpower.com**) have become ubiquitous in the ham shack. Almost all the station accessories you purchase use them or can be purchased with them as an option. I even go as far as putting a Powerpole connector on the end of OEM supplied power cables with a new rig. As good as Powerpoles are, there are two things I dislike about them. First, there are those tiny, hard to insert, easily lost, roll pins. The other is that a slight tug can disconnect them.

I have found a solution to both those problems, and I even went a step further. Anderson offers a small retention clip (pn 110G68) to insert into the holes where the roll pin would normally go. The clip keeps two connectors together; no longer will a tug on a cable separate the connector and there is no need for roll pins.

The ones shipping today have a small hole, which is a good place for a small piece of twine to tie it to the cable. I prefer to tie the clip to the power source side of the wire. Since I also use dual wall heat-shrinkable tubing to form a strain relief on the connector, I tied the twine to it and, with a small dab of glue, I can keep everything neat and in place. — *73, James Philopena, KB1NXE, 265 Frost Hill Rd, Marlborough, NH 03455,* **kb1nxe@arrl.net**

Mobile Mic Holder

I have tried everything I can think of to keep my mobile microphone off the floor. As you are probably aware, many microphones no longer have a button on their back for a holder. I needed a convenient way to mount the microphone that didn't require me to drill holes in the dash of a pickup truck that cost as much as my house.

What I came up with made my wife chuckle but works quite well. I purchased a 3 inch piece of ¾-inch PVC pipe, a pipe cap and a small brass screw in hook. The pipe is 0.848 inches and the cigarette lighter is 0.818 so a little sanding on the end of the pipe was needed. I inserted the finished product into my cigarette lighter socket (see **Figure 30**). It works quite well for parts from the junk drawer. — *73, Ted Pirolli, KD8HYT, 143 E Walnut St, Petersburg, MI 49270,* **tedpirolli@ yahoo.com**

Off the Shelf Expander

Belkin's RockStar, marketed to "connect up to five pairs of headphones," functions well for connecting bugs, keys, and paddles to the transceiver. I have a Begali Sculpture paddle (**www.i2rtf.com**), which I use most of the time, a Vibroplex Original Standard (**www. vibroplex.com**) and a Begali Blade straight key, which are in place and ready to immediately be put in the game. An extra 3.5 mm plug to 3.5 mm plug cable is included in the package (see **Figure 31**). — *73, Bob Mayo, W2TAC, 113 Taconic Lake Way, Petersburgh, NY 12138,* **bobmayo@outlook.com**

Fuse Audit

Power problems are a fact of ham life. Whether it is a winter storm causing your mains to sag, surge, or just plain disappear, or perhaps a summer lightning strike that sends a zap through your radio, system power is never 100 percent reliable. On the front lines, protecting your rig, is the lowly fuse.

Here's a great station maintenance project that is worth the trouble. First, inventory each piece of station gear containing a fuse. Note

Figure 31 — The Belkin RockStar splitter is an inexpensive and neat method for connecting multiple accessories in your shack. [Bob Mayo, W2TAC, photo]

that some contain internal circuit protection fuses, as well as a power fuse on the rear panel. Operating manuals often omit the exact specifications on these fuses so it's best to pull them, take a close look, and make note of the type, rating, and size of each fuse for each piece of equipment. [It's very important to check that second-hand purchases have correct fuses. — *Ed.*] The variety of fuses you find will, no doubt, surprise you.

Next, go to your junk box and find any fuses that match your equipment's needs. Not many spares found, right? That brings us to the fun part. Start hunting the flea markets, parts stores, and catalogs for the missing spares. Once you find them, buy a parts storage box with enough bins for each piece of equipment on your list.

Put one or two spare fuses for each piece of fused gear into each bin and add a label with the equipment information. Label the box and put it near your station where you won't forget it. Then, when one of those fuses is lost defending your rig, you'll be troubleshoot-ready. — *73, Robert Griffin, K6YR, 1436 Johnson Ave, San Luis Obispo, CA 93401,* **k6yr@arrl.org**

Induction Coupling of Transceiver Audio

Hearing loss with age often results in a 20 dB loss in certain frequencies. For those of us with more than a "normal" loss, the inability to distinguish between Bs, Cs, Ds, and many other letters becomes the norm. This inability to distinguish such sounds is why we use phonetics when we need to get a call correct.

The addition of hearing aids usually increases our ability to understand what has been said and communicate more effectively. Unfortunately, hearing aids have a down side. Used in conjunction with a headset, they often produce feedback. Using a speaker instead of a headset usually requires one to turn up the volume, filling your room with noise, annoying others in your household, and driving your hearing aids to distortion.

Fortunately, the solution is simple; most hearing aids come with a telephone setting (M or M+T) which allows the hearing aid to pick up the sound from the telephone by magnetic induction. To better interface your hearing aids with your radio, switch the hearing aid to the telephone mode and replace the headset with a magnetic loop pickup. Magnetic loop pickups are small induction coils that you can either hook behind your ear, or loop around your neck. (Search for "hearing aid magnetic loop pickup" to locate sources; they run around $40.) Plug the loop pickup into your transceiver's **PHONES** jack using a stereo to mono converter. The audio output of your transceiver will be routed to the loop pickup and then to your hearing aid

by magnetic induction.

Now all of the technology you acquired with your hearing aids can work directly with your rig's noise reduction and signal enhancement capabilities to optimize your perceived audio. As a bonus there is no feedback, distortion is minimized and the nonhams in your household don't have to listen to your contact. — *73, Lee Yackey, AI6G, 12715 Goethe Pl, Granada Hills, CA 91344-1420,* **yackeyfamily@cs.com**

Paper Clip Paddle

I have been homebrewing for many years and have built a number of the hacksaw blade-type paddles. I recently decided to design and build the simplest and cheapest paddle imaginable, which led to this idea.

I used one large paper clip for the paddle arm and two smaller ones for the contacts. After putting the first version together and trying it out, I found it was far too springy. I lowered the contact point between the paddle arm and the two contacts almost to the bottom. Then I modified the paddle finger piece moving it close to the contact clips for the least amount of spring in the arm. For a further refinement, I tied the two contact clips together at the top with a plastic-insulated tie wrap to reduce contact clip movement. This resulted in a very workable paddle.

This key is not for the heavy-handed operator, as it takes only a light touch to work. I find it comfortable for an easy, long CW chat at up to 20 WPM.

Bend the clips to the shapes shown and secure them as shown in **Figure 32**. Screw the clips down to a wood board and make any necessary adjustments. — *73, Larry Block, N2OH, 206 Deanne Dr, Lakewood, NJ 08701-7315,* **n2oh@optonline.net**

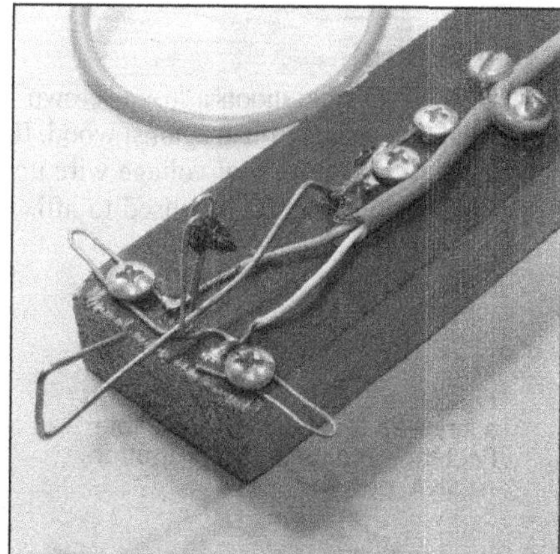

Figure 32 — **The simplest homebrew paddle ever devised will keep the code flying up to 20 WPM.** [Larry Block, N2OH, photo]

The Buzz on CW

The quality of CW has been improved through the use of filtering, IF band shifting, and notching together with electronic processing of the received signal. The sound that results is a very pure sine wave tone. The question is, is this pure tone the most recognizable for our human brains? I have noticed that a buzzing sound makes CW easier to understand.

A Little History

The original landline telegraph used the up and down clicking of the sounder to convey a message to the operator. The sounder often included an arch that helped improve the "tone" of the clacking arm. When I visited the Antique Wireless Association (AWA) museum (**www.antiquewireless.org**) in Bloomfield, New York, I was amazed by an old telegraph sounder that was mounted in a triangular enclosure. The operator had put a small cigar can with its lid half open over the sounder to modify the tone. Bob Hobday, a member of the AWA Board, explained that the can was used to produce a different sounder tone to distinguish between the signals coming from Buffalo and Syracuse. This enabled the operators to copy the different signals when both sounders were clicking simultaneously. The cigar can would vibrate to produce a unique tone — a rattling or buzzing.

CW Filter

I have tried several active audio filters, but I found the extra battery, connector, and speaker were inconvenient. I tried listening to W1AW's code practice with a portable general coverage receiver that I modified by covering its speaker with a small piece of cardboard, which served to dampen the higher frequency tones. I found it easier to copy the CW signal using the resulting low tone. The cardboard acted like a low-pass filter, but the sound vibrations tended to make it shift position.

To combat this, I put a small screwdriver on it. The combination of the cardboard

Figure 33 — **An old CD, a bottle cap, and a fishing weight can be used to give your CW a buzz.** [Hwang Nam Chang, KE2HF, photo]

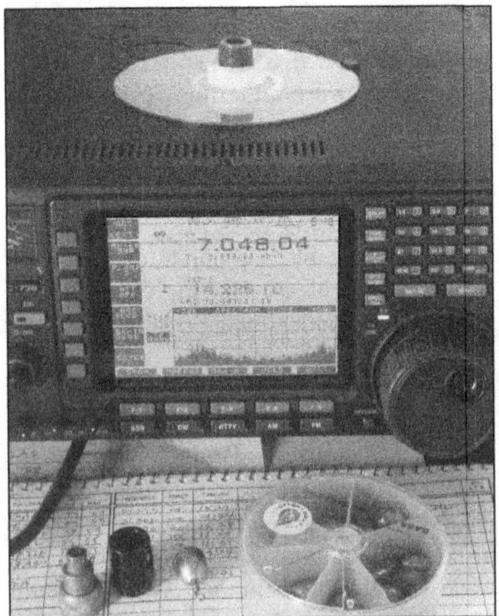

Figure 34 — The addition of a CD and a small weight can be used to adjust the CW tone of any transceiver with a top-mounted speaker. [Hwang Nam Chang, KE2HF,

with the weight of the screwdriver produced a kazoo-like buzzing sound. I replaced the screwdriver with some fishing weights and found they had the same effect.

When I tuned to a CW signal and gradually lowered the tone to the range of 300 – 600 Hz, the buzzing sound was generated, which sounded the same as the active audio filters I had tried. The buzzing sound is not a clean sine wave but contains some harmonics that give it a "deep" sound. It sounded more distinct, as if it was amplified, and I found it was easier to copy.

CW Muffler

After some experimentation, I found that I could replace the cardboard with a CD whose center hole I plugged by gluing a water bottle cap to the opening. It made a good muffler and yielded the best buzzing sound. I also found that I could add a small fishing weight to the bottle cap to change the tone (see **Figure 33**). When I added a lighter weight the sound became lower, and adding a heavier weight produced a higher tone. I did find that to copy very weak signals, a smaller weight worked better.

Next, I tried this CD muffler on my IC-756 transceiver speaker, which is located on the top cover of the rig (see **Figure 34**). I attached it with a couple of magnets to prevent the muffler from sliding around. I believe this technique will work on other radios that have top-mounted speakers. However, it does not work with speakers that have a fabric cover because the fabric dampens the vibration effect.

The plastic CD cuts off the high frequency audio in the same way a low-pass filter

would. The deep buzzing sound produced is easier to separate from any background noise and there is no ringing effect. The big plus is that it's basically free and there is no need for a battery or separate speaker. — *73, Hwang Nam Chang, MD, KE2HF, 109 Franklin St, Newark, NY 14513-1526,* **hchang2@rochester.rr.com**

Stabilizing a Portable Key

I frequently operate CW portable, sometimes hiking to a Summits on the Air (SOTA) activation or just taking the portable rig with me on vacation to play radio from a new location. In either case, I want to travel light, which means traveling without a heavy base. This left me trying to figure out how to stabilize a small portable paddle and a straight key, when there often isn't a flat surface available.

I devised a solution that fits my needs while being lightweight and fairly versatile. I fashioned a 6 × 2.5 inch piece of $\frac{3}{16}$ inch acrylic sheet and attached a piece of 3M Dual Lock Reclosable Fastener on it and on each key bottom (see **Figure 35**).

The acrylic is lightweight and long enough to grip with one hand while the other operates the key. This thickness is rigid enough for me, although ¼ inch thickness might be a bit better for a little extra weight. The 3M Dual Lock Reclosable Fastener material has a distinct advantage over standard hook and loop fasteners because it holds the key to the acrylic more rigidly, yet still allows you to easily reposition the key as needed. When you push the key against the Dual Lock fastener it pops into place with a satisfying snap. I also put some low profile adhesive feet on the bottom of the plate to raise it a bit, which makes it easier to grip when used on a flat surface. — *73, Chip Stratton, AE5KA, 1728 S Detroit Ave, Tulsa, OK 74120,* **lightdazzled@gmail.com**

Cable Control

The T18 staple gun shoots a "round crown" staple that does not lay flush against wood. It is specifically made for low-voltage wire up to $\frac{3}{16}$-inch in diameter. If it is used to affix,

Figure 36 — The rounded shape of the T-18 size staple is perfect for anchoring cable ties to a wall, floor, or desk to keep wire and cable runs under control. [Joe Morse, AD4W, photo]

say, 6-inch cable ties instead, its usefulness is greatly improved (see **Figure 36**). It can now be used to neatly route bundles of coax and wire around your shack, attic, basement, or crawlspace. When another cable needs to be added to a bundle, there is no need to pull out the old staple. Just clip and remove the existing cable tie and slide a new one under the protruding crown of the old staple. For very large wire/coax bundles, the T-25 $\frac{9}{16}$-inch staple and gun can be used with larger size cable ties. — *73, Joe Morse, AD4W, 317 Westlawn Rd, Columbia, SC 29210-5622,* **ad4w@sc.rr.com**

An Improved Ground Connection

The need for a common ground connection in ham radio stations has been well-documented. A good ground is necessary for safety and good radio operation. Several publications suggest using ½-inch copper pipe for a common buss and small hose clamps to secure wires to it. This can prove difficult, especially with heavy wires or braid. I have found that using round-head brass screws and nuts soldered to the pipe can provide a strong, easy-to-use means of attaching wires to the ground buss. All materials for this project (see **Figure 37**) can be obtained at any hardware or big box store. Construction is easy, requiring only simple hand tools.

First, decide where along the length of the pipe you want to place the connection points; these locations will be where you install the

Figure 35 — The acrylic plate and a key with Dual Lock Reclosable Fastener attached. [Chip Stratton, AE5KA, photo]

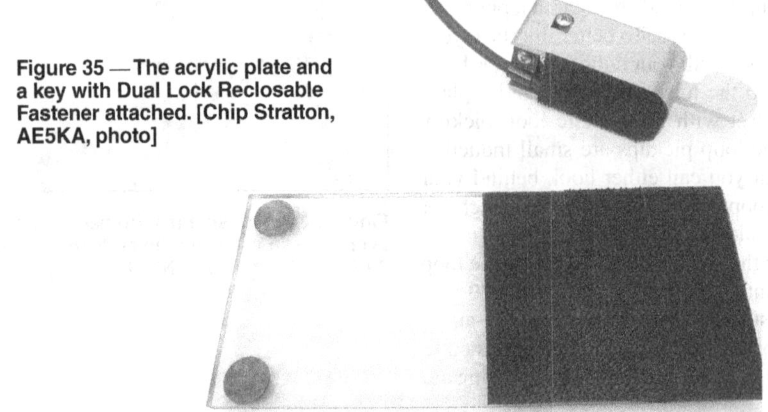

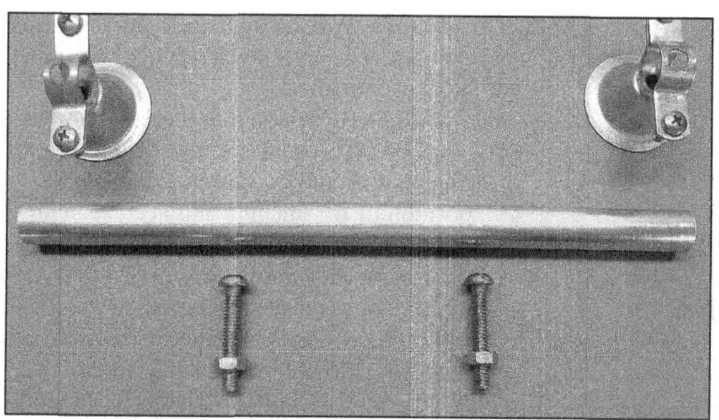

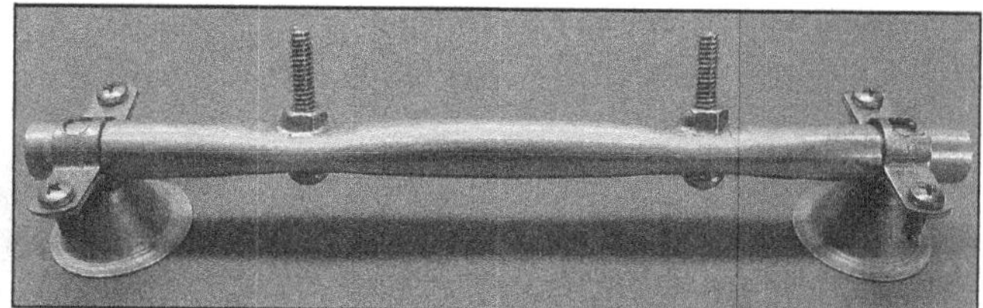

Figure 37 — These components are all that are necessary to make a solid ground connection system for your shack. [Jason Mueller, AE3JM, photo]

Figure 38 — The completed buss. Note how the pipe is flattened slightly where the screws pass through and that the screws are soldered to the pipe. [Jason Mueller, AE3JM, photo]

screws. Then compress the pipe at the screw locations to provide a flat surface for drilling the holes. This compressed surface also makes soldering the bolt head and jam nut to the pipe easier. I used a large **C**-clamp to flatten the pipe but a vise or hammer will also do. Do not flatten it completely as this will reduce its strength.

Once you have the pipe mounting points flattened, drill a hole at the center of the connection point that is wide enough to accommodate the type of screw you will be using. I used 1½-inch long, round-head brass screws and nuts; however, you should use a bolt size that will match the size of the ground connectors you will be using.

Once the screws and nuts are installed and tight, solder them to the pipe using plumber's soldering techniques. As with all soldering, all parts to be soldered must be cleaned "bright." Fine steel wool can be used to clean the parts. Soldering the nuts and bolts requires more heat than most soldering guns can produce; paste flux and a torch are called for. As with any soldering, use just enough solder to bond the bolt head and jam nut to the pipe. Any solder that wicks up the bolt can be removed by careful filing and running a steel nut up and down the bolt for rethreading. **Figure 38** shows the completed buss.

Use lugs to connect wires to the buss. A second or third nut can be used on each bolt to stack connections. A fair amount of torque can be applied to the bolts, assuring a good mechanical and electrical ground connection. — *73, Art Mueller, WA3BKD, 1532 Millers Run Rd, McDonald, PA 15057,* **n3dq@hky.com**

Dog Chow Rig Protection

My hobbies are ham radio and camping. I needed a way to protect my gear during outings. After some searching, I found a tough, waterproof solution for safeguarding my gear — a large dog food bag.

To use the bag, first dust off the outside with a damp cloth and then turn it inside out. Dry pet food will leave an oily residue on the inside of the bag, which is now the outside. To keep this residue from attracting insects,

I fold the bag flat and wipe each side of the "outside-inside" with a mild kitchen cleaner or Windex.

That done, slide your rig or anything else that needs waterproof protection into the bag and seal it. After folding over the end, you can use binder clips or wrap the bag with twine. If the bag isn't too full, you can twist the end and use twist ties or zip ties to hold it closed. — *73, Ralph Phillips, KE5HDF, 9102 Weymouth, Houston, TX 77031,* **ke5hdf@ sbcglobal.net**

Homebrew Microphone Stand

Desk microphones are very handy while taking traffic or filling in log sheets (for us non-computer types). But they can be very expensive and lack the useful control buttons often incorporated in the factory-provided ones.

While taking traffic on the Wyoming Cowboy Net, I found myself holding down paper with an elbow while trying to write. This was hardly efficient or comfortable. It occurred to me that the factory mic had a built-in holder on the back. If I could mount it on a stand then, using VOX, both hands would be free. The result was the mic stand shown in **Figure 39**.

I have access to welding and cutting equipment so I made mine out of steel. But the same result could be done with a variety of materials, such as a tuna tin filled with plaster, with wood for the upright.

Referring to **Figure 40**, the bottom is a piece of 1 × 2½ × 3 inch steel drilled for

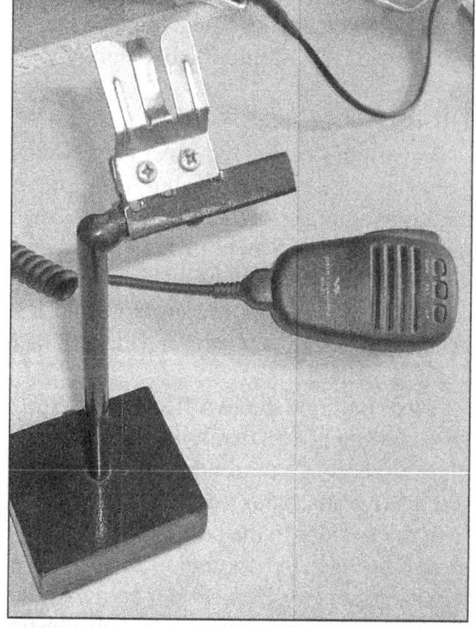

Figure 39 — This homebrewed stand allows the typical factory hand microphone to double as a desk mic. [Leonard Gordon, KD7CLO, photo]

a ⅜-inch round rod, 7 inches long, for the upright. The upright height can be any length to match the needs of the individual. The base is drilled and tapped for a set screw to hold the upright rod. A 3-inch piece of ⅜-inch rod was welded at a 90° angle to the upright. A piece of ½-inch square tube 2 inches long with a set screw is placed over this piece of ⅜ inch rod.

A simple tab is welded and drilled to

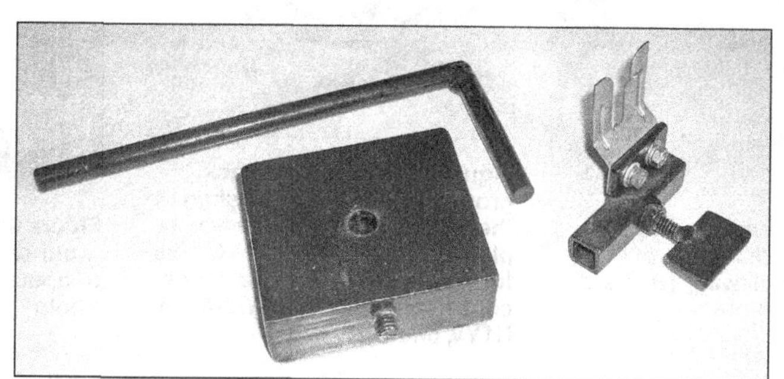

Figure 40 — While this stand is made from steel, wood could be used for simpler construction and a more pleasing look. [Leonard Gordon, KD7CLO, photo]

this square tube to accommodate the screws for the microphone mounting clip. A bit of spray paint was applied for a professional appearance. Finally, I glued a piece of craft rubber on the bottom to prevent slipping and possible scratches on the desk. Slip the microphone into the clip and you have a desk mic with full features. The total height of my stand is 9¾ inches, but it can be whatever meets your individual needs.

As a bonus, when we have one of our numerous summer afternoon thunderstorms, I simply lay the mic on the desk to indicate that the antennas are grounded. When I return to the shack there is no possibility of forgetting to reconnect the antennas because the mic on the desk is a visual reminder. — *73, Leonard Gordon, KD7CLO, 2112 Rooks Ave, Cheyenne, WY 82007,* **samicat@bresnan.net**

Tilt-up Feet

Most transceivers and other ham equipment have a tilt-up leg feature to improve the viewing angle of the controls and meters. I purchased a new power supply and discovered that it didn't have this feature. To adjust the viewing angle I created front leg extensions from a few parts and pieces I had in the workshop.

First, unscrew the two front rubber feet. Drill a center hole in two ½-inch electric box plugs, to accept the rubber foot's mounting screw. Now attach the box plug to the bottom with the flat side to the bottom of the equipment case using the original screws from the removed feet (see **Figure 41**). To finish the modification, screw on two ½-inch plastic threaded electrical tubing adapters. You now have front feet extensions about an inch and a half long (see **Figure 42**). The viewing angle is similar to that of the other equipment and, for equipment with cooling vents underneath it also allows extra cooling air in. I spray

Figure 42 — With the addition of some short "legs," the power supply meters are now much more visible and more cooling air can flow underneath. [Richard Russo, KB3VZL, photo]

painted mine flat black to improve their appearance. The total cost of parts is less than $3. — *73, Richard Russo, KB3VZL, 105 Colonial Ave, Norristown, PA 19403,* **kb3vzl@arrl.net**

Code Key Covers

For an inexpensive way to display or keep the dust off of your shiny code key, consider picking up a clear plastic "model car" display case from your local hobby shop. They can be had for under $10 and come in different sizes ranging from those suitable for the larger mechanical "bugs" down to the smaller electronic keyer paddles. If you like, you can use the shaft of a soldering pencil to easily melt a slot in the lower back wall of these cases to accommodate wiring. — *73, Joe Morse, AD4W, 317 Westlawn Rd, Columbia, SC 29210-5622,* **ad4w@sc.rr.com**

K400 Mount Problem

During the installation of a new Yaesu FT-857D transceiver and a Yaesu ATAS-120A screwdriver antenna on my friend Joe Goodell's, W1GC, vehicle, a problem was encountered. We found a solution that could be of use to others.

The problem was that the radio and antenna combo would not tune on 15 meters. The radio would pass through the low SWR position of the antenna and keep searching for a 15 meter low SWR, until eventually stopping at an incorrect position.

The hood-mounted Diamond K400 mount had been previously installed for an existing dual-band FM radio. A separate ground from the barrel hex screw used for the mount's vertical adjustment (see **Figure 43**) to the chassis along with multiple braided ground straps from the hood to chassis ground (see **Figure 44**) were added for the new antenna installation. The Diamond mount was retrofitted with a new connector to match the ATAS-120A. The base of the mount, on the lip of the hood, was not altered.

After attempting a factory radio reset, per Yaesu Technical Support's direction, the problem persisted. Vehicle engine noise was also higher than expected on several HF bands, in spite of all of the grounding.

The Diamond installation sheet states that some antennas may require that the setscrews under the mount (see **Figure 45**) be in direct contact with the metal of the vehicle. This was never an issue with the previous dual-band FM radio but proved critical for the new Yaesu HF rig. We scraped paint away so the setscrews were in contact with metal. As a result, the antenna tuned correctly and the engine noise was reduced as well.

I made a follow-up call to Yaesu to let them know what was discovered. The Yaesu technician cautioned that the setscrews under the SO-239 receiving mount on the Diamond K400 can loosen over time and this has caused eventual intermittent tuning issues with the ATAS-120A antenna. An inspection of your mobile mount is something that should be

Figure 41 — A few inexpensive components can replace the low feet of a piece of equipment to allow for easier viewing. [Richard Russo, KB3VZL, photo]

Figure 43 — The thick black ground wire was added when the original VHF antenna was replaced with the ATAS-120A. This functioned well on all bands except 15 meters. [John Wittmann, N1YV, photo]

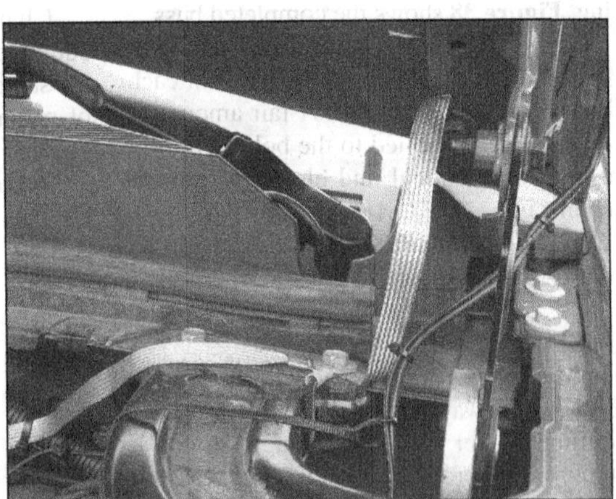

Figure 44 — The black ground wire and additional grounding braids were needed for the ATAS-120A to operate on the HF bands. [John Wittmann, N1YV, photo]

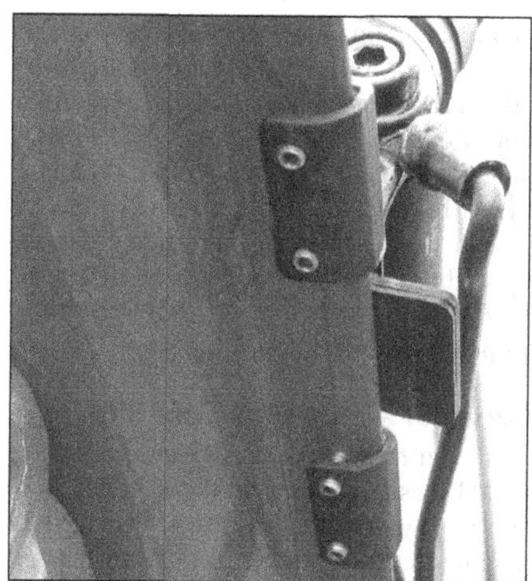

Figure 45 — In addition to all the added ground wiring, the setscrews of the K400 mount must be in contact with bare metal for the ATAS-120A to tune across all the HF bands. [John Wittmann, N1YV, photo]

added to your regular maintenance list. — *73, John F. Wittmann, N1YV, 1 Brentwood Rd, New Milford, CT 06776-2603, **n1yv@arrl. net** and Joe Goodell, W1GC, 16 Carriage Dr, New Milford, CT 06776, **joegoodell@ yahoo.com.***

Junk Box Porta-Paddle

I needed a lightweight CW paddle for portable use. My trusty old Vibroplex paddle weighs about 2¾ pounds. That's great for stability on my operating desk, but far too heavy for travel.

My goal was to build a portable paddle using junk box parts. The paddle I built weighs in at only 5 ounces, has adjustable gaps, and fills the need nicely.

I began by searching my junk box and found a pair of microswitches that could be actuated with very light pressure. I decided to experiment a little and ordered several similar-looking microswitches from a surplus electronic catalog, to see if I could find a pair with a slightly lighter touch. Predictably, the original units were the best for this application. They combine sensitivity with good tactile feedback and a quick return.

I fashioned a paddle from a piece of scrap plastic and a base from some scrap wood. I made a **U**-shaped bracket from a strip of steel and found some 6-32 screws and nuts to form the adjustable gaps. Going outside the "box," a hardware store provided a small brass-plated hinge.

I have always believed that a homebrewed project should not only work well, but should look good. Therefore, I sanded and stained the wood, and coated it with a two-part epoxy. The resulting glossy finish looks like many coats of varnish (see **Figure 46**).

The switches I found in my junk box were originally single-pole double-throw, but only the normally open terminal is used. I drilled small holes in the wood base and added four adhesive-backed feet to provide space to run the wiring underneath. The feet also provide some stability. Of course, with the light weight of this paddle, I have to hold it in place when operating. I may try attaching it to the radio to solve this problem. — *73, Sumner Weisman, W1VIV, 43 Agnes Dr, Framingham, MA 01701, **w1viv@ arrl.net***

S-Biner Mic Hook

My son, Caleb, KK4UWJ, needed a microphone mounting clip for his van installation and came up with this idea. He used a Nite Ize brand, S-Biner #2 clip on his Yaesu mobile microphone (see **Figure 47**). It is a good fit on the clip button on the back of the mic and can probably be adapted to other microphones as well.

For a mount, you can use either a hook or a loop as the S-Biner can open up and clip into the loop. As a matter of fact, I like this better than the button style, as you can clip it on anything with a hook or loop. They sell for under $2 at outdoor stores. — *73, Cameron, KT3A, and Caleb, KK4UWJ, Bailey, PO Box 173, Mount Wolf, PA 17347, **kt3a@arrl.net***

Vinegar Corrosion Control

I bought a scanner at a garage sale. Upon examination, I discovered that batteries had been left in a removable battery holder. The batteries had leaked, leaving a chemical residue all over the interior of the holder. I asked around to see if anyone had a good way to remove the chemical coating without damaging the box. One of my friends, Steve Brichta, WB9ZNL, suggested using something acidic. He recommended freshly squeezed lemon or orange juice, because these were more potent than bottled juice.

Because strength seemed to be a requirement, I decided to buy white vinegar, which has 5% acidity. I put the battery holder into a sealable plastic bag, covered it with the white vinegar, and put on some safety glasses, just in case. After a few moments of swishing the battery holder around in the sealed bag, the box was as good as new. — *73, Klaus Spies, WB9YBM, 8502 N Oketo Ave, Niles, IL 60714, **wb9ybm@juno.com***

Keep Your Paddle Fresh

I was preparing to operate portable and needed to find some way to carry my Bencher paddle without having to disassemble it — visions of an expensive hard-sided case came to mind. I started prowling around the house looking for something to hold and protect my paddle during my portable operation.

My wife, Lyndel, asked me what I was looking for, and I explained my problem. She

Figure 46 — Take a few junk box parts, add a bit of ingenuity, and you too can have a no-cost, portable paddle. [Sumner Weisman, W1VIV, photo]

Figure 47 — A common S-biner clip used in hiking and backpacking can be repurposed to keep your mobile microphone stowed while not in use. [Cameron Bailey, KT3A, photo]

Figure 48 — The feet are mounted on the outside of the lid to provide scratch-free support and stability for the key when operating. [Richard Zolnowski, W0RPZ, photo]

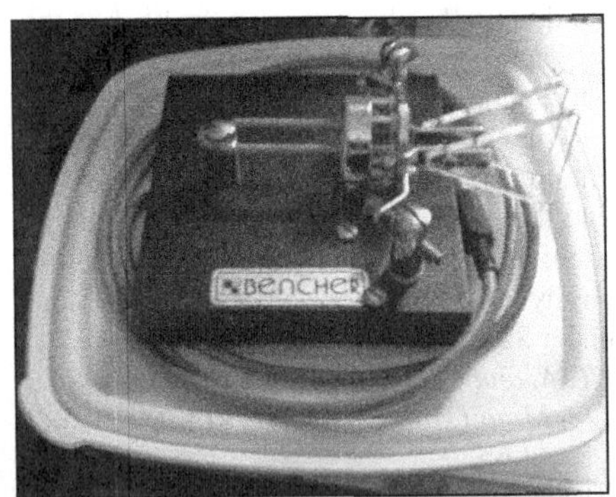

Figure 49 — The plastic spacers between the base of the paddle and the lid provide enough space for the cable, and the lid is large enough to hold the entire unit inside the container when closed. [Richard Zolnowski, WØRPZ, photo]

left the room and returned with a plastic food storage container.

"Can you use this upside down and put the paddle on the lid?" she asked.

"What a great idea," I thought.

I put spacers under the key in place of the feet and positioned the key on the inside of the lid. Then I put the feet on the outside of the lid and screwed them into the spacers through the lid (see **Figure 48**). The threading was the same on the spacers and the screws in the feet. This leaves room for the wires under the key (see **Figure 49**) and places the feet under the lid so the key stays still when in use, plus there are no screw heads extending out to scratch something. The feet also act like washers, keeping the screws from pulling through the lid.

The paddles are close enough to the edge of the lid so that there is no interference during use, but there is still ½ inch between the end of the paddles and the container when it is closed up. The bottom of the container is stiff, with an outward arch, so that nothing will warp it enough to hit the paddles. Finally, the case is water-resistant to boot!

My Bencher isn't tied to my bench any more. — *73, Richard Zolnowski, Jr, WØRPZ, 8706 Kentucky Ave, Raytown, MO 64138,* **w0rpz@arrl.net**

Compact and Inexpensive Connectors

I bought a good supply of breakaway machined socket pins to make IC sockets in perforated board prototype circuits. I epoxy them into holes I drill in the perforated board to make them rigid. Because they stack, I started using them to make cable and wire connections as well. **Figure 50** shows how conveniently these breakaway pins connect and fit in a very small space. The two five-pin connectors are programming interfaces for the Atmel ATmega328P ICs, the CPU chip in the Arduino Uno. I now add this interface to any board I build with that chip to enable remote programming.

Figure 51 shows construction details of the top and center stacking connectors. **Figure 52** shows a prototype circuit with two ATmega328P chips, each with a programming interface and an array of connectors for various sensor inputs and control outputs that were part of the original design. After it was built, my requirements changed and I needed to add several outputs to control additional LEDs and an input from a pyroelectric infrared (PIR) motion sensor. It proved relatively convenient to add more socket pins for connectors

Figure 51 — Breakaway connectors can be stacked to increase the number of connections to allow for configuration changes. [Sam Green, WØPCE, photo]

in the scant remaining space. Then I decided I could improve performance by adding another PIR sensor. I added diodes to the PIR sensor outputs so I could wire-OR them and plugged the cables from both into the same socket by stacking them. Consider this approach when you need more interconnects in a tight space. — *73, Sam Green, WØPCE, 10951 Pem Rd, St Louis, MO 63146,* **w0pce@arrl.net**

Surveyor's Tape for a Safe Site

If you have tried to navigate through a Field Day site, you may have found that hazards lie in wait. Tent stakes, dipoles with ladder line feeds, and power cords strewn about on the ground make for many safety hazards. In the dark, it gets even worse, especially as you try to navigate to the coffee pot. Then there are public service events where you take your go-kit and find that tucking coax and power cords away from the surrounding foot traffic usually ends up being a tangled mess.

There is a tool you can use to help keep your coax, power cords, and antenna leads neat and orderly, as well as easy to tie up and store in the go-kit. It is called surveyor or flagging tape, a 1-inch-wide ribbon of colored polyvinyl that is non-conductive and non-sticky. It costs about $2 for a roll that is 200 feet in length and is available in a variety of bright colors. You can get it at home improvement and hardware stores.

It is great for marking, locating, and identifying, and can be used as you might use light string or small zip ties. Coil up that 50-foot length of RG-8 and tie it with the surveyor's tape. Use a different color for different types or lengths of coax. Coil up the 16-foot coax from your ¼-wave antenna to easily store in your go-kit.

Use your imagination. Two hundred feet of tape will go a long way. Because the tape is an insulator, you can hang your 2 meter antenna from a ceiling hook or tree limb

Figure 50 — Breakaway connectors are stackable and fit in small spaces, making them very useful when adapting a prototype to changing requirements. [Sam Green, WØPCE, photo]

Figure 52 — The need for a second sensor is easily accommodated by stacking a second connector onto the first. [Sam Green, WØPCE, photo]

using it. The roll is lightweight and does not take up much space in the go-kit. It is both cheaper and easier to use than zip ties. I hope it will help you keep your gear neat, orderly, and easy to identify. — *73, Stephen Bellner, W8TER, 6853 Deer Ridge Rd, Apt 6, Maumee, OH 43537,* **w8ter@arrl.net**

Headphone Holder

I've been thinking for months about what to use to hang up my headphones. My first thought was cup hooks, but they are too small. Then it came to me — a shaker peg! Pegs are available at most hardware stores as well as woodworking suppliers and hobby stores. I just drilled a hole in the side of my desk where I wanted to hang the headset and glued in a peg. — *73, Hal Welch, WA1ZJL, 521 Winn Rd, Lee, ME 04455,* **wa1zjl@ arrl.net**

Moldable Plastic PTT

For a few minutes of work and a few dollars you can make a push-to-talk switch (PTT) that fits your hand perfectly. RadioShack carries a marvelous product called U Mold Plastic (2770161). At room temperature, this plastic is very rigid but has a pleasant waxy feel. When submersed in very hot water for a short time it becomes a soft, transparent, sticky material that can be shaped easily.

To make a custom PTT, connect a momentary single-pole-single-throw (SPST) switch (2750609) to a jack appropriate for your rig with a twisted wire and form the warm plastic around the assembly, incorporating your own finger impressions for a comfy hold (see **Figure 53**). When the switch eventually fails, merely immerse the assembly in hot water and reuse the plastic. — *73, Chuck MacCluer, W8MQW, 1390 Haslett Rd, Williamston, MI 48895,* **w8mqw@arrl.net**

Handheld Hanger

I have many handheld scanners and radios and I wanted to keep them out of the way, yet readily available. I realized that a flat-style curtain rod would work very well to hold a number of handheld transceivers or scanners. The rod can be mounted to a wall, keeping all your handhelds neat and in one place. To top it off, it's a very inexpensive solution. — *73, Rusty Hack, NM1K, 21 Montano, Enfield, CT 06082,* **nm1k@amsat.org**

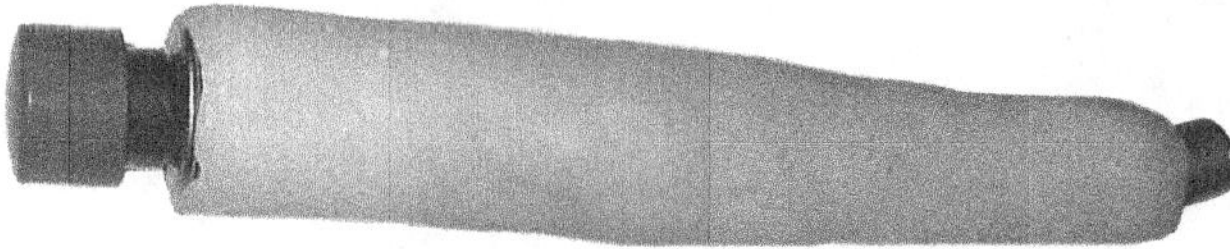

Figure 53 — Moldable plastic can be used for a variety of repairs and projects, such as this custom-fit PTT switch. [Chuck MacCluer, W8MQW, photo]

A Simple Dimple

My Kenwood TS-440SAT is compact and packed with features. One feature it lacks, like many other rigs, is a tuning knob dimple — that small concave spot on the face of the main tuning dial that lets you use one finger to quickly and easily change frequencies across the band. Trying to duplicate that function by using a finger on the knurled circumference of the main tuning knob is a pain (literally), and not too precise either!

I found a solution in a jar of old clothes buttons. Sorting through the collection, I found a button that was the perfect size and shape — flat on one side and concave on the other. It was even close in color to the rig's main tuning knob (see **Figure 54**). If you don't have one that fits your need around the house, a department or sewing/knitting store will have a great selection. I chose to use a small piece of carpet tape (heavy-duty, double-sided adhesive tape) cut to the proper size to mount it. This gives me the option to remove it if I want but you could use super glue or epoxy to make it permanent. Put the tape on the flat part of the button and push it in place so the outer rim of the button is just a bit back from the edge of face of knob (on the TS-440 there is a decorative metal band on the knob's face and the edge of the button rests against that). This simple dimple looks and works great. —*73, Mike Berlin, WU1M, 63 Clifford Dr, West Hartford, CT 06107,* **wu1m@arrl.net**

Figure 54 — A spare button makes an effective tuning dimple for rigs whose big knob lacks that useful asset. [Mike Berlin, WU1M, photo]

RJ45 Mic Extension

Most manufacturers now favor RJ45 connectors for microphone cable connections to their mobile radios. I have operated mobile since 1984 and, with the exception of my Yaesu FTM-10R, have not had any radio that offered a good remote cable extension for the microphone. I have used plastic RJ45 straight-through connectors and Category 5 (CAT 5) cables to remote the microphone with some success, but securing the straight-through connector has always been problematic.

I found a solution in the Switchcraft EHRJ45P5ES RJ45 CAT 5e shielded, panel mount jack (which costs about $9). This jack, along with a straight-through (not crossover) CAT 6a shielded cable allows for a clean, mechanically secure way to remotely mount the microphone. I mounted mine to a lower panel in my Prius. You could also mount the jack in a metal bracket. You should take care to mount the jack in a location that avoids side and vertical loads on the relatively weak microphone cable connector.

Also, remember to trace the wiring to ensure that the cabling is a direct match, wire for wire, from your microphone to the radio. This was the case with my IC-7000 but may not be the case with your radio. An incorrect match may damage your microphone or circuits in your radio. Both the connector and the use of a CAT 6a extension cable should prevent noise and crosstalk. — *73, Joe Scalet, WKØG, 36287 West 231st St, Edgerton, KS 66021,* **wk0g@arrl.org**

Phone Patch Revisited

Phone patches have a long history in ham radio. In their heyday, during the Korean and Vietnam war eras, many of our servicemen and women overseas stayed in touch with family and friends this way.

Currently, with cell phones and the Internet, many of us have forgotten about the disaster-related possibilities of using phone patches. I feel we need to maintain that capability, because in a disaster one of the first things to fail is the phone network, be it cellular or landline.

However, the phone network doesn't fail everywhere. Outside the disaster area it will be operating normally. Just because we are hams doesn't mean we are chained to communicating by radio. In a disaster situation, the cellular/landline phone system can have advantages. By placing a station outside the disaster area and linking it by phone patch to working cellular or landline systems, a reliable alternative communications channel is made available.

With this in mind, I've combined the capability of cellular and landline phone patching in one system. It's possible, with some experimenting, to be able to add Skype,

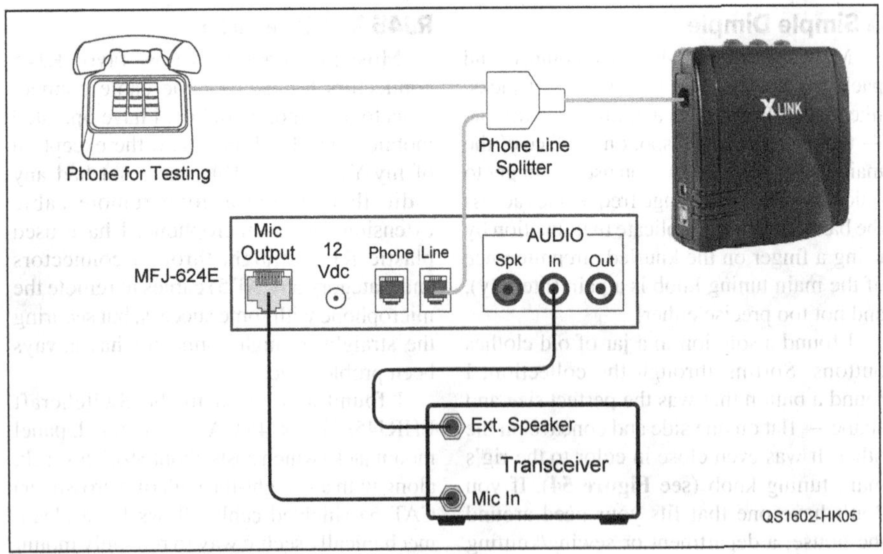

Figure 55 — Diagram of the XLINK phone patch setup.

Vonage, or other phone services as well.

In order to build this radio/landline system, you will need the following equipment:

- an XLINK BTTN (**www.myxlink.com/index.aspx**)

- an MFJ-624E Deluxe Hybrid Phone Patch

- a single to double phone line adapter

- assorted connectors to fit your rig

Begin by getting the patch working. Follow the instructions to set it up for your radio. Configure it to allow you to use your microphone to talk over your radio in bypass mode. The true test is to see if you can hear yourself on a second radio.

Once the patch is working, set up the XLINK. You will need a cell phone and a standard desk-type phone (this will be for testing purposes). Using the instructions in the XLINK manual, link your cell phone to the XLINK, which, in turn, will link to the desk phone. If you're successful, when you pick up the desk phone you should hear a dial tone. If the setup is working properly, you should be able to both receive and make cellular calls via your desk phone.

Next, you will need to add one more line to the phone connection on the XLINK. This is accomplished with a single to double jack modular adapter, which you plug into the top connector on the XLINK. The desk phone is connected to one side of the adapter and the line input of the phone patch is connected to the other.

Complete the setup by connecting the external speaker output of the radio to the **AUDIO IN** connector of the patch. When all is done, your setup should look like **Figure 55**.

Test the system either with a friend on the air or with a handheld transceiver locally.

Start by keying up the transmitter to see if you can hear it in the phone. Then call yourself on another phone to check that the audio is coming through. You will need to adjust the level controls to match your specific equipment; some trial and error is required. Also, if you intend to use both a cell phone and a land line, the levels will be different for each.

In order to monitor the audio coming into the patch, you can plug a speaker into the **AUDIO SPK** jack on the patch's rear panel. This will allow you to listen to the phone side of the conversation. You can add a phone switch at the point where the XLINK connects to the patch to allow you to have both cellular and landline capability.

With this setup, you can connect by radio to a station in a disaster area and link them into the landline/cellular network to provide another communications option. — *73, Dennis Smith, KA6GSE, 19631 Soldon Ct, Santa Clarita, CA 91351,* **thepres@thevine.net**

Coiled Cord Cable Control

The remote installation of a Yaesu FT-857 control head requires at least three cables: a six-conductor cable for the control head, an eight-conductor cable for the microphone, and a two-conductor cable for the speaker.

There are a couple of additional functions that I like to remote as well, such as the ability to switch to low power and to key the transmitter in CW regardless of the mode selected. These require another three wires. All these wires can lead to a mess in a hurry. In the past, I have used cable ties to bundle the cables, but the rough edges can be tough on the hands and will snag on anything close.

While building a remote cable set for use with my HF go-box, I was looking through my collection of cables when I spotted a new 25-foot coiled telephone handset cord. I was able to slide the microphone, control, and speaker cables into it and used the coiled cord for the other functions I needed. The 25-foot length worked well to contain the 10-foot cables I am using (see **Figure 56**), allows good flexibility, is less prone to snagging, and is easy on the hands.

As the use of wired phones decreases, old handset cords should be easy to find and can be put to use as cable minders. If an installation requires only a small number of wires or cables, just slide them through the center of the coiled cord, then shift the coils to distribute them evenly along the length. If the number or size of the wires or cables is large, the coiled cord can be spiraled around the wires at an appropriate pitch. To keep the coiled cord from creeping away from the ends over time, use tape or wire ties at each end.

As mentioned above, I used the coiled cord to carry signals as well as to contain the cables. I use standard surface mount telephone jacks as breakout boxes at each end. If you do this, remember that the connectors on the coiled handset cords are not the same as those used to connect the phone base to the wall. I found it easy to replace the handset connectors with standard RJ11/14 connectors. — *73, Albert Reynolds, N4AGG, 11104 Coachmans Way, Raleigh, NC 27614,* **n4agg@arrl.net**

Sound Director

If you're have difficulty understanding voices or subtle sounds on your radio even though your hearing is good, your receiver's audio might be the culprit. The fact is, many radio speakers direct the sound upward, and some even direct it downward, muffling the subtle sounds on your receiver. The common solution is to increase the volume, but that can disturb others and high volume can create

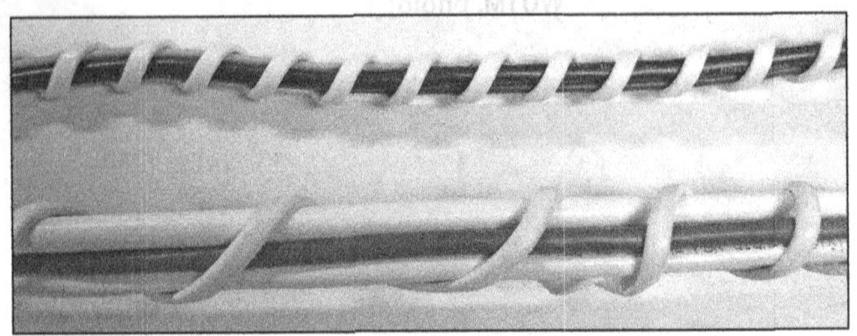

Figure 56 — Discarded coiled telephone cords will keep those bundles of cables under control. [Albert Reynolds, N4AGG, photo]

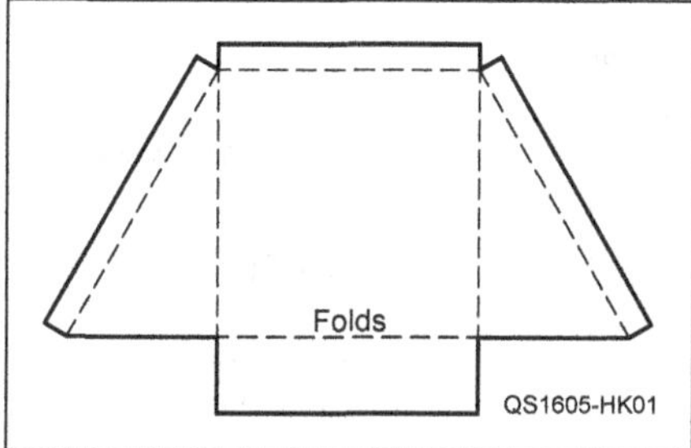

Figure 57 — This diagram shows the method for cutting and bending cardboard to create a sound director.

Figure 58 — A cardboard sound director formed using the drawing in Figure 1. [Hal Rogers, K8CMD, photo]

distortion. Outdoor operations, such as Field Day and public service deployments, are a particular challenge. Boosting the volume can interfere with other operators. Headphones or an extension speaker are an option, but a much easier way to improve your receiver's sound is a "sound director." This easy-to-make device redirects the sound toward you.

The simple layout in **Figure 57** can be cut from a piece of thin cardboard. I used a shoe box, but a cereal box will work as well. The finished sound director can be seen in **Figure 58**. There's nothing critical about the dimensions. Just make sure it covers the speaker grill, but doesn't block the ventilation holes. Finally, either tape the three flaps down to the transceiver case or, if your radio is steel, glue strips made from an old magnetic business card to the flaps.

Another option is to cut a small plastic food bowl in half, or fabricate your sound director from a milk carton. With the receiver's sound directed toward you, you'll enjoy clarity at a much lower volume. — *73, Hal Rogers, K8CMD, 7811 Dogwood Ln, Parma, Ohio 44130,* **k8cmd@arrl.net**

Stand-up Radio

Have you ever placed your 2 meter handheld transceiver on the breakfast table and had it tip over and spill your orange juice? The loud bang will surely wake everyone up. Because my radio isn't meant to be an alarm clock, I built the stand shown in **Figure 59**.

All the parts are available at your local building supply store. You'll need a Simpson Strong Tie A24, which costs about $3.

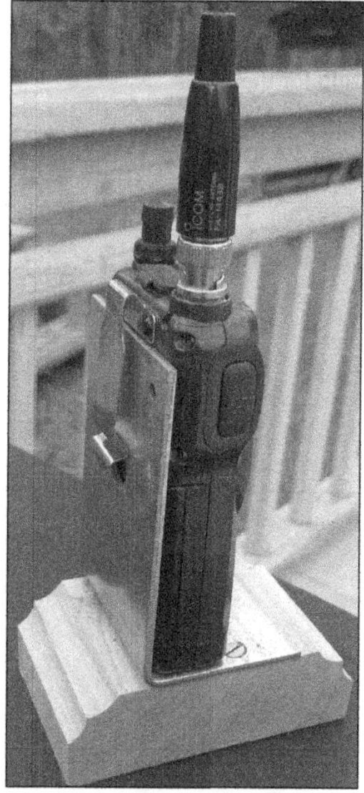

Figure 59 — Some scrap wood and an appropriately sized steel angle bracket prevent impromptu shock testing. [Phill Grant, N1YPS, photo]

The Strong Tie is a piece of 12 gauge steel about 4 inches tall with a 2-inch bend that I use as a base. The width is 2½ inches. Two flat-head #6 ¾-inch screws fasten the Tie to the wood. I did countersink the two holes on the bottom part of the Tie so the radio would sit firm and not wobble on the screw heads. The wood block was a piece of deck railing found in my kindling box; any appropriate piece of scrap will do. A piece of weather stripping on top of the base would give your radio a softer seat. — *73, Phill Grant, N1YPS, 119 Hoe Shop Rd, Bernardston, MA 01337,* **phill112643@verizon.net**

Mic Cord Protector

I live in a tropical country and have experienced a problem with my microphone cords deteriorating after a year of use. The cable's outer insulator disintegrates, especially when they are installed in my mobile. The heat and humidity probably play a role in the cable's deterioration.

I have tried replacing the cables, but the result is the same. The cable's outer insulation can't stand up to my tropical environment and the cables require periodic replacement, which gets expensive.

The solution I found is called a curly wrap cable protector, which I purchased at a store where they sell cellular phone accessories (see **Figure 60**). The protectors are miniature versions of the spiral wrap protectors used on large cables. These are designed to protect cell phone power cords. — *73, Eduardo Victor J. Valdez, DU1EV, 9 Guiho St, Monte Vista Subdivi-sion, 1802 Marikina City, Philippines,* **du1ev@hotmail.com**

Powerpole Fuse Holder

While breadboarding a circuit, I found myself without a fuse holder. I did, however, have a Powerpole-to-alligator-clip adapter that I had made for another project. The Powerpole connectors were mounted side by side and I noticed that some automotive blade fuses I had fit into the Powerpoles perfectly. Putting them together gave me an instant fuse holder (see **Figure 61**). — *73, Norm Beausoleil, N1RFK, 99 Smith St, Putnam, CT 06260,* **n1rfk@arrl.net**

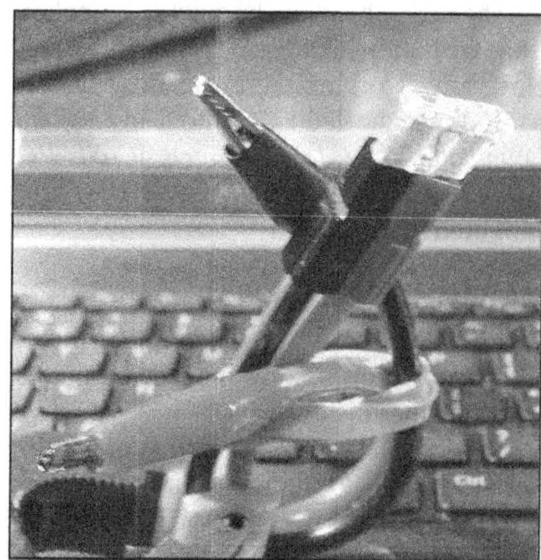

Figure 61 — A pair of Powerpole connectors mounted side by side will form a socket that will fit automotive blade fuses. [Norm Beausoleil, N1RFK, photo]

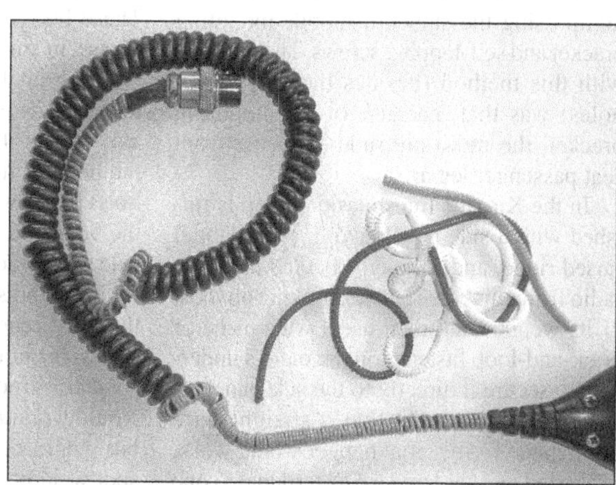

Figure 60 — Cell phone cable protectors work well to increase the longevity of microphone cords in difficult environments. [Eduardo Victor J. Valdez, DU1EV, photo]

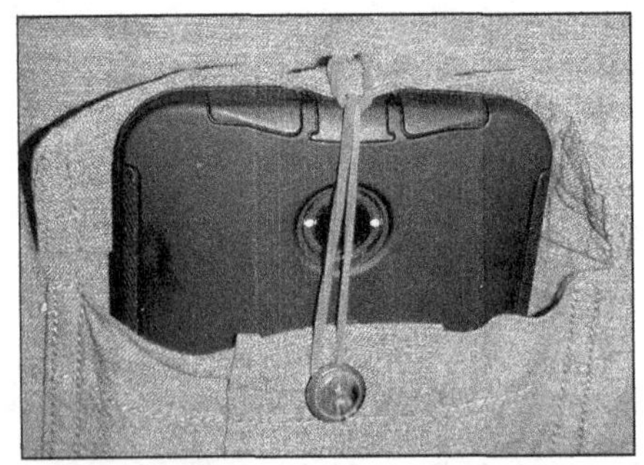

Figure 62 — This rubber band gimmick will keep your expensive smartphone safe in your pocket during your Field Day set-up and take-down activities. [John Nogatch, AC6SL, photo]

Cell Phone Safety Belt

A rubber band extends my shirt pocket button closure, making a convenient safety belt to retain a large cell phone (see **Figure 62**). The rubber band is inserted through the buttonhole, then through itself, and then looped over the button, in order to keep the cell phone from falling out.

The rubber band can be quickly unhooked from the button in order to access the phone. Now, after working on antennas, I can get back down the ladder without my phone falling out! — *73, John Nogatch, AC6SL, PO Box 1129, Boulder Creek, CA 95006-1129,* **jnogatch@gmail.com**

Mobile Hook Mount

In a past issue of *QST*, I described a method of providing a convenient microphone hook on the dashboard of some vehicles. Here I describe a different way of mounting a mobile radio to the transmission "hump" of some vehicles.

When I purchased a Kia Soul, topmost in my mind was how to mount several of my mobile rigs without drilling holes all over. The first radio I installed was an Alinco DR-235 that, in my previous vehicle, had been mounted on the right side of the transmission hump using the standard mobile mounting bracket and self-tapping screws. The problem with this method (besides the ugly screw holes) was that, because of the mounting bracket, the radio protruded into the front seat passenger leg area.

In the Kia, the transmission hump is finished with a side panel having a substantial raised ridge, and I thought I could hang the radio from this ridge (thereby taking up most of its weight) and then use a wide piece of hook-and-loop fastener on the radio's underside to secure it directly to the side panel.

At first I thought that a straight wire hook made from a coat hanger would work. I crimped and soldered a ring terminal to one end of a 2-inch piece of coat-hanger wire. The ring terminal was secured under the radio's side mounting screw. I bent a hook on the

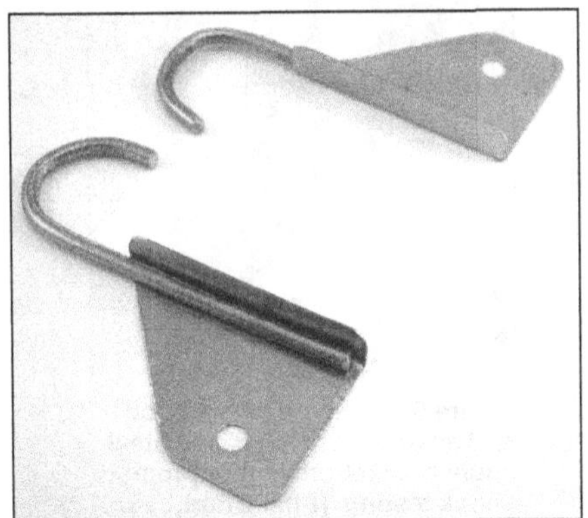

other end and hung it on the ridge. This did not work well at all, partly because the coat hanger wire was not stiff enough, but mainly because the angle of the hook wire caused the radio to not hang flat against the hump side panel. What was needed was a small bracket to keep the mounting hooks parallel to the hump mounting surface.

Figure 63 shows the dimensions of the hook/bracket I designed. The hooks were made from 1/8-inch brass rod, bent in a vise around a piece of 5/8-inch wooden dowel. These hooks were then soldered along the crease in the brass plate (see **Figure 64**). Note the small bent "lip" across the width of the plate. This helps to stiffen the plate, and also makes it easier to position and solder the hook to it. Be sure not to make the two brass plates with the same bend direction, as the 90° bend must be opposite on the other side of the radio.

I used brass rod and sheet brass, because they are readily available at most hobby stores, the brass takes solder readily, and the solid brass rod results in a strong hook. (Steel would work too for both the hook and plate, but would be harder to solder to, although spot welding would work, if available. Aluminum was not considered, as I had no easy way to fasten the hook/plate parts together using that material.)

Figure 64 — The completed hook/bracket assemblies. [D. Dorward, VA3DDN, photo]

I pretinned the solder area on the hooks and the plate using a Hakko 60 W solder station with a 1/8-inch wide solder tip. This was marginal in terms of heat, but got the job done. (A 100 W chisel-tipped iron might have been a better choice.)

The soldering was done with the plate lying flat on a small brick. The solder wire I used was 60/40 rosin core. After soldering, all rosin residues were wiped off using isopropyl alcohol.

You can see in Figure 62 that the plate has a 5/32-inch mounting hole that is positioned to match the location of the first mounting screw on each side of the radio. Dimension A is the horizontal distance from the radio face plate to the first mounting screw. Dimension B is the vertical distance from the radio bottom to the first mounting screw. If the mounting screws on your radio are located differently, you can adjust the overall length and width of the brass plate according to your measurements for A and B.

Figure 64 shows the finished left and right mounting hook assemblies before I painted them black. The plate was de-burred, its sharp corners rounded off, and the top right corner was trimmed off at approximately 45°, for appearance. **Figure 65** shows the hooks in use.

To make sure the radio will stay in place even if your vehicle is in an accident, add hook-and-loop mounting tape. I recommend using a 2-inch-wide strip of Velcro #90593 (or other industrial-strength fastener with a high-temperature adhesive, if possible) across the width of the radio. If you are not comfortable with relying on the hook-and-loop fastener tapes to secure the body of the radio, I suggest you add a metal clamp attached with a self-tapping screw into the mounting area.

If you are using fasteners, it is essential to first clean the areas where the fasteners will be applied. Use isopropyl alcohol on a cloth, to remove any surface oils that will prevent the fasteners from sticking. Allow the surfaces to dry thoroughly before applying the fastener and allow the fastener's glue to set for an hour before mating the tape closures. [Caution: Don't be tempted to forgo

Figure 65 — The completed mount shown attached to the radio, hanging from the center console. [D. Dorward, VA3DDN, photo]

using the hook-and-loop fastener — it is essential for a safe installation. — *Ed.*]

Hook-and-Loop Fastener Considerations

These fasteners have been around since the mid-1950s. Today there are many variants with an astounding number of specification choices, just a few of which are strength of closure, rated number of closures, type of adhesive used, and temperature resistance. They are also used in numerous applications, ranging from clothing to aerospace.

Although the commonly known name is "Velcro," according to the manufacturer, that name relates to the company and not the product. Velcro (the company) offers eight different adhesive choices.

3M Corporation also manufactures competing hook-and-loop fasteners in at least eight different product groups and also sells products under their trademarked "Scotch" brand. [3M also sells a line of "Dual Lock Reclosable Fasteners" that they advertise as being five times stronger than hook-and-loop fasteners. — *Ed.*]

To add to the confusion, there are a host of house brands where the product may actually originate from Velcro or 3M, but are packed and branded by the industrial supply houses under their own name, and with minimal specifications provided. I have found it is nearly impossible to learn the specifics of the product lines offered through retail suppliers. Sadly, specific information for the consumer products consists of generic descriptions like, "superior holding power" or "not recommended for *x* or *y* application." This situation is further complicated by the fact that stores limit our choices to the one or two best-selling variants. Choose carefully when buying these products. — *73, Don Dorward, VA3DDN, 1363 Brands Crt, Pickering, ON L1V 2T2, Canada,* **va3ddn@arrl.net**

Moldable Mount

I own the Icom 5100 mobile transceiver and was never satisfied with the mounting arrangement. I had tried metal plates on the dash, flex mounts, and other available options, but it was never a clean-looking install. As I searched for a solution, I came across something called Sugru (**sugru.com**), a silicone-based moldable glue.

It is available in a kit with magnets, from which I was able to create the perfect mounting system. It is completely heat and cold resistant, unlike double-sided tape, and it can be cleanly removed when and if I ever sell my Jeep. I'm even considering molding a hook for the microphone. Sugru beats drilling holes in the dashboard. — *73, Jeffrey Baer, WT4K, 4150 Praline Ct, Marietta, GA 30066-2394,* **wt4k@arrl.net**

Safe and Charged

Like most hams, I have several handhelds that use rechargeable batteries or battery packs, as well as battery-powered hand tools and personal electronics. These power packs can be severely damaged or even cause a fire if they happen to short out. Also, it is often necessary to recharge these packs before use, but it is not good to overcharge them.

To avoid these two pitfalls, overcharging and short-circuit danger, I use electrical marker tape or clear, write-on, cellophane tape to put the date charged on the case and to cover the terminals to prevent shorting out (obviously, remove this before using or recharging the pack). This allows me to come back several weeks later and know which packs or cells might need recharging before use, and prevents the possibility of a short as the packs bounce around in a go-box.

This idea can help you be better prepared by keeping you aware of the charge status of your batteries, and it could prevent the serious damage possible from a shorted-out battery pack. — *73, William Cleveland, WD5IBY, 501 E Adoue St, Baytown, TX, 77520-4933,* **wcleveland2@yahoo.com**

Another Powerpole Adapter

The September 2015 issue of *QST* had a fine article by Phil Salas, AD5X, about adding a Powerpole adapter to the banana jacks of his station power supply.

While the concept is a great idea, it is not adaptable to all power supplies. On my Pyramid Phase III Model PS-35 supply, the center hole in the banana jacks (probably more correctly called "binding posts") is only

Figure 66 — The adapter is mounted to the power supply by tapping the binding posts to accept screws, upon which the adapter board is mounted. [Ken Hollenbeck, K9CAX, photo]

about ⅜ of an inch deep, not deep enough to insert banana plugs to a proper depth for a secure connection.

I have devised a relatively simple and effective modification that allows this adapter to be used on any power supply with binding posts that do not accept banana plugs.

Prepare the PC board as shown in the original *QST* article, except that the holes, designed to mount the banana plugs, should be drilled out to accept a 10-32 screw.

The center holes in the binding posts are almost the exact diameter to accept a 10-32 tap, so I tapped them full depth, ending with a bottoming tap to ensure full seating of a 10-32 screw.

When tapping the holes, try to keep the tap level and square with the front of the power supply, so that when installing the studs they will be as parallel as possible.

Take a 10-32 × 1½ inch brass screw and run a 10-32 brass nut halfway down its length. Screw the brass screw into the previously tapped binding post until it hits bottom and is tight, then run the nut down until it is snug. Cut the screw off to a length of ⁷⁄₁₆ inch as measured out from the face of the nut (see **Figure 66**).

An easy way to cut the screw off is to use a Dremel tool equipped with a narrow abrasive cutoff disc. An alternative method is to use a fine-toothed hacksaw blade. Regardless of the method used, make sure a 10-32 nut will start easily on the stud. If not, clean up the threads with a small file.

I wired the module with #10 AWG stranded wire. The ends that fasten to the PC board were terminated in a size 10-12 uninsulated eye terminal where they were crimped and soldered securely (see **Figure 67**). The eye terminals were temporarily mounted to the PC board with short 10-32 screws and nuts to assist in determining proper lead length and routing.

To assemble the unit, be sure to place a size 10 brass washer over the studs previously secured to the binding posts. Next, slide the eye terminals over the studs, followed by the PC board. Complete the installation by add-

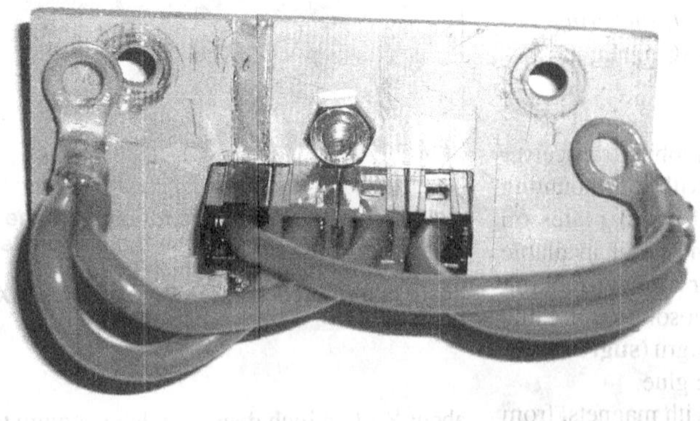

Figure 67 — The Powerpoles connect to the power supply with ring connectors on #10 AWG stranded wire. [Ken Hollenbeck, K9CAX, photo]

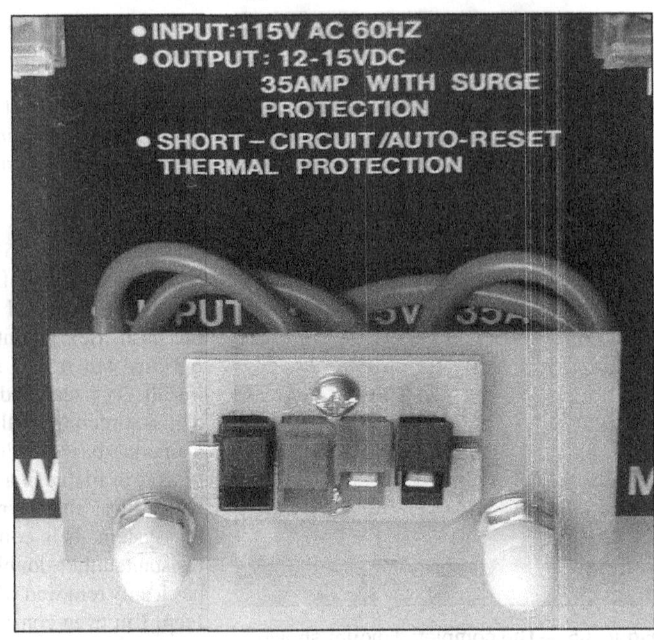

Figure 68 — The completed adapter mounted on the power supply. Note the nylon caps on the ends of the screws to prevent accidental shorts. [Ken Hollenbeck, K9CAX, photo]

ing a size 10 brass washer and a 10-32 brass nut on the outside of the PC board.

Tighten the nut securely to ensure a solid mechanical and electrical connection, which is very important in high-current applications. The installation is finished by adding nylon 10-32 acorn nuts to the ends of the studs. This helps prevent an accidental short from occurring (see **Figure 68**). Note that in Figure 68 the lower corners of the mounting plate have been beveled to provide extra clearance for the brass mounting nuts. Also, I highly recommend that the connections to the Powerpole contacts be made with an appropriate crimper.

One word of caution: be careful not to overheat the PC board when removing the narrow strip separating the plus and minus side, as it is possible to burn or char the bare board. — *73, Ken Hollenbeck, K9CAX, 2433 Cherrywood Ln, Sister Bay, WI 54234,* **marken52@charter.net**

Easy Homebrew PTT

I wanted a very light headset for portable use with my Icom IC-706MKIIG transceiver. The Heil BM-10-iC headset with the Icom OPC-589 microphone and modular plug adapter was just the ticket. For push-to-talk (PTT), the adapter includes a ¼-inch connector for either a foot switch or a hand switch.

The BM-10 and the adapter were expensive, so I decided to build my own thumb-operated PTT switch in order to operate using my left hand for the PTT switch and my right for logging. After a little deliberation, I came up with the solution shown in **Figure 69**. The only parts needed were:

- Pushbutton switch (RadioShack part # 2750609)

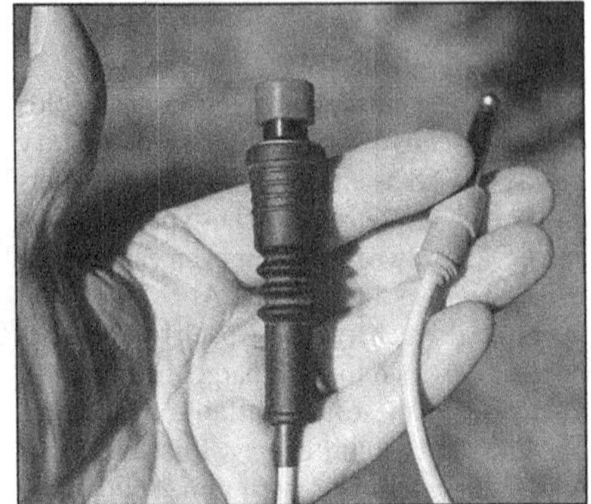

Figure 69 — A sparkplug boot can be used as a convenient grip for a handheld PTT switch. [Chuck Greeno, WA7BRL, photo]

- An automobile sparkplug boot
- About 2 feet of shielded audio cable
- One ¼-inch standard phone plug
- 1-inch piece of heat-shrink tubing for strain relief, as needed.

Most of the parts I had in my junk box; the RadioShack switch was my only purchase. The hole in the sparkplug boot needed to be opened up a little to fit the switch body. Once that was done, I placed some heat shrink over the cable, and then slid on the boot. Next, I removed the nut and washer from the switch body and soldered the cable to the switch. With the cable connected, I pushed and twisted the switch into the boot so the threads on the switch body would hold it snugly in place. Mine feels very comfortable in my hand, looks nice, works fine, and I saved the $30 price of the commercial switch. — *73, Chuck Greeno, WA7BRL,* **wa7brl@arrl.net**

Locking Powerpoles

Powerpole connectors have become the standard for power connections in ham radio, displacing Molex connectors. Powerpoles do have one disadvantage — they are more difficult to lock together than the Molex-type connectors. Facing such a problem, I found that 0.095-inch serrated weed trimmer line works nicely to lock Powerpole connectors together. It is inexpensive and non-conductive.

To lock the connectors, press the line into the Powerpole connector's pin hole until the opposite end is almost flush and trim it with a flush cut diagonal cutter (see **Figure 70**). After it is trimmed, press the "pin" into the connectors with the flat side of the cutters. Be sure to buy line with a serrated edge; the points of the serrations collapse slightly and secure the pin in place. — *73, Lynn Burlingame, N7CFO,* **n7cfo@arrl.net**

Mic Clip Fix

The plastic clip on my Yaesu handheld transceiver's speaker-mic disintegrated on its

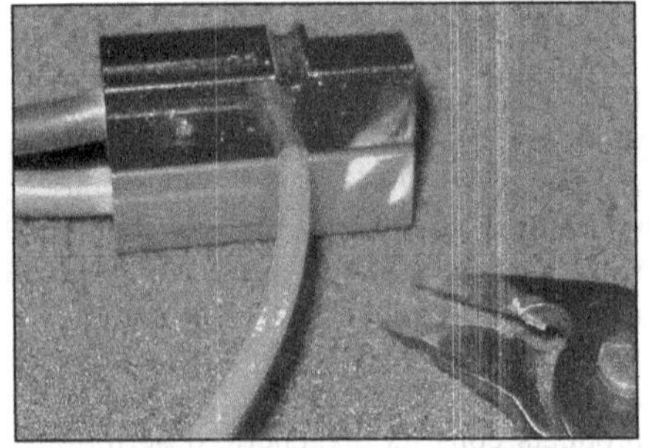

Figure 70 — Serrated weed trimmer line makes an effective and inexpensive locking pin for Powerpole connectors. [Lynn Burlingame, N7CFO, photo]

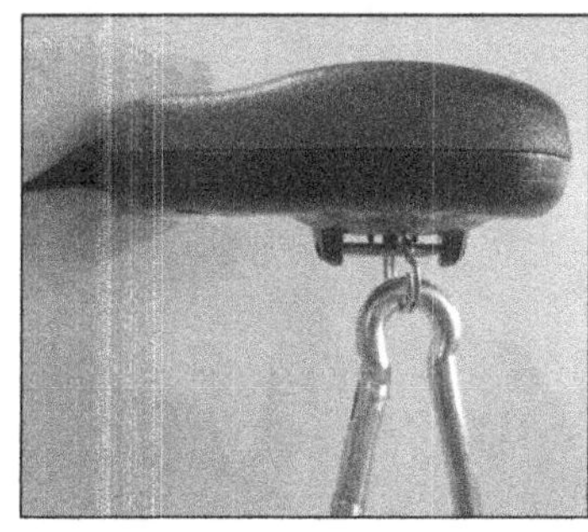

Figure 71 — A small key ring and a carabiner can be used to repair a broken clip on a Yaesu speaker-mic. [Mary Duval, K1MTD, photo]

Figure 72 — The carabiner provides a flexible attaching device that permits a variety of attachment options for your speaker-mic. [Mary Duval, K1MTD, photo]

third deployment. Being an innovative ham, I jury-rigged the mic to my Community Emergency Response Team (CERT) vest and used it for a half-marathon event. Later, I devised this simple fix (**see Figure 71**). Remove the remaining clip parts (if any), keeping the rotating base with the metal pin. Thread a small key ring onto the pin, centering it in the slot on the base. This keeps the key ring from sliding back and forth on the pin. Attach a small carabiner-type clamp to the key ring, and you'll have a strong new clamp for your speaker-mic (see **Figure 72**). — *73, Mary Duval, K1MTD,* **k1mtd@arrl.net**

Mobile Mounting and Modern Auto Safety Devices

Statistics from the National Highway Traffic Safety Administration prove that driver distraction is the leading cause of vehicle crashes. This fact prompted ARRL® to adopt a *Mobile Amateur Radio Operation* policy statement, which can be viewed by entering "Mobile ARS Policy" in the "Website Search" box at the top of the ARRL home page and clicking on the "Policy Statement" link. As mobile operators, we must do our part, by installing our radio gear in a safe manner, while also minimizing distraction to the driver.

By any measure, it is becoming more difficult to install Amateur Radio gear in and on modern vehicles. There are a myriad of reasons for this, not the least of which are the safety devices installed in new vehicles.

Airbags are the major mitigating factor; they deploy at speeds as fast as 200 mph. As many as a dozen are installed in an average vehicle, all designed to protect the passengers from front, rear, and side impacts. Collapsible steering columns, inflatable and/or power retracting seat belts, and automatic adjusting headrests add to the mix of safety devices.

Other innovations like Engine Idle Shut-off (EIS) require the inclusion of a battery monitoring system. This fact complicates power wiring, as discussed in my previous article. Shock-absorbing bumpers and aerodynamic body designs add complications to antenna mounting. Predictably, prospective mobile operators look for workarounds, but all too often end up decreasing the effectiveness of these modern vehicle innovations, especially airbags!

No doubt the greatest infraction is mounting control heads atop the dashboard. This is especially true when magnets, spring clamps, suction cups, and bungee cords are used for attaching them. Mounting control heads and whole transceivers in overhead storage bins where dangling cables interfere with the driver's vision is a close second.

Whatever means are used, gear should be placed to minimize distraction — the greatest single cause of vehicle crashes. Interference to vehicle controls should also be avoided. This includes dangling microphone and power cords.

The difficulty in mounting extends to the outside of the vehicle as well. Nowadays, there is a great reluctance to drill holes in a car's body to properly mount antennas. As a result, magnetic and trailer hitch mounts are often employed as a workaround. Modern vehicles are designed to absorb the energy of a crash by collapsing in a controlled manner. Arguably, these workarounds can produce a reduction in this collision protection.

All of these issues point out the need to plan every installation as if your life (and the lives of your family members) depends upon it — because it does! This author cannot describe the necessary steps for every vehicle make and model. However, there are some salient points that can be made regarding any mobile installation:

1. Plan your installation. Write down everything you think you'll need, and have it on hand before starting. This includes wire terminals and fuses, coax and connectors, ground strapping, cable ties, and proper tools.

2. Don't hurry! An average transceiver/antenna installation may take well over a day to complete.

3. Admittedly, finding a safe, convenient, and distraction-free interior mounting location isn't easy. The best place is low in the center console, well away from exploding airbags. Footwells, cubby holes, and storage bins should not be used, as they may not provide sufficient ventilation. Do not use any mounting style that doesn't use bolts and/or screws to hold gear solidly in place.

4. Microphone cords and interconnect cables must not interfere with vehicle controls and should be securely restrained.

5. Proper wiring is essential. Never use existing vehicle wiring — especially accessory sockets — to power radio gear. This includes using sockets to charge secondary batteries. After all, there is always a way to route wiring through the firewall when necessary.

6. For both safety and efficiency, antennas should be permanently through-hole mounted. Improper mounting, particularly on HF, can result in common mode current flow, which causes both receive and transmit RFI issues. If you're reluctant to punch holes, seek professional help. Two-way radio or car stereo shops are good resources.

Obviously, these are only a few of the important installation parameters. But if it isn't apparent, the single attribute that should be kept in mind at all costs is the value of your life, and the lives of your passengers. In other words, we all need to exercise due prudence when undertaking any mobile installation. In short, be an operator, not a statistic. — *73, Alan Applegate, KØBG, 3202 Notting Hill, Roswell, NM 88201-0403,* **k0bg@arrl.net**

Batteries and Other Power Sources

Simple Crowbar Circuit

Does your power supply have a "crowbar"? Most low voltage (12 V) power supplies use high current pass transistors to control output voltage. The voltage sensing components feed control signals to the base of the pass transistors holding the output voltage to 13.8 V. Ahead of the pass transistors is a brute power supply of 18-24 V. If one of the pass transistors failed it will most likely short emitter to collector. This will apply 18-24 V directly to your transceiver, which could be bad news and it's back to the factory for an expensive repair.

Figure 1 is a simple and effective crowbar circuit I have used for many years. As the supply voltage rises above 14 V, the Zener diode draws current causing a voltage across R1. At 15 V the silicon control rectifier (SCR), which has a 100 A surge capacity, fires shorting the output, blowing the primary fuse and protecting your transceiver.

Check to make sure that flea market special power supply has a crowbar. If it doesn't, you can install one inside or outside in a minibox or a RadioShack project box. Simply connect it across the output of your power supply using a short piece of #12 AWG wire. — *73, Roger Snowdall, WØKWJ, 8405 Everett St, Raytown, MO 64138-3132,* **w0kwj@att.net**

A Battery Supply for the IC-703

I use my IC-703 as a bedside radio, for low-power contests and for Field Day. I power it from a 12 V, 7.5 Ah, sealed-lead-acid (SLA) battery. I needed a simple, cheap means of recharging the battery without the bother or expense of a commercial three-stage charger. This solution fits my needs perfectly and may interest others as well.

I solder an inexpensive male dc chassis mount connector to the negative blade of the battery, doing so in a way that does not preclude the use of wires with slip-on wire terminals. A 2 W, 10 Ω resistor connects the center pin of the connector to the positive terminal of the battery. Across the resistor I solder a 3 V, miniature, wire-lead lamp, which serves as a charge indicator. This lamp is the same used in the Littelfuse low-voltage indicator fuse post.

The power source is a 12 V dc wall wart, which plugs into the connector and charges the battery (see **Figure 2**). Wall warts come in many types, some regulated and others not. The unregulated type generating full-wave rectified ac having peaks reaching about 18 V is the type needed for this application. The regulated type will not work because the voltage is limited at too low a value to charge the battery.

The open circuit voltage of this type runs anywhere from 17-19 V peak and the waveform is full-wave, rectified ac. With this type of wall wart as a source, the current it supplies to the battery will start at about 150 mA with the battery at low charge and slowly diminish to a few milliamperes as the battery attains full charge.

The lamp is moderately bright at the onset, but diminishes to a very slight glow as the battery reaches full charge. We have a simple, automatic tapering charger. These very low charge currents produce no outgassing or heating of the battery but be advised, this is not a quick charge. It may take 24 hours or more to bring the battery up to full charge.

I've used this circuit for more than 3 years on batteries with excellent success. I can leave the IC-703 connected to the battery without danger of over-voltage during the charge cycle and the charger produces zero hash noise as might be present with the typical pulse width modulation type charger. Try it — it's a cheap and easy solution. — *73, David Gauger, W9CJS, 3900 Bluebird Ln, Rolling Meadows, IL 60008-2907,* **w9cjs@arrl.net**

B+ From a Battery

Lately there seems to be a renewed interest in using vacuum tubes in small CW transmitters. The main problem in working with tubes in this SMT era is how to obtain the high plate voltage (B+) needed to operate them. Depending on the operating power level, tubes require from 100 V for low power operation, up to perhaps 200-300 V for a powerhouse like the old 6L6 tubes, which are still readily available.

B+ plate transformers are not an everyday shelf item these days. Wiring a B+ power supply directly from the ac mains (without a transformer) is *very dangerous* and also makes portable operation impossible. The trendy, and more elegant, option is to provide for operation from a 12 V battery source, which might, in turn, be charged by a solar panel. Recent weather events have certainly born out the need for portable, self-sufficient readiness.

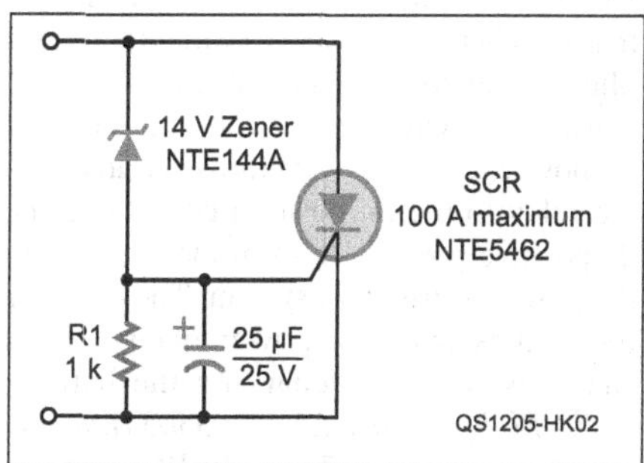

Figure 1 — This crowbar circuit is a simple and reliable method for preventing an overvoltage that could damage your equipment.

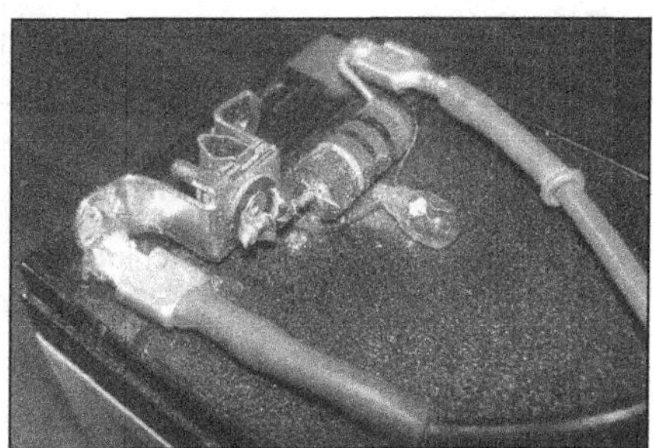

Figure 2 — This simple arrangement combined with an unfiltered wall wart works well as a trickle charger for SLA batteries. [David Gauger, W9CJS, photo]

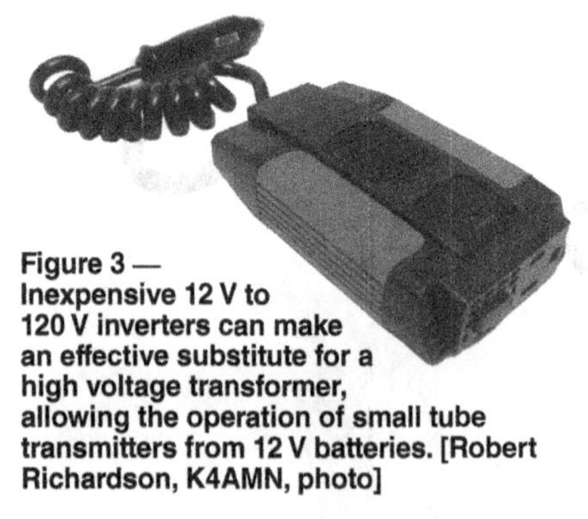

Figure 3 —
Inexpensive 12 V to
120 V inverters can make
an effective substitute for a
high voltage transformer,
allowing the operation of small tube
transmitters from 12 V batteries. [Robert
Richardson, K4AMN, photo]

I have thought up a simple solution. The small 12 V dc to 120 V ac inverters that are available can substitute for the B+ transformer in a power supply circuit. If you don't have one already, you can purchase one very inexpensively. The Model 66944 inverter shown in **Figure 3** is available at Harbor Freight for $17 and is good for 80 W.

You can use the inverter like a transformer to feed a simple power supply circuit and convert the inverter's 120 V ac to a higher level dc voltage. In the "Power Supplies" chapter of the *ARRL® Handbook* are half- and full-wave rectifier and voltage multiplier designs that will supply B+ range voltages.[2] Using an inverter in place of the usual high voltage transformer is inherently safer and well suited to portable emergency operation especially considering that tube equipment is much more resistant to EMP effects. — *73, Robert Richardson, K4AMN, 12418 Colby Dr, Woodbridge, VA 22192-2105,* **inoue@att.net**

Emergency Power Switchover Relay

I have an emergency power system at home that includes two 75 Ah liquid acid batteries, which I originally had charging from an Astron 20 A linear power supply.

It was a simple setup, but it worked flawlessly. The linear supply was turned on to supply power to the station, the dc powered lighting, and a small laptop. The linear supply was connected in parallel with the battery and then to the power distribution system so when the ac mains failed, the battery was already in line and took over to power the system. After the ac mains failed, I would start turning off all nonessential items that were powered by the backup battery.

Eventually I replaced the Astron with a switching power supply that is much smaller, but it included a noisy fan. Besides the fan,

[2]Available from your ARRL dealer or from the ARRL Store. **www.arrl.org/shop.**

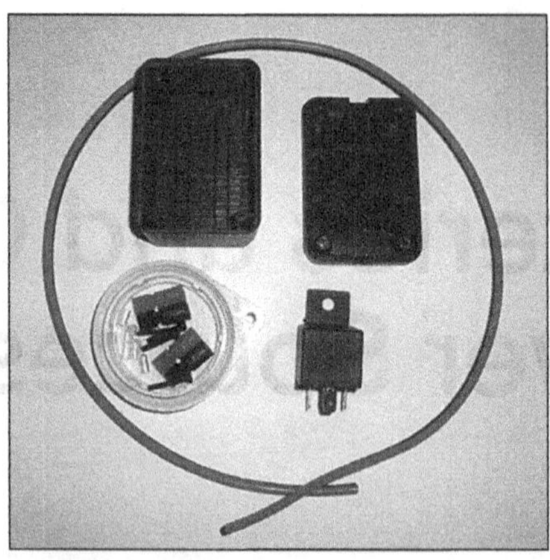

Figure 4 — These are all the parts needed to make the relay switch: a salvaged wall wart case, a small white 12 V wall wart (this one is not a switching type but a linear wall wart), about 3 feet of red/black wire, two sets of Anderson Powerpoles, the case screws, and the 40 A relay. [Philip Karras, KE3FL, photo]

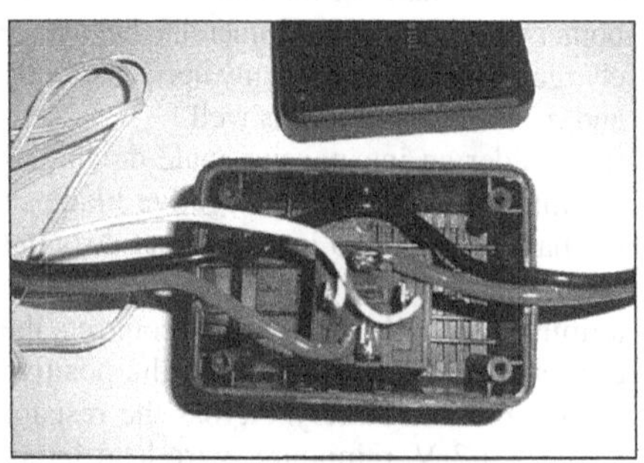

Figure 5 — The relay is mounted in the case. The red power lines connect to the relay contacts and the wall wart lines to the relay coil connectors. [Philip Karras, KE3FL, photo]

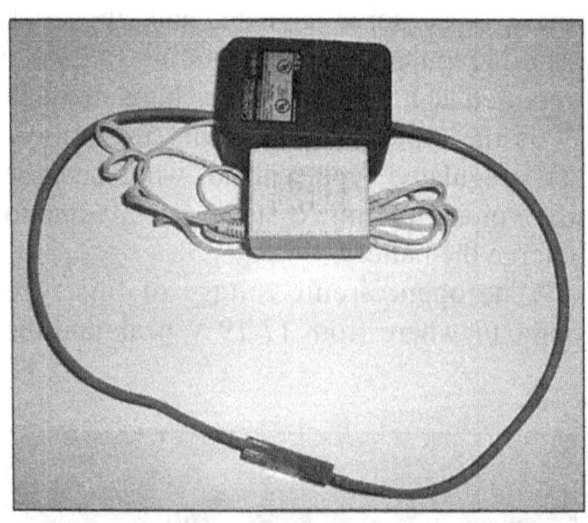

Figure 6 — The finished project, after the Anderson Powerpoles were attached to the wire. I just need to insert the relay in the positive dc line between the switching supply and the battery. [Philip Karras, KE3FL, photo]

the switcher supply had one major difference I had not considered when I decided to change from the linear supply.

When I first substituted the switching supply for the linear supply I got a rude surprise. With the ac to the switching supply turned off, the supply's fan and lights were still on. Upon reviewing the switcher's schematic, I found that even with the ac main's power turned off (the same condition if the ac power had failed), the supply was still drawing power.

I quickly realized that the battery was draining though the switcher supply. I had to actually disconnect the switcher from the power distribution system. That meant that when the power came back on, I had to remember to plug it back into the system to recharge the battery. It was not an acceptable emergency power system any longer.

It turns out that the switcher's fan and light both run on the output dc, not off the input ac. The linear supply didn't have a fan or any internal lights and the full-wave bridge in the linear supply kept the battery from draining through the secondary of the transformer, so there was no reason to protect the battery from discharging through the linear supply.

I had to come up with a way to protect the battery from discharging through the switcher supply. It couldn't be left connected to the power distribution system because the running fan and light would discharge the battery.

My first method was to use a 40 A diode. This worked fine, but I never liked the voltage drop, so I came up with my present system. I connected a 40 A relay that I obtained from All Electronics Corporation (**www. allelectronics.com**, part # RLY-351) in series with the switcher's positive dc output lead. I used a small 12 V wall wart to hold the relay on, passing dc from the switcher to the power system (see **Figures 4** and **5**). When the ac fails or is turned off, the relay will then return to its off state, disconnecting the switcher from the power system. I used a small ac splitter to power both the switcher supply and the wall wart, and this splitter is plugged into a switched ac socket.

Using this arrangement (see **Figure 6**), I can leave the switcher turned on and ready. When ac power is available, it supplies power to both the switcher and wall wart. The switcher turns on and supplies dc power; the wall wart turns on and flips the relay to the on position connecting the dc positive power line from the switcher to the power distribution system. The switcher then takes over the power needs of the shack as well as recharging the battery. — *73, Philip Karras, KE3FL, 3305 Hampton Ct, Mount Airy, MD 21771,* **ke3fl@arrl.net**

Simple Solar

The students of the Central Middle School Amateur Radio Club (sponsored by the Midland Amateur Radio Club, W8KEA) were building AM radio sets that operate on a 9 V battery. During the meeting, one of the adult volunteers suggested the idea of using a 9 V solar panel to operate the radios. His comment sparked an idea.

I had the solar cells and associated circuitry from six solar yard lights. Each of them had an AA rechargeable battery to store energy during the day and supply it to an LED at night. I decided to make a 9 V supply from these discarded yard lights.

Each light contains a solar cell, a control circuit, an AA battery, and an LED. First I had to understand how the control circuit worked. During the day, the sun shines on the solar cell and recharges the battery. When the sun goes down, the LED lights come on. In order for these units to work as a 9 V supply, I needed to be able to draw current for the radios and charge the batteries at the same time. I also didn't want the LED to drain the battery when the sun wasn't shining.

With no schematic for the circuit available, I used a circuit drawing program called *ExpressSCH* (a free download at **www.ex-presspcb.com**) to reproduce the circuit as it was laid out on the controller board.

Once completed, I dragged the components around to rearrange the schematic and make it easier to understand how the circuit worked (see **Figure 7**). The diodes and transistors were unmarked so their exact identities are not known. For the purpose of the circuit analysis I assumed that Q1 and Q2 were bipolar transistors.

The AA battery charges directly from the solar cell through D2. The LED requires about 2.6 V just to turn on, and about 3.5 V for full output, but neither the battery nor the solar cell can supply that much voltage. The voltage is generated by a simple stepup converter (boost converter). While the sun is shining, the base of Q1 is biased off and so is Q2 as a result. As soon as the sun stops shining Q1 is turned on and thus Q2 as well. This causes current to flow through L1 and Q2 to ground. The LED does not light because the voltage is still below 2.4 V. At this point, C1 will briefly pick up a charge and the voltage at the junction of C1 and R1 will cause Q1 and subsequently Q2 to turn off. However, the energy stored in the magnetic field of L1 will cause the voltage at D3 to rise high enough for the LED to turn on and shine. Meanwhile, the voltage at the junction between C1 and R1 will fall to the point where Q1 turns on again, causing the cycle to repeat itself. The frequency of this cycle will be influenced by the R1C1 time constant. This cycle will continue until the battery voltage drops too low to operate the LED or the sun comes back up.

I disconnected one side of L1 to disable the step-converter — preventing the LED from draining the battery when the solar cell was in the dark. Furthermore, I could access the solar panel's output at the battery terminals, whether a battery was installed or not.

So with six of these units hooked up in series and mounted in a nice oak and black walnut case (see **Figures 8** and **9**) I had my 9 V supply ready to go. With the rechargeable AA batteries installed, I can leave it all day in the sunshine and listen to my AM radio at night. — *73, Dennis Klipa, N8ERF, 644 E Whitethorn Dr, Midland, MI 48640,* **n8erf@arrl.net**

Ryobi Power Adapter

As the ARES® Director of the Foothills Amateur Radio Society, I have long advocated for better battery packs to keep a handheld running in a pinch. These can be AA- or D-cell belt packs, Anderson Powerpole outlets, or cigarette lighter plug adapters for vehicles.

Professionally, I'm in the construction industry and I use Ryobi 18 V drills and impact drivers. I upgraded to the new lithium batteries, which, according to the label, have 48 Wh of power. Also the lithium batteries retain their charge at low temperatures and for long periods. With my handheld transceiver connected to my four-element Yagi, it takes ½ W to hit the local repeater. If we don't consider conversion losses, that's 96 hours of transmit power! Doing the math, the drill

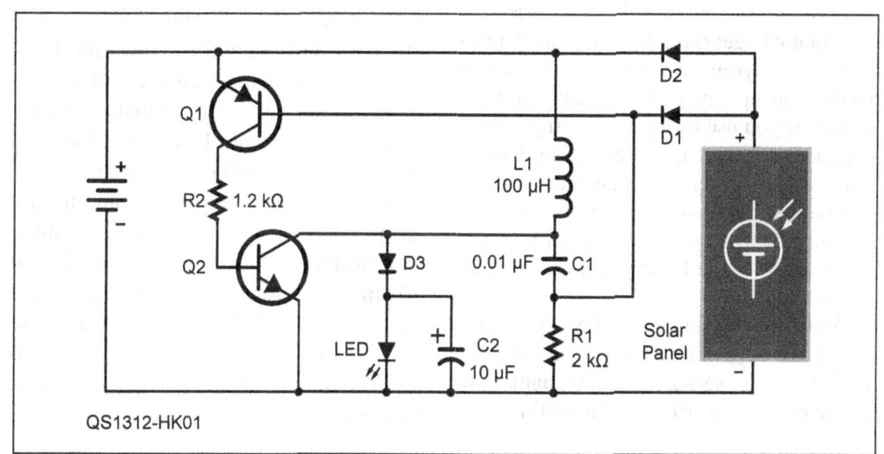

Figure 7 — Schematic of solar LED yard light control circuit. [Dennis Klipa, N8ERF, photo]

Figure 8 — The completed 9 V solar power supply in an oak walnut case. [Dennis Klipa, N8ERF, photo]

Figure 9 — Bottom view showing the six controllers mounted in the case with their batteries installed. [Dennis Klipa, N8ERF, photo]

Figure 10 — An accessory battery tester is modified with the addition of a regulator and Powerpole connectors to convert electric tool battery packs for use with portable radios. [Wally Gardiner, VE6BGL, photo]

battery pack would outperform the AA batteries, hands down.

The problem is how to convert the typical 18 V tool battery to 12 V. Ryobi integrated a voltage tester into their new line of batteries, leaving the older battery tester on the sale table for $10. This provided the solution for connecting to the battery pack. With a little modification, a voltage regulator was mounted, two Anderson Powerpoles attached (see **Figure 10**), and my venerable IC-02AT handheld transceiver was operational on the local High River 146.7 MHz repeater.

For the voltage regulator, I used the new 78H12, which, according to specs, has a continuous current output of 5 A with a 7 A peak current output. You could, in theory, run your 25 W mobile, but the heat sink on the 78H12 gets pretty warm at 5 W.

I left the factory voltage indicator circuit in place and added a 100 pF capacitor to filter any induced RF on the power leads. There are several circuit variations available on the Internet. A fuse or two and possibly a crowbar circuit could also be added depending on the size of the enclosure. The hardest part of the project is finding a connector for your brand of power tool battery. A little ingenuity may be required, but the concept seems to work just fine. — *73, Wally Gardiner, VE6BGL, 390 Ellis Crescent, High River, AB Canada T1V 1J1*, **gardinercomputer@shaw.ca**

Karo Corrosion Cure

Early in my broadcast engineering career I did maintenance on mountaintop microwave relay stations, which had large banks of lead-acid batteries. The maintenance exercises at these stations consisted primarily of cleaning corrosion from battery terminals until one of the technicians came up with what was a stroke of genius — using clear Karo syrup for corrosion protection.

At the time I learned this trick, the battery in my Jeep had top terminals. I cleaned the whole battery and the clamps, then re-assembled everything, including the wire I'd attached to the positive clamp for my rig. I carefully poured clear Karo syrup over each terminal. I used just enough to cover the battery post and most of the terminal, finally using a small screwdriver to encourage the syrup to migrate beneath the clamps. I kept that car for another 2 years and never saw a spot of corrosion on either battery terminal again.

It is necessary in using this technique to be careful not to get syrup into the cells. I never applied the syrup anywhere near the battery ventilation openings. Also, make sure the container you use to apply the syrup is non-conductive. Your car or emergency battery may be low voltage but lead-acid batteries can deliver hundreds of amperes. Shorting the hot side to ground will cause very rapid overheating of the battery that could result in fire and explosion.

With the presence of fiber in today's world, the days of microwave relay stations with their house-sized battery banks are pretty much over. But this interesting trick of using small amounts of Karo syrup to neutralize battery acid may still be useful today. — *73, Tom Norman, KQ7T, 452 N Section Ext, South Lebanon, Ohio 45065*, **tom1norman@live.com**.

Plugging the Power Leak

I built a six-switch outlet box where I could plug in all the wall warts, also known as small plug-in power supplies that convert 120 ac to some lower voltage.

Recently I've noticed that this box, while still helpful in reducing the power wasted by these devices, is not as helpful as it once was. This is because today's wall warts are being made with switching power supplies that use only 10 – 15% more power over that needed for the load device. This means that when your device is fully charged and turned off, the switching wall wart uses much less power than the older linear versions. I decided to investigate further.

I have five similar radios plugged in around the house. These are AM/FM/shortwave radios with wakeup timers and clocks. They came with 300 – 450 mA, 9 V linear wall warts. When I measured three of the radios while they were turned off, with only memory, clock, and LCD display being maintained, the Kill-A-Watt meter read 0.3 W, which equals 33 mA.

I had noticed that a 12 V, 750 mA switching wall wart would use almost no power when it wasn't driving a load and used it in an incandescent to LED lamp conversion project. The LEDs required 12 V at 160 mA and used just 2 W, much less than the original 40 W lamp. I decided to use either 12 or 9 V switching wall warts to replace the standard transformer based wall warts.

Figure 11 — The three radios operating at a reasonable volume only draw 3 W using the newer, switching-type wall wart, a power savings of more than 50%. [Phil Karras, KE3FL, photo]

I bought enough of the 9 V switching wall warts and swapped the plugs for ones that fit the radios. [Switching type power supplies can be strong RFI generators. Buy only quality devices and check for RFI when they are first installed. — *Ed.*] These were used for the test. Powering the same three radios with only the clock and memory running took so much less power that the Kill-A-Watt meter measured 0.0 W.

I checked one radio. When off, it used 3.2 mA at 9 V, to keep the memory, clock, and clock display working. Efficient switching supplies use only 10 percent more power over the load requirements to operate. Let's assume it needs an additional 15% more power over the load device requirements. That would be 1.15 times the 3.2 mA for a total of 3.68 mA. Then for three radios, the total needed becomes about 11 mA for a total of only 0.099 W or 99 mW, while the older linear supplies used 0.3 W (300 mW) to supply the needed 0.086 W (86 mW) the three radios use. This represents a power savings of about 71%.

When operating the radios at a comfortable volume, the total power used with the original wall warts was 7 W and with the replacement switching supplies was only 3 W (see **Figure 11**), the same power as the transformer supplies when the three radios were off. A summary of my results is shown in **Table 1**.

Considering the number of wall warts you probably use, changing to the switching-type to replace your existing transformer-based linear power supplies could make a decent dent in your electrical usage.

Using switching supplies, the power used by the radios while only maintaining memory and displaying the time, was too small to measure with the Kill-A-Watt meter. I didn't expect to save as much when the radios were on, but with the new switching supplies it looks like a savings of more than 50% compared to the older transformer-based supplies and about 70% when off. — *73, Phil Karras, KE3FL, 3305 Hampton Ct, Mount Airy, MD 21771,* **ke3fl@arrl.net**

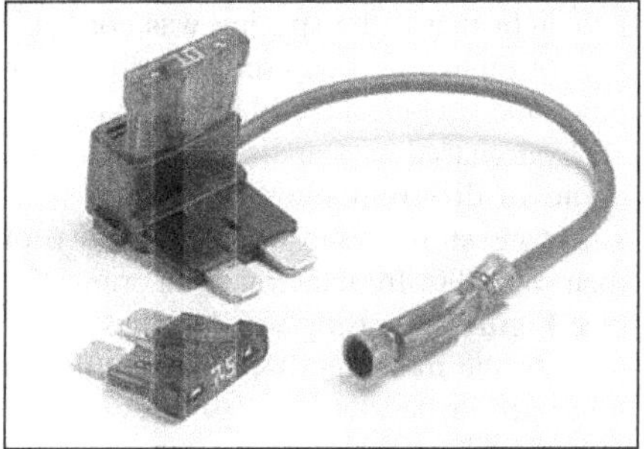

Figure 12 — The Fuse Tap is an auto accessory that plugs into the fuse panel and splits out a +12 V line for a small transceiver or other accessory. [Richard Kriss, AA5VU, photo]

Quick Auto Power Tap

The Fuse Tap shown in **Figure 12** is a fast and easy-to-use device that lets you split and tap one of your car's existing +12 V accessory fuses. The cool thing about this device is that you can tap +12 V by removing the factory accessory fuse, inserting the tap, and adding another fuse for your rig.

I use the Fuse Tap in a C6 Corvette and the +12 V is keyed like other auto accessories in that the +12 V is muted while the auto starter is engaged — just like the auto sound system and other vehicle accessories. Note that this is a +12 V tap, so users will have to find a good common ground. I found one just outside of the fuse panel. I tapped the Corvette's heated seats fuse and have no problem running both the heated seats and my Yaesu FT-8800 50 W rig. I do not suggest or recommend the Fuse Tap for a HF rig. [Caution — the Tap *doesn't increase* the current capacity of the 12 V branch circuit you install it in. The rating of the fuse you remove to install the Tap is the maximum current for that branch. *Don't overload the branch. — Ed.*]

Yes, I know the preferred method is to run a direct line from the battery, but that involves tapping the battery and running wires through the firewall. The Fuse Tap is an easy way to feed a VHF or UHF handheld transceiver, mobile rig, or other low-power accessory. The Fuse Tap is available from various retailers and is usually supplied with several fuses. It sure beats trying to splice wires to add an inline fuse. — *73, Richard Kriss, AA5VU, 904 Dartmoor Dr, Austin, Texas 78746,* **aa5vu@arrl.net**

Storing Your Battery Pack

Some of us have alkaline battery cases for our handheld radios so that we are prepared when our rechargeable battery packs become discharged. Beware of how the case is stored. Some battery cases boost the voltage of the alkaline batteries in order to provide sufficient voltage for the radio to operate. If such a case is stored with the batteries installed, the booster circuitry causes a constant drain; the batteries could be discharged when you try to use them.

For this reason, and to prevent ruining the case from possible battery leakage, do not store the case with the batteries installed, even if it is not connected to the radio. — *73, Art Samuelson, W6VV, 440 Davis Ct, Apt 611, San Francisco, CA 94111,* **w6vv@ arrl.net**

12 V DC Power Strip

While reconfiguring my test bench, I realized I never had enough 12 V dc connections available. Since many pieces of test equipment and homemade devices run on 12 V, having a convenient distribution panel would be helpful. After some thought, my design goals became something that would have Powerpoles, a car power socket, a "power on" indicator, and the ability to directly hard wire pieces of test equipment.

I remembered reading an article where someone had used circuit breaker box neutral bus bars for holding ground radials.

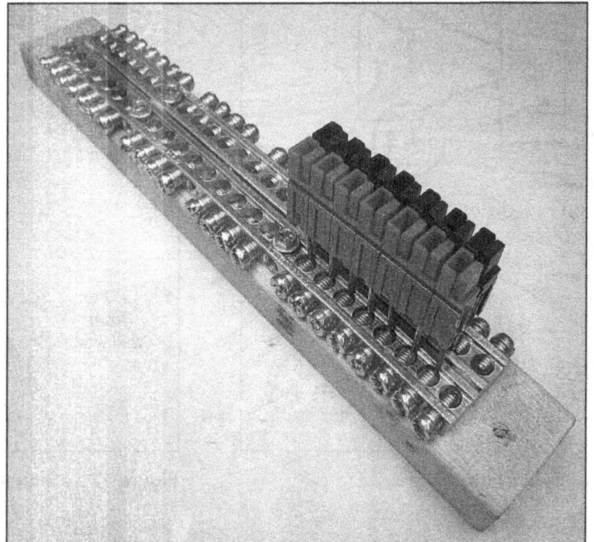

Figure 13 — The basic power strip setup. The bus bars are mounted on their sides, screwed to a wooden base. The Powerpole adapters are mounted in some of the positions while others are left available for hard wire connections. [Allen Wolff, KC7O, photo]

Table 1

Power Usages for Three Radios

Radios	Power Supply	Watts
Off	Old	3
On	Old	7
Off	New	0 (~0.1)
On	New	3

Figure 14 — The cut-off nylon screw fits into the holes in the bar and carries a nylon washer for spacing and insulation. [Allen Wolff, KC7O, photo]

Figure 15 — Powerpole adapters are made with 12 AWG solid wire and are mounted to the lower section of the bus bar to provide convenient connection points for Powerpole adapted equipment. [Allen Wolff, KC7O, photo]

I thought that if the spacing was correct, a pair of them could be used to hold Powerpoles and become a distribution device.

I purchased two neutral bus bars (they come in different lengths) from a local hardware store. I screwed them down on their sides, 90° from their normal orientation (see **Figure 13**), using wood screws passed through the mounting holes. The mounting holes are drilled through, so I also placed nylon washers as insulating spacers at the mounting points. I cut off a section of an #8-32 nylon screw, added a nylon washer, and fitted it into the holes in the bar at the mounting points (see **Figure 14**).

To attach the Powerpoles, I soldered (not crimped) short lengths of solid #12 AWG house wire to each contact (see **Figure 15**). Since house wire is exactly what is required by code, it should hold fine under the bus bar screws. The 12 V supply is fused with a 20 A blade fuse and the supply wires are soldered to short pieces of #12 AWG house wire. I added an LED with series 470 Ω resistors soldered to each lead [the resistance will vary depending on the type of LED you use. — *Ed.*] and insulated with heat shrink at the top end of the strip to indicate when the power is on.

The distribution strip is mounted vertically under a shelf where a short is unlikely. Plastic protective shrouds could be added to prevent shorts, if desired. — *73, Allen Wolff, KC7O, 57 W Grand View Ave, Sierra Madre, CA 91024,* **ajwolff@earthlink.net**

Wall Wart Battery Backup Improvements

I took particular interest in Paul Danzer's, N1II, *QST* article, "Done in One: Battery Backup For Your Wall Wart," as I deal with these kinds of circuits and applications in the commercial world as part of my job.[2] I made some quick simulations with *LTspice* and while the circuit seems to be generally headed in the right direction, there are some key areas where it can be improved and simplified (see **Figure 16**).

I replaced the LM741 with one half of an LM393 comparator. This is a very common and inexpensive part that has two main advantages over the LM741. First, its supply range is much greater. It can operate from as little as 2 V to as much as 30 V. It also has an input voltage range that goes from its own negative supply pin to 2 V below the positive supply. To keep the inputs below the supply rail, I scaled down both the inputs to roughly half the supply voltage using R4/R5 and R7/R8. Note that the slight imbalance of the two dividers is intended to fine tune the switchover point. In most cases, we can make the two dividers equal using only 10 kΩ resistors, but if you wish to minimize the drop on switchover, this is the way to implement it.

The change of the op amp and the reconfiguration of the inputs are the key changes. In analyzing the original circuit, I found that the inherent body diode of the FET (not shown in the original schematic) along with D4 in the original circuit also perform the "OR" function of the two input rails, so we can power the op amp directly from the output and delete the original components D1 and D2.

The analysis also indicates that D3 and R6 of the original circuit have minimal effect. As a consequence, D3 has been removed and R6 has been re-purposed as a pull-up resistor on the output of the LM393.

---

[2]Danzer, N1II, "Done in One: Battery Backup For Your Wall Wart," *QST*, Jan 2016, pp 49 – 50.

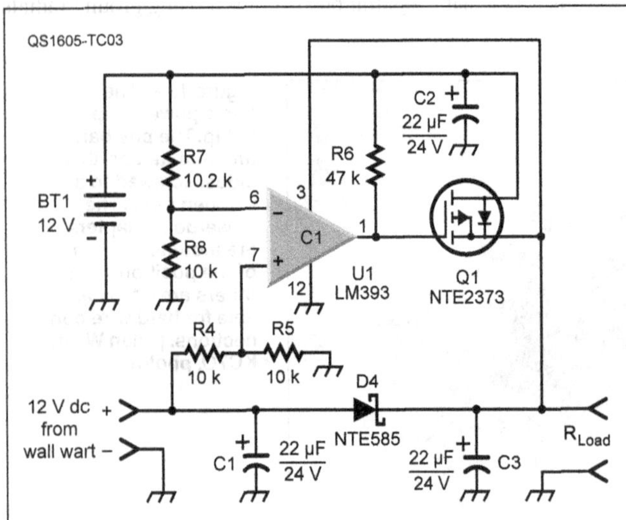

Figure 16 — Redesign of the wall wart backup circuit.

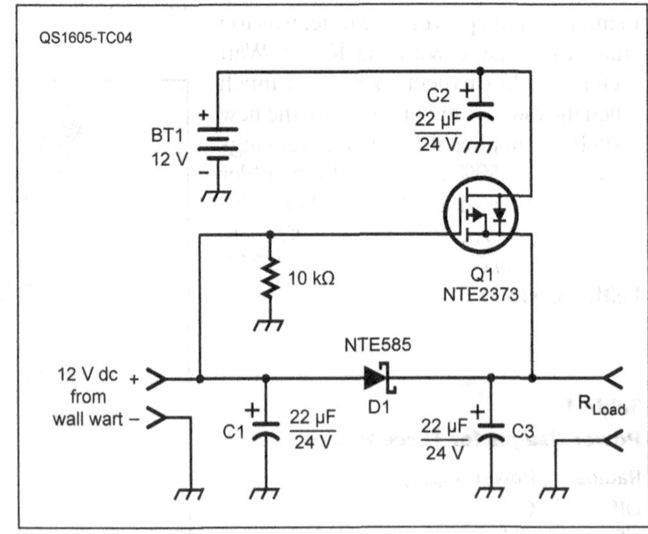

Figure 17 — Simplified circuit that operates without a comparator.

Overall, Figure 16 works well. In redesigning it, however, I came up with a much simpler circuit that, for 12 V applications, is good enough for most uses. While the absolute simplest approach to OR two power sources would be a pair of diodes, replacing one diode with a FET, as shown in **Figure 17**, is a time-tested design that is very useful. This circuit lacks the precise FET turn-on threshold that the comparator gave us in the more complex circuit. Instead, the FET's inherent turn-on threshold of approximately 2 V is used.

When applied to our wall wart case, the FET would only turn on when the 12 V wall wart dropped to approximately 10 V. We still have the battery feeding the load through the body diode of the FET so the load voltage would not drop out completely. It would only briefly drop to 11.3 V before the FET turned on and brought it back up to 12.0 V. The only advantage of the more complex circuit is the elimination of the brief period where the load is fed through the 0.7 V drop of the FET body diode. Even this can be minimized by paralleling the FET with a Schottky diode, which typically have 0.35 V forward drops. — *73, Dave Lundquist, N2DJE, 36 Annandale Rd, Stony Brook, NY 11790,* **n2dje@arrl.net**

Two-Range AC Power Supply

This power supply circuit configuration (see **Figure 18**) has been in use at K2AOP for many decades. It uses two capacitors with voltage ratings of half the voltage that is required when a single capacitor is used and the percentage of ripple on the low range at full load is half as much as with one capacitor. This arrangement provides the ideal result; equal percentage ripple on both ranges with the lowest voltage rating on the capacitors.

In the low-voltage switch position, you have two capacitors in parallel following a bridge rectifier. In this circuit, the diodes are rated 2 A continuous with a 100 V peak reverse rating. In the high-voltage position, you have a voltage doubler with the capacitors in series. When in the doubler mode, the two diodes on the right side of the bridge are in the reverse (back) biased condition and

are inactive. Cover up the rightmost diodes and you will see the typical full-wave voltage doubler circuit when the switch is in the high-voltage position.

With a transformer of a different voltage, the diode reverse voltage rating should be at least four times the transformer's secondary RMS voltage, and the capacitors should be rated for at least two times that voltage. — *73, John Clark, K2AOP, 6226 E Carolina Dr, Scottsdale, AZ 85254-1930,* **k2aop@ arrl.net**

Diode Voltage Drops Raise Battery Power Drain

The May/June issue of *Elektor* magazine (**www.elektormagazine.com**) reminds us that it is important to consider the forward voltage drop (V_f) of rectifier diodes used for reverse-polarity protection in series with the power source. This is a common technique for battery-powered equipment that is regularly disconnected and reconnected to power sources. Because power dissipation is P = $V_f \times I_{avg}$, that simple diode can eat up a lot of a battery's stored energy. For example, a typical Instantaneous Forward Voltage Characteristics curve for a 1N4001 diode (see **Figure 19**) shows that at 250 mA of average current, the garden variety 1N4001 (50 V, 1 A rating) dissipates 0.8 V × 0.25 A = 0.2 W. Although manufacturers' data for Schottky diodes, such as the 1N5819 (40 V, 1 A rating), show them to have a lower forward voltage drop (0.3 V_f) than silicon junction diodes (see **Figure 20**), they still dissipate a significant amount of power (0.075 W). Also, while it may be counterintuitive, smaller diodes of any type often have a higher V_f for the same current as larger ones.

Regardless of whether you use silicon junction or Schottky diodes, the voltage drop can be reduced by wiring several diodes in parallel to keep all of them operating at a point where V_f does not increase as steeply with current. For example, if you use four 1N4001 diodes in parallel, this will reduce the average current in our example to 62.5 mA per diode, which, in turn, reduces V_f from

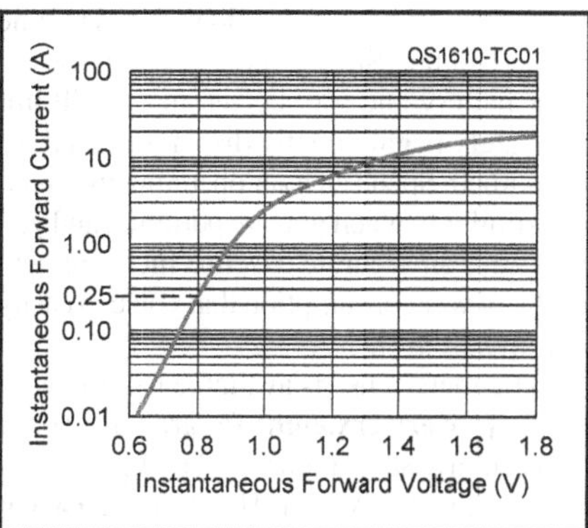

Figure 19 — This typical I-V characteristic curve for a 1N4001 diode shows that at 250 mA, the V_f will be about 0.8V. While 0.8 V is low, it still represents a significant loss of battery energy as heat.

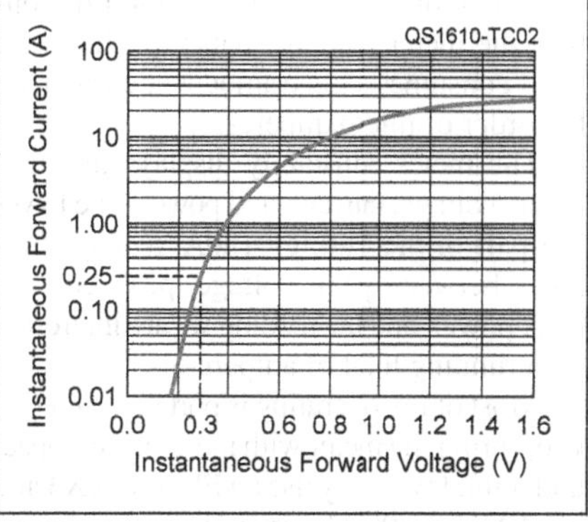

Figure 20 — This typical I-V characteristic curve for a 1N5819 diode shows that at 250 mA, the V_f will be about 0.3 V, much lower than the 1N4001 diode, but still dissipating battery energy as heat.

0.8 V to about 0.7 V, a power savings of 10% at a cost of pennies per diode. This assumes an equal division of current between the diodes, so it is best to use diodes all purchased from the same manufacturing batch — so buy a large number all at once from one vendor. You also need to have a reasonably good idea of the maximum current to be drawn so you can pick the right number of diodes.

At currents above an ampere or two, consider using a MOSFET with its much lower **ON** resistance as described in *The ARRL Handbook*'s "Reverse-Polarity Protection Circuits" section of the "Power Sources" chapter. The chapter also includes information on other polarity protection techniques that may work better for your application. — *73, H. Ward Silver, NØAX,* **n0ax@arrl.org**

Measure Battery Power Demand

Battery endurance during portable or emergency operations can be difficult to predict. Instead of attempting to predict how much power you will require, measure

Figure 18 — Schematic diagram of the dual-voltage power supply.

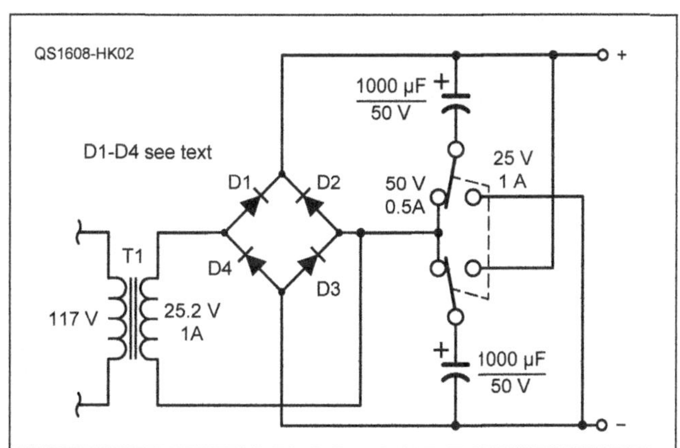

it directly. This is easy to do with a dc inline power meter. Simply connect it between your battery and transceiver and it will immediately begin directly displaying total Ah consumed. Spend an hour operating in a manner similar to when you are portable, and you will have an accurate baseline measurement of the power consumption that you can easily extrapolate from.

The meter I use is available from Powerwerx (**powerwerx.com**) for around $50, or you can save some money by buying the similar Turnigy 180A from Hobby King (**www.hobbyking.com**). (I chose the latter, but had to install the Powerpole connectors myself.) The meter is simple to use — you just plug it in and it continuously monitors your battery. Keep in mind that these meters are not self-powered and they have no memory, so do not remove the meter from the power source until your measurements are complete. (You can turn the radio on and off, just don't disconnect the meter from the battery.) The meter itself consumes a trace amount of current, on the order of microamperes.

The meter constantly displays instantaneous voltage, current, and power. The lower left part of the Turnigy 180A display alternates between peak voltage, peak current, peak power, and the reading we are interested in, a running total of ampere-hours.

Most of my operating is portable, and I always bring the meter with me — it's compact and light. I typically use an Elecraft KX3 and a 4.2 Ah LiFePO4 battery. Using the meter, I have discovered that this battery is more than adequate for a typical weekend backpacking trip, as I typically find that I have 1.5 – 2 Ah "left in the tank." The meter is also extremely useful when I set up a solar-powered station for Field Day. It tells me how much power my panel is delivering under different sky conditions, and how much current is being delivered to my battery. I never leave home without it. — *73, Mark Volstad, AI4BJ, 6098 Tosha Dr, Burlington, KY 41005-9303,* **mvolstad@twc.com**

Float Supply Fire Danger

Spring storms here in Texas are anything but predictable. Because of such storms, we like to be prepared for power outages. We do that by placing batteries in parallel with our power source. Once installed, these batteries have to be kept charged. At my station, I use a power supply to float them at 13.8 V.

Inside most quality power supplies is a crowbar circuit. The crowbar circuit is a silicon control rectifier (SCR) tied across the plus and minus outputs. Its job is to turn on in the event that the output voltage rises above the set voltage. If the output voltage gets too high, the SCR turns on, shorting plus to minus, which should blow the fuse in the power supply's input circuit. The crowbar's job is to protect your downstream equipment from excessive voltages in the event the regulator or some other part of that circuit fails.

In my Astron RS-35VM, there is such an SCR crowbar circuit. A lightning strike to a disconnected antenna not only blew the antenna into pieces but somehow managed to find its way into the power supply, taking out the LM723 chip — and the SCR.

Remember the backup battery I spoke of? Well, it was hooked up to the power supply. Because I had installed some safety precautions, there was no smoke or fire — just a dead station and many questions.

Performing a root cause analysis, the first issue was to repair the power supply, carefully analyzing what failed. I determined that the output was shorted and a look at the schematic quickly led me to the crowbar circuit. Cutting the SCR out of the line removed the short, enabling me to repair the supply. Replacing the SCR, the regulator chip, and the metal-oxide varistor (MOV) in the ac line put the Astron back into service. While testing the transistors, I noted that one of them was installed incorrectly from the factory, in that one of its legs was between the metal frame and the socket, instead of through the pin as it should be.

I had installed a 30 A automotive fuse in the line between the battery and the power supply. This precaution saved the house from going up in smoke, as the batteries are 90 Ah units. That much power would be enough to melt something, the weakest link being the SCR, but the wires could get hot enough to start a fire. [Caution: A 90 Ah battery's output is *not* limited to 90 A. High amp-hour capacity batteries, when shorted, can deliver many hundreds of amperes *instantly* and represent a significant risk of fire, explosion, and/or personal injury. — *Ed.*]

Astron and other power supply manufacturers sell supplies designed for battery charging. If, however, you are using a standard supply to float charge high capacity storage batteries, check your power supply out and make sure you have the battery connections fused to prevent a power supply output side short from becoming a calamity. — *73, Scott Taylor, N5MJQ, 1022 Nottingham Dr, Carrollton, TX 75007,* **Dok@TimeDok.com**

4

Construction and Maintenance

Soldering Iron Antioxidant

I have been soldering things for 66 years. During that time the one thing about soldering that irritated me was the constant battle to keep the soldering iron tip clean of oxidation (that hard black scale that builds up). I keep a wet sponge handy and was constantly cleaning the tip.

When building with larger components, it was fairly easy to find a clean spot on the tip to heat the component and solder simultaneously. Working with SMD (Surface Mount Devices) and small printed circuit boards is quite different. The component terminals and solder pads are small. These small components, especially the SMDs, are amazing devices and can withstand quite a bit of heat — for a very short time. Soldering such devices requires that the heat transfer from the soldering iron to the component or pad happen very quickly. A small, clean and properly tinned tip is a must, which was a problem with my old 12 and 25 W irons.

Keeping Out the Air

Oxidation is a result of the hot metal coming into contact with oxygen in the air. It occurred to me that oxidation could be reduced if the tip was not exposed to the air while in the holder. I thought, "Why not keep the tip buried in a pool of molten solder?" As this idea developed, some questions arose.

First, what to use as a cup to hold the molten solder? It needs to be small, shallow, cheap and able to withstand substantial heat. I started walking the aisles of my home improvement store and found a solution. A ⅜ inch Axle Cap Nut (Lowe's part #008236768541). They are used on kids' toys to keep the wheel on the axle. I bought a package of two.

Next, I started to wonder if the soldering iron would generate enough heat to keep the solder molten? And what would I use to hold the cup? Whatever was used had to withstand the heat from the molten solder.

I found a piece of oak board about an inch thick. I cut it to size and bored a shallow hole to hold the cup and another hole to support a soldering iron holder formed from

a coat hanger. A couple of screws through the 25 W soldering-iron base into the wooden block and my new soldering station was complete (see **Figure 1**).

First Heat

As my 25 W soldering iron was heating up, I used my 300 W soldering gun to melt enough solder to fill the cup. I cleaned the tip of the soldering iron and stuck it in the homemade holder. Then I withdrew the iron and the bottom ⅛ inch of the tip was clean and perfectly tinned — no hard scale. I grabbed a few pieces of wire and a PC board and soldered and desoldered a bunch of connections without ever touching the cleaning sponge.

The original holder for my 12 W iron was a small piece of angled aluminum. It didn't have an existing base to work with, so that holder was built completely

from scratch. A scrap piece of wood, the other cup, a formed section of coat hanger, a few minutes time and this project reached a successful conclusion (see **Figure 2**).

Safety

Remember, even low-wattage soldering irons are *hot*! So too the liquid solder in the cup. Don't use this around children or others who could be burned by the molten solder or iron. Even though the amount of solder in the

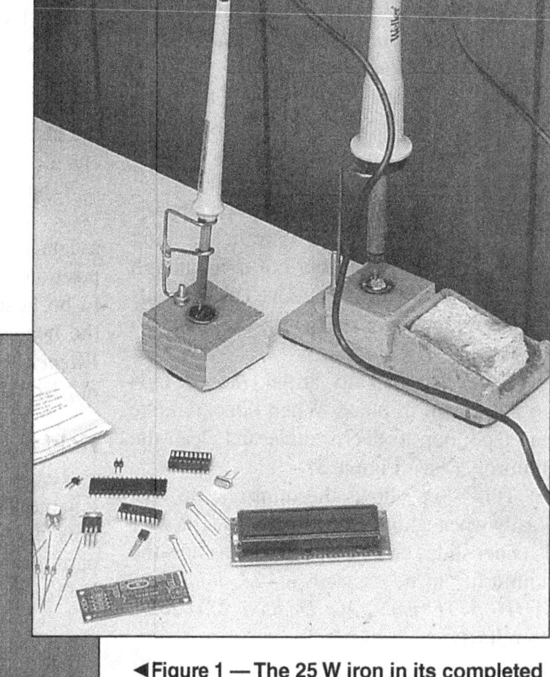

◄Figure 1 — The 25 W iron in its completed antioxidant holder. [Wayne Smith, WA4WZP]

▲Figure 2 — Some metal cups, a piece of wood and a coat hanger are all you need to keep your soldering tip shiny and ready for action.

cup is small, it's enough to cause a serious burn if spilled. The base of my 25 W iron is heavy enough to resist tipping over. Not so for the 12 W iron. For safety purposes, it is attached to the workbench with industrial strength hook and loop fastener. Always use soldering equipment in a proper work area and take steps to keep hot items away from yourself and others.

The solder cups are held in the wooden blocks with epoxy, as well as the lower coathanger support of the 12 W iron. Be careful to use an epoxy that can tolerate the heat, such as J-B Weld, which can withstand 600°F.

From a cold start it takes about 10-12 minutes for the solder in the cup to liquefy. A dark layer of rosin will form on the top, but it does not stick to the tip of the iron. — *73, Wayne Smith, WA4WZP, 224 Saint John's St, Arden, NC 28704, **wa4wzp@arrl.net***

Mini Paint Roller

Occasionally, I have a radio project that needs painting. In cases where the surface area is small and I want a smooth finish, a miniroller is very handy. Most paint departments sell 3 inch wide rollers and matching paint pans. In cases where the 3 inch roller is too large, I have made my own microroller as follows:

1. Cut a 3 inch paint roller brush to the desired length (1 inch for my projects).

2. Slide the paint roller onto a nail (mine is about 4 inches long) with a flat head to keep the miniroller from falling off.

3. Cut a piece of #20 AWG solid insulated hookup wire to approximately 6 inches.

4. Twist a loop at the midpoint of the wire next to the bottom of the 1 inch roller.

5. Run the two hookup wire tails parallel to the nail. Make the wire holder snug against the miniroller, but not tight enough to interfere with rolling action. This prevents the miniroller from sliding off the other end of the nail.

6. Run a loop of tape around the wire tails to hold them in place. When done, remove the tape loop to disassemble and clean the miniroller (see **Figure 3**).

This setup allows the miniroller to turn easily when adding or rolling paint. The wire retainer slides on and off the nail to allow the miniroller to be cleaned. — *73, Tom Hart, AD1B, 54 Hermaine Ave, Dedham, MA 02026, **tom.hart@verizon.net***

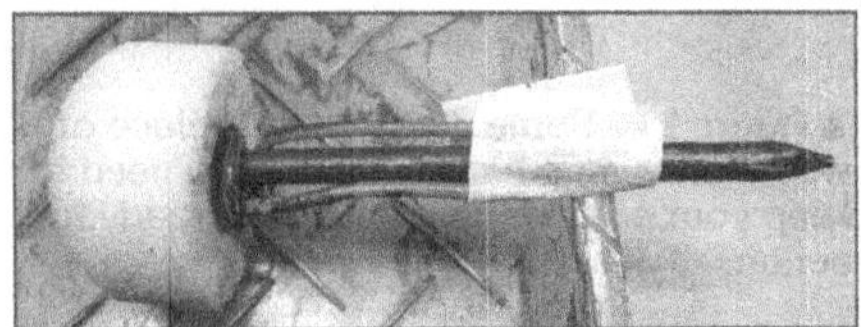

Figure 3 — The completed miniroller ready to get into those tight spaces. [Tom Hart, AD1B]

Switch Shaft Extension

I wanted to add a long extension shaft to my antenna selector switch. I found a perfect 36 inch length of ¼ inch aluminum rod stock for only about a $1.50 at our local home improvement store. Then it occurred to me that I needed a way to join my new shaft to the short switch shaft. Wandering around the plumbing department, a little compression union for ¼ inch copper pipe caught my eye (see **Figure 4**). It cost about twice as much as the aluminum shaft but I was in a hurry. I slipped the extension shaft in one end of the union and the short switch shaft in the other and tightened the nuts. So, if you are in a hurry, give the coupling a try. By the way, they look cool too! — *73, Martin Huyett, KØBXB, 7735 Big Pine Ln, Burlington, WI 53105, **huyettmeh@tds.net***

Hair Dryers in the Shack

The August 2011 hint on PA module repair reminded me how often I use a hair dryer in my shack.[2] I use it to preheat the heat sink prior to working on a PA module or other device that has a heat sink or large amounts of printed circuit board copper attached. With the copper or heat sink hot, the heat from the iron is concentrated on the work area when soldering.

I also use the hair dryer to heat up PL259s before I solder them, for heat shrink and to assist in finding intermittent issues in radios. A 1200 W hair dryer is fairly hot. It can actually melt plastic or solder if given the chance but with careful use I have not encountered issues.

I have a Revlon Iconic, which has a "nub" designed to sit on the table as well as a detachable curved spout. It allows me to position it on the bench and move the item to be heated around the air flow or point the heated air where needed. — *73, Ron Wagner, WD8SBB, 5065 S Kessler-Frederick Rd, Troy, OH 45373, **wd8sbb@arrl.net***

Hold the Solder

I suffered a stroke 10 years ago, which left me with only one working hand. This presented a problem. How could I solder parts in place to a circuit board where you hold the soldering iron with one hand and

[2]Larkin, W8RVT, "PA Module Repair," *QST*, Aug 2011, p 57.

Figure 4 — A ¼ inch pipe compression coupling repurposed as a switch shaft extension. [Martin Huyett, KØBXB]

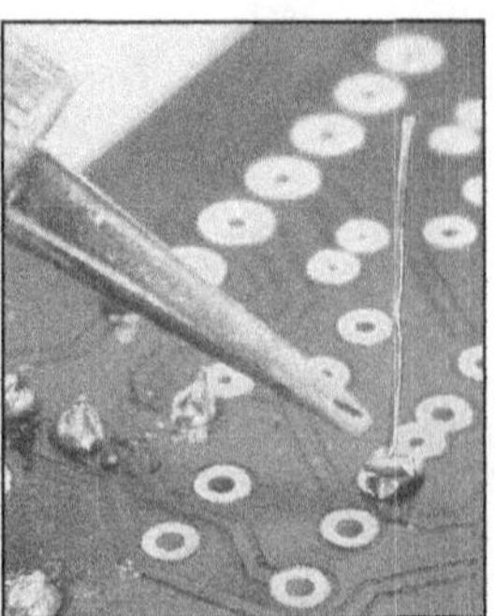

Figure 5 — By preforming the solder into a coil that can be slipped over the component lead, soldering can be a one-hand operation. [Jim Bull, KE7OW, photo]

the solder with the other? I thought of at least two ways.

First, get at thin nail, pin or other piece of metal with a diameter slightly larger than the leads of the resistor, capacitor or other component you intend to solder. Clamp the nail with a vise, letting about 1-1½ inches extend above the vise. Then take your solder and wrap it around the nail one or two times, leaving the end of solder as part of the coil. Remove the coil from the nail and slip it over the wire until it touches the board. At this point, you may take your wire clippers and cut the solder, so that only the solder coil is left (**Figure 5**) or leave the coil and let the heat from the soldering iron separate it.

Another way to manage the solder is to clamp the solder with an alligator clip attached to a small portable vise or other holder. Leave 2-6 inches of the solder free, which you can then place on the board at the point requiring the solder. I found this method good where the components had short leads.

What if there is too much solder and some needs to be removed? I have a desoldering tool consisting of a small vacuum bulb that is put over molten solder to remove it. Here, again, you need two hands: one holds the bulb while the other guides the soldering iron. I solved this problem by building a stand for the soldering iron. I cut a piece of aluminum to about 4×1 inch. The aluminum is wrapped around the handle of the soldering iron and held in place by a vise (I use an X-acto mini vise). It's then positioned so it heats the area to be desoldered while I use my hand to work the bulb. — *73, Jim Bull, KE7OW, 26441 161st Ave SE, Covington, WA 98042, **ke7ow@comcast.net***

Airdux Coil Cutter

A homebrew project I was working on required cutting a few turns from an Airdux coil form. The coil was 1½ inches in diameter with very close, fine-wire turns. I couldn't cut the coil form with any cutters

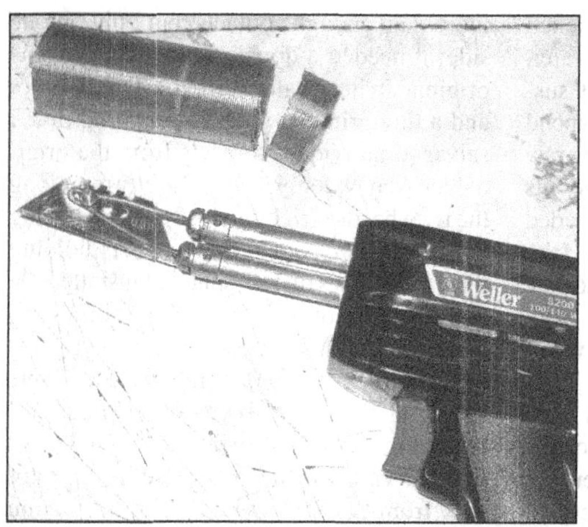

Figure 6 — This tool made from a box cutter blade and a soldering iron will cleanly cut the plastic forms on an Airdux coil. [Whitey Doherty, K1VV, photo]

or razor knives I had. I solved the problem by attaching a box cutter blade to the tip of my Weller soldering gun with a machine screw (see **Figure 6**). When heated, it easily cut through the four plastic form sections to remove the required number of coil turns. — *73, Whitey Doherty, K1VV, PO Box 1193, Lakeville, MA 02347-1193,* **k1vv@comcast. net**

Cure T-90A Wobbles

My ICOM IC-T90A has given me good service for many years. But the large triband antenna flexing on the small SMA antenna connector has been the Achilles' heel of this radio. I have repaired it several times now and had to replace the SMA connector once. It is not too hard to get into the radio but desoldering is required on the function pot and two small chassis-to-circuit-board shields. When that is done, you can remove the circuit board to access the solder point for the SMA connector. The center pin of the SMA connector is "blob" soldered to the circuit board and you will frequently find this joint broken. Cleaning out the solder joint and soldering several small wire strands to the circuit board and the SMA center pin make a flexible connection that allows any antenna move-

ment to flex the wire instead of the circuit board.

When my flexible connection failed, I decided to investigate the SMA connector motion. There is a flexible weather seal inside the radio on the antenna connector that will not allow it to be tightened securely, no matter how much you tighten the jam nut on the outside. The real problem seems to be the antenna itself. It will screw down on the SMA connector only so far, making the slim SMA nut the sole security. Also, the SP/MIC connector weather seal, which is very close to the antenna base, may contribute to the problem (see **Figure 7**).

The fix was simple. Add a washer or washers as necessary between the antenna base and the SP/ MIC jack seal to allow the antenna to compress more tightly. This will eliminate the antenna wobble that shakes the SMA connector, breaking the solder joint. If you repair your radio, I would remove the internal weather seal so the SMA nut on the outside will pull the connector base tightly against the inside of the chassis. [Note that changing the factory weather seals will affect the "splash resistance" rating of the T-90A. — *Ed.*] — *73, Charlie Liberto, W4MEC, 619 Hideaway Cove, Hendersonville, NC 28739,* **w4mec@arrl.net**

Paper Roll Coil Forms

Powdered iron toroids are very much a part of modern equipment construction and have contributed enormously to the miniaturization of much of today's gear. As useful as these items are, they do have some drawbacks.

First of all, they are not commonly available at the corner hobby store. Secondly, they do not lend themselves to adjustment of coupling when auxiliary windings are added. Finally, to those of us of frugal sensibility, they cost money!

Cylindrical plug-in coil forms, once a mainstay of the ham's parts pantry, have be-

come hard to find. Searching for a substitute for a vintage project, I happened upon some discarded cores from paper rolls used in credit card terminals.

One variety is made from a milky plastic similar to polyethylene. These rigid cores are of a lightweight honeycomb construction and are very easily worked with a small drill or even a sharp awl or pin. Much less material is contained within the field than would be the case with the popular PVC piping and mine passed the "microwave oven" test (30 seconds in the microwave with a cup of water) with flying colors. Sharp peaks at resonance suggest reasonable Q, at least at HF. A second type is black. I suspect these may contain carbon — perhaps useful for winding chokes but not for high Q requirements. As a final bonus, either type core can be end mounted using coarse-thread self-tapping screws or with hobby cement.

Figure 8 shows the white cores in use in a vintage regenerative receiver, one as a coil form in a regenerative circuit with a "tickler" winding (center). It also illustrates two different mounting possibilities. Not plug-in, but perhaps some clever experimenter will solve this dilemma as well. — *73, Bill Higgs, NT4C, 12304 Old Henry Rd, Louisville, KY 40223-2232,* **bill@lizcurtishiggs.com**

Repairing Stuck Pots

If you pick up used gear at hamfests or from the various online sites, you will eventually get one that has a potentiometer or rotary switch that King Kong leaning on a 36 inch Stillson wrench can't turn. The problem will not be the pot's wiper or the detent mechanism of a rotary switch, it will be where the shaft passes through the body that attaches it to the panel. I've tried several methods to free them up, but this one works the best.

The first step is to remove the stuck control or at least get it back from the panel. Looking closely at the mounting area of the control, you will see that just past the threaded mount, there will be a boss slightly bigger than the threads. This is the area that

Figure 7 — The addition of a washer to the SMA connector can help provide mechanical support to the T-90A antenna connector. [Charlie Liberto, W4MEC, photo]

Figure 8 — The author's two-tube regenerative receiver using three paper roll coil forms. [Bill Higgs, NT4C, photo]

Figure 9 — To cure a stuck potentiometer drill a small hole at the base of the its mounting threads and apply a small amount of light oil or WD-40. [Charlie Liberto, W4MEC, photo]

contacts the back of the panel so none of the stress of the tightened control nut is transferred to the body of the control, distorting it. Usually, there is a gap between this boss and the threads.

Using a slow speed drill with a number 50 or smaller drill bit, drill a hole into the control (see **Figure 9**), just enough to break through to the shaft. If there is no gap between the threads and the boss, drill as close as you can. On a pot, you want to drill opposite from the terminal area so chips won't get into the pot. If you can rotate the control at all, you can verify you have drilled deep enough as you can see the shaft turn. If you can't rotate it, use your best judgment as to depth.

Use penetrating oil or WD-40 sparingly; don't get it on the contacts of a rotary switch or the innards of the pot. Begin working the shaft and it should slowly start to give way and turn like new. After it is free, use a light machine oil to lube the control, especially if you used WD-40. If you did get oil in the control or contacts, use a good contact cleaner like De-Ox-It, don't leave WD-40 on any important surface.

If the control is just too involved to remove, take off the mounting nut and any washers, then drill as close to the panel as you can. This will keep the area of the threads that are roughened up by the drill, in an area where the nut will not need to reach, or the nut will act as a die and clean the threads when you put it back on.

In the picture, I was able to drill the pot easily just by holding it in one hand and using a battery drill. You don't need much force, the smaller bits will really cut quickly without much effort. This procedure just may keep those log taper pots doing their job with silky smoothness. — *73, Charlie Liberto, W4MEC, 619 Hidaway Cove, Hendersonville, NC 28739,* **w4mec@arrl.net**

Rebuilding a Broken Wafer Switch

Okay, we all know hams are cheap. I was facing a repair-or-replace decision involving a double-wafer ceramic band switch in a Tokyo Hy-Power HC-2000 antenna tuner I'd acquired (on the cheap) at a hamfest. Some contacts on the original switch had literally melted for some reason. A previous owner already had started rebuilding the damaged wafer and I completed the job by replacing the remaining burned contacts and securing them with scrounged machine hardware. I also polished away some burned spots on the rotor and replated that section in silver. [Replated areas should be inspected periodically for wear. — *Ed.*] As I was trying to install and rewire the switch, the second "good" wafer broke! Tokyo Hy-Power offered to sell me a new switch from its old stock for about $110. The original switch was rather flimsy so the decision to repair was a no-brainer.

As it so happened when I first opened the tuner's case I found a huskier two-wafer, multiposition ceramic wafer switch (I suspect the original owner didn't want to spend $110 either) and it provided the requisite raw material. I cannibalized that switch for its ceramic wafer and rotor disk, but I needed more very small machine hardware (the previous do-it-yourselfer had used some nice small brass screws I couldn't find locally). [A good source for unusual hardware is **www.amazonsupply.com** — *Ed.*]

The tuner switch had wipers on both sides of each wafer and was of the shorting variety. Tiny rivets would have been the preferable fastener, but these and the gear to install them were beyond the scope of my workshop and supply cabinet. A hardware store yielded some 1 mm stainless screws and nuts, although the pieces cost nearly 25 cents each! Even so, it was far less expensive than a new switch.

The switch wafer I was repurposing had

Figure 10 — The rebuilt wafer, with burnished solder "caps" on each of the nuts securing the contact lugs and a drop of thread-locking compound (still drying) on each of the nuts securing the rotor components. [Rick Lindquist, WW3DE, photo]

more than enough contacts, but only on one side; it needed a double set to replicate the original switch wafer. Using a Dremel tool and a tiny grinding point or a cutting disk I salvaged individual contacts from the original, broken wafer by *very carefully* removing the rivet heads (see **Figure 10**). After shaving down the flanged end of the rivets holding the new wafer's disks in place, I installed the salvaged lugs opposite each of the originals, using the 1 mm stainless hardware.

In addition, I had removed the rivets securing the two silver-plated pieces of the rotor, since that piece of ceramic also had cracked. I reattached these to the rotor disk from the new switch, using the 1 mm stainless hardware, making sure to align the rebuilt rotor on the common shaft with the original rotor, to maintain correct switching sequence (*you must install the rotor before the lugs!*). The cutting wheel trimmed the excess screw threads.

When all was done, I had a rebuilt switch wafer with five sets of contact lugs (see **Figure 11**). Drops of thread-locking compound secured the stainless nuts on the rotor; while solder held the nuts on each contact lug (soldering stainless can be a challenge). I swished some Tarn-X over the silver plated components, put the entire switch assembly back together and ran it through the ultrasonic cleaner to eliminate any grinding debris and dirt. Once it was clean, I replaced it in the tuner and the switch has performed flawlessly even at 600 W. — *73, Rick Lindquist, WW3DE, 25483 Jamie Ct, Seaford, DE 19973-8310,* **ww3de@arrl.net**

Soldering Iron Tips

The solder cup technique shown in Wayne's, WA4WZP, hint for preventing oxidation is indeed an interesting solution.[1] But for anyone serious about soldering, the first place to look is at the tool itself. An iron of 25 W or less is not suitable for general soldering. Such irons warm up slowly, get

Figure 11 — The reassembled two-wafer band switch, prior to ultrasonic cleaning: The original (and flimsier) wafer is on the left, the entirely rebuilt wafer on the right. [Rick Lindquist, WW3DE, photo]

too hot while on their stand and then too cool during heavy use.

Most irons used for soldering in professional environments have heating elements rated about 90 W combined with a temperature sensing feedback circuit that cycles the element on and off to maintain the correct temperature. They heat about five times as fast, don't oxidize nearly as much (because they don't overheat) and can maintain their temperature while in use. Even though the power rating is much higher, they are actually safer for sensitive components, because the temperature is tightly controlled and they have the heating capacity to solder quickly before overheating sensitive devices.

Even among temperature controlled irons there are significant differences. The most important difference is the tip material. I have worked in a number of labs with a variety of temperature controlled irons. The cheap temperature controlled irons are constantly plagued with tip oxidation problems. The high quality ones use different plating on the tips that resists oxidizing. Combined with minimal regular cleaning, I have seen these tips function with no problems for a year or more of daily use, including extended idling periods. Also, beware of products that have a power control but no temperature sensor. They will not maintain a constant temperature.

The most popular irons in assembly houses are made by Metcal (**www.metcal.com**). Unfortunately, they are quite pricey, even for commercial use. A close second is the Loner series by Edsyn (**www.edsyn.com**), which are more affordable. Third place goes to the Weller/Unger products. These cover a range of prices from under $100 to $300 or more. Online auction sites are a good place for a hobbyist to look for any of these. While I don't personally have experience with them, Hakko products (**www.hakko.com**) may fit in this range somewhere as well.

There are a number of (mostly Asian import) products that are temperature controlled and frequently available for prices that look quite attractive compared to the above. Avoid them. In my experience, they have tips that oxidize quickly, leading to poor quality soldering ability only hours after being taken from their box. Between frequent tip replacement and lost time, they simply aren't worth the price. Spending $100 or more for a soldering iron may seem extravagant, but the difference in ease of use and quality of joints is night and day. Compared to our radios — well, they're pretty cheap. A good iron is essential for anyone who is serious about building electronics. — *73, Wilton Helm, WT6C, 320 Old Y Rd, Golden, CO 80401-9563,* **wt6c@arrl.net**

[1]Smith, WA4WZP, "Soldering Iron Antioxidant," *QST*, Jan 2012, p 63.

Figure 12 — These Gessobord™ artist panels are designed as hardwood boxes that can easily be repurposed to contain electrical projects. [Sherry Goeller, VE3DCU, photo]

Project Boxes

I stumbled upon a product called "painting panels" at the local art supply store. They are white on top and have a wood frame with the top part being similar to a hardwood panel. They come in a wide range of sizes. I bought an 8 × 8 inch box that is 2 inches deep (see **Figure 12**). It's perfect for some projects that I have in mind, and the price was reasonable at $14. The boxes are made in the USA and are perfect for tube type projects and active component type projects. I can imagine dials, knobs and meters on the front panel already. The painting panel boxes are called Gessobord™ and are manufactured by Ampersand (**www.ampersand. com**). — *73, Sherry Goeller, VE3DCU, 58 Jones St, Hamilton, ON, L8R 1Y1, Canada,* **sgoeller@cogeco.ca**

Make Before Break Can Break What You Make

I toiled for hours crafting a new chassis for an oscillator module to act as a precision time base. It required several voltages to operate and since it was mounted on a shelf attached to a 19 inch rack mount faceplate, there was room for some of the power cubes in my junk box. There was ample space on the faceplate to add a voltmeter to monitor the supply voltages. Next to the voltmeter I mounted a wafer switch to switch the meter between the 15 V and 5 V dc supplies. I tested all of the equipment prior to assembly and the wiring was checked before powering up the chassis.

The meter came to life when I turned on the power and it measured both voltages accurately. Unfortunately, I heard a small arc-like "pop" when I rotated the switch between the two voltages. Then the oscillator module failed to generate a signal and I realized that something had gone wrong.

I thought back to the "pop" I heard and unplugged the oscillator module from the chassis. While rotating the wafer switch very slowly I noticed the voltage change abruptly from one value to the other with no zero voltage indication between. I realized the wafer switch had a "make before break" wiper. Such switching has its place in keeping a circuit loaded while selecting various outputs (switching speakers), but not in switching between voltages to be measured or supplied.

As I switched from 5 V to 15 V there was a momentary connection of the two separate power sources placing 15 V on the 5 V line. This damaged all the digital logic. Don't repeat my mistake. Check those rotary switches before use and remember that they come in "make before break" and "break before make" varieties. My belated solution was to move the voltage sources to nonadjacent positions placing a dead zone between them. — *73, Den Nendza, W7KMV, 4219 E Oxford Dr, Tucson, AZ 85711,* **w7kmv@arrl.net**

High Voltage Parts

High voltage (HV) power tubes are still with us and probably will be for many years to come. I have found that discarded microwave ovens are a good source for some of the expensive high voltage parts that they require.

The transformer output voltages of microwave oven transformers vary, but they output high voltage, are powerful and are free for the taking on trash day. The secondary is also relatively easy to remove and rewind for other voltages. These ovens also contain a high voltage diode (which is often the reason for the oven's failure) and a high voltage capacitor, which contains a bleeder resistor. I find that the capacitance value is too low for use as a filter, but perhaps it's enough for other applications.

In addition, the oven is a source of some useful hardware such as a fan, a small "stirring" motor, sheet metal or an almost finished cabinet and some electronic parts on the control board. — *73, Jim Wallace, KB5MT, 111 Deer Island Rd, Mabank, TX 75156-6816,* **kb5mt@arrl.net**

Going Straight

I have several microphones with cords whose ends have been cut off or have had connectors soldered onto them. I wanted to change a connector but the coiled cord end presented a problem. There was only a short length of straight cable before the coils of the cord. I used a heat gun on the coiled area and straightened enough cable to be able to remove and replace the connector. I generally straighten out about 2-4 inches (7-10 cm). After straightening the cable, the connectors are easy to replace and it looks better too. — *73, Tony Fonseca, VA7TF, 44 Fulmar St, Kitimat, BC, V8C 1T4, Canada,* **va7tf@arrl.net**

Drill Chuck Helping Hand

Most of us have a battery powered drill. When the batteries finally die, we dispose of it and buy a new one. I had a couple of these drills and I was about discard them when I decided to open them up and see what I could salvage. The main part is the chuck. I thought it might be handy for holding things when soldering or brazing.

I modified the chuck by grinding off the main gear until it was flat and would stand by itself. I bent some wire, soldered alligator clips to the ends and now have a "third hand" to help with holding parts or wires (see **Figure 13**). Two of them are even better (see **Figure 14**). The chucks could be glued to a piece of wood or metal to create a support fixture. They are helpful in my construction projects and they aren't adding to the local landfill.
— *73, Van Johnson, KH6UX, 4567 S Mulford Rd, Rockford, IL 61109,* **copperking12@ comcast.net**

Coil Wire Supplier

I was never proficient at winding coils, but I thought I could wind one coil. I have plenty of wire and coil forms from 2 inch diameter and up, so I ordered a kit I'd had my eye on

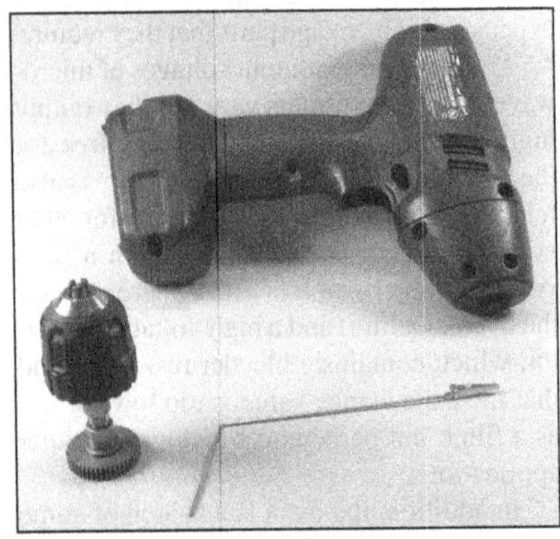

Figure 13 — Before discarding that worn-out drill, remove the chuck; it will make a great "third hand." [Van Johnson, KH6UX, photo]

that only required winding one coil.

When it arrived, the wire was not included but the coil form was. When I looked at it I was stunned! I had forgotten that I was no longer back in the old tube days when coils were *coils.*

The form looked like a small black rubber grommet. The instructions said to put 24 turns of wire on it. My wire wouldn't even make two turns before completely filling the hole. Rereading the instructions, I discovered that this form is called a ferrite core and the wire required is 28 gauge magnet wire. What is magnet wire? Time for an Internet search!

After a little research I found that magnet wire is nothing more than enameled wire. Plain old copper wire with a coating of enamel, which is very hard and will withstand a lot of abuse. The next question was where to get this baby-sized wire? Rolls were available online but I only need a piece about 2 feet long.

One day, I went into a craft store for another item. The store was crowded and, as I maneuvered around the other shoppers, I found myself in the bead aisle. There, hanging on the wall, were small rolls of wire in a variety of gauges and colors. They called it "beading wire." It was available in 28 gauge. Just what I needed! I did wonder if this was really copper wire with enameled coating? For a few dollars I bought the smallest roll, which contained 72 feet. Removing it from the package, I scraped off the enamel and sure enough, it looked like copper wire. I looked up the resistance per foot for 28 gauge wire and multiplied it by 72 feet. An ohmmeter measurement settled the question.

So, if you need short lengths of magnet wire, stop by your local craft store and pay a visit to the bead aisle. Be careful though, some beading "wire" might be steel or other light metals, or even colored string.

Now I have to wind that confounded coil!
— *73, Wayne Smith, WA4WZP, 224 Saint John's St, Arden, NC 28704,* **myrepwayne smith@gmail.com**

Cutting Circular Pads

While visiting a local Harbor Freight Tools Store (**www.harborfreight.com**) I found a small rotary cutter that I thought might be useful for producing circular pads on copper clad boards. I picked one up and was very pleased with the results. Isolated pads, cut into printed circuit board material, provide a quick way to produce circuit boards suitable for projects of small to medium complexity. With this technique the board develops along with the project. It produces boards similar to those using "Manhattan-style," "dead-bug" or "ugly" construction. In some cases, this method can be used to modify existing printed circuit boards to handle a few extra components. The only requirement is that the existing board has sufficient copper area to accommodate the new pads. Pads can usually be added without removing the PC board from its mounts.

The tool is an electric drill attachment called a "Rotary Spot Weld Cutter," (see **Figure 15**). I purchased it from Harbor Freight, item # 95343. Its intended purpose is to remove spot welds from sheet metal. When used to make isolated pads on copper clad material, it produces a pad approximately 8 millimeters in diameter with a 1 millimeter wide gap around each pad (see **Figure 16**).

If used in a drill press, the depth of the cut will be uniform all around the circumference of the pad, but perfectly good pads

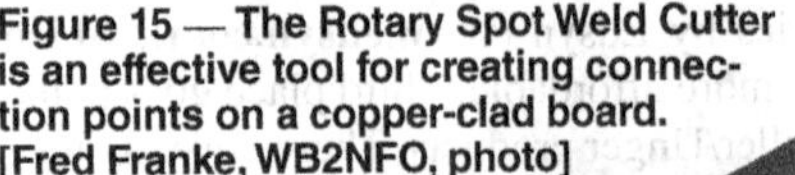

Figure 15 — The Rotary Spot Weld Cutter is an effective tool for creating connection points on a copper-clad board. [Fred Franke, WB2NFO, photo]

Figure 16 — The cutter makes 8 millimeter diameter connection "pads" in copper-clad boards allowing for quick and easy circuit construction. [Fred Franke, WB2NFO, photo]

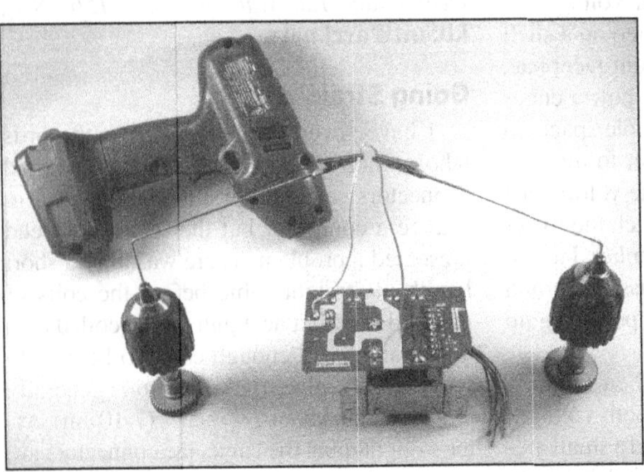

Figure 14 — The chuck from a worn-out drill and a stiff wire with an alligator clip can be a big help on the work bench. [Van Johnson, KH6UX, photo]

can be produced using a handheld electric drill. Just use a light touch so you don't cut completely through the board. The tool has a spring loaded center punch that keeps the cutter from wandering from its intended position. Since the cutter is designed to cut into steel it should be able to produce a great many pads in copper before becoming dull. If the teeth eventually do wear down, the cutter head is double sided and can be flipped 180° introducing a new set of teeth.

The tool costs $5, but Harbor Freight usually has "20% off" coupons and catalog specials so you probably will pay even less. Despite not having the best manual dexterity, I have had excellent success producing and modifying several circuit boards using just this tool and a ⅜ inch electric drill. — *73, Fred Franke, WB2NFO, 26 Dogwood Rd, Kings Park, NY 11754, **ffranke@suffolk. lib.ny.us***

Inexpensive SMT Microscope

In the March 2012 issue of *QST* Wayne Smith, WA4WZP, wrote an excellent article on using a USB webcam as a digital microscope for surface mount work.[1] That got me thinking about possible alternatives. In the period between the use of overhead projectors and the transition to laptops/tablets, a device called the document presenter (aka document camera) was sold widely. Its purpose was to digitize written documents and overheads and transfer them to a projector. These devices often had a high resolution camera with excellent optics, autofocus, lighting and other great features. Although still sold for classroom use, their popularity waned so they are currently available, used, for drastically reduced prices. I have found them to be an excellent magnifier for SMT work.

When shopping for a unit the important characteristics are:

■ Camera resolution; the higher the better

■ The optics; generally the higher the zoom the better for SMT work

■ The type of video output; In general, an XGA output is best, followed by VGA with NTSC video having the lowest quality. [SXGA outputs are also available as are some other high resolution video outputs but the aspect ratios of these vary and they may not display correctly on a standard monitor. — *Ed.*]

■ Refresh rate; the higher the frames per second, the better for soldering

Other great features that are sometimes available are the ability to fold up into a small

Figure 17 — Document cameras are available, inexpensive on the surplus market and make excellent magnifiers for SMT or other fine work. [Mat Breton, AB8VJ, photo]

space, backlighting and the ability to control and capture images through a USB interface.

Because the surfaces are non-conductive plastic, it is important to use a ground mat or other anti-static procedures when working with sensitive electronics. A sheet of glass may suffice if using the backlight. Either will protect the soft plastic surface from stray solder drippings or scratches.

I was able to pick up an excellent unit for less than $1 on eBay [not including shipping — *Ed.*]. It has LED lighting, XGA resolution, a 22X optical zoom, both XGA and USB outputs as well as other great features (see **Figure 17**). I added an old flat panel monitor and now have an excellent digital microscope for next to nothing in terms of both effort and cash. — *73, Mat Breton, AB8VJ, 35229 Rosslyn St, Westland, MI 48185, **ab8vj@arrl.net***

Desktop to Kilowatt

The cost of cabinetry for ham gear is daunting and some of us are not able to pay big bucks for new equipment enclosures. In many cases, they are the highest dollar item on the component list. Computer cabinets are inexpensive, reliable enclosures. Both the desktop and bench top style cabinets are plentiful and sufficiently RFI shielded for most projects, plus, they are real space savers. **Figure 18** shows an RF linear amplifier I built using a desktop case I bought new on eBay for less than $30. Of course, there are

Figure 18 — This amplifier uses two 4-400A tubes to generate 600+ W on 80-10 meters. It uses an external power supply to provide 2500 V. The filament transformer is installed next to the tubes. [Richard Calhoun, W6DZT, photo]

[1]Smith, WA4WZP, "Webcam Microscope for the Radio Amateur," *QST*, Mar 2012, pp 38-39.

many similar cases available used from thrift shops, garage sales and such. In fact, you may have an old computer yourself that could be recycled in this way. — *73, Richard Calhoun, W6DZT, PO Box 77313, Corona, CA 92877,* **richardcalhoun@hotmail.com**

Taking the Heat

Frequently, I find myself replacing final transistors in a rig in which they've been replaced before. I usually find far too much heat sink compound (HSC) between the transistor and the heat sink. I can only assume that the person who replaced the parts saw no visible sign of HSC and so made sure that the newly installed set had more than enough to do the job. Unfortunately, the extra compound acts as a thermal insulator, inhibiting heat transfer.

How is that possible, you ask? There are three reasons:

■ Metal to metal contact transfers heat far better than any HSC.

■ The HSC is *only* needed to fill in the air gaps between the component and the metal heat sink.

■ The air gaps we're trying to fill are small, invisible imperfections in the metal surfaces.

The HSC is not supposed to take the place of metal to metal contact. Its purpose is to fill the air gaps between the metal surfaces with something that transfers heat better than air does.

In a perfect world, if we placed two pieces of metal together, they would have no air gaps between their surfaces. This level of smoothness can be achieved, but machining and polishing surfaces to such a high level of smoothness is expensive. An interesting side note is that, when two surfaces are machined that smooth and attach together, they almost become one. The surfaces, over time, begin to share electrons forming a bond that is extremely difficult to separate.

In our real world of electronics, this never happens because transistors and heat sinks aren't machined that precisely so there are limited points of metal to metal contact between any transistor (or resistor) and the metal heat sink surface and this limits heat

transfer. The other areas are air-filled gaps that form a thermal barrier, similar to double and triple glazed windows.

The best way to increase the heat transfer in these areas is by filling them in with HSC. Metal to metal contact points are far better at transferring heat than the HSC so it is very important to keep the use of the compound to a minimum. Use just enough HSC to fill the gaps but *not* so much that it interferes with the metal to metal contact. When HSC is applied correctly, the best possible heat transfer between the two metal surfaces will occur.

Using too much compound "floats" the transistor or resistor above the metal heat sink. While better at transferring heat than the same distance of air, it is far worse than contact points actually touching, metal to metal.

The best way I've found to apply HSC is to put a small amount of it on a finger and swipe it across both the component and the heat sink. If I see ridges I've applied too much so I clean my finger and wipe the surface again until I see no excess compound (see **Figure 19**). At that point, the transistor/resistor is ready to mount to the heat sink.

When tightening the parts, if any HSC is forced out from between the two surfaces, there's too much HSC. The parts should be removed, cleaned and a thinner coating of HSC applied again.

A good rule of thumb is, since you can't see the air gaps you're trying to fill you shouldn't be able to see the HSC filling them either! In other words, using the minimum amount of HSC, and not a big blob, will provide better heat transfer performance. — *73, Phil Karras, KE3FL, 3305 Hampton Ct, Mount Airy, MD 21771-7201,* **ke3fl@ arrl.net.**

Flame Soldering

Recently, I was soldering some wire and wire lugs and my soldering iron was not getting hot enough. I couldn't finish the job properly. I thought about what I had available that might help. I ended up using a small candle. It took less than a minute to complete the job. The wire and wire lug both became hot and the solder flowed smoothly forming a good solder joint. Some of the

Table 1
Conduit Punch Sizes

Punch (inches)	Hole Size (inches)
½	0.880 (⅞)
¾	1.115
1	1.36
1¼	1.68
1½	1.933
2	2.42

plastic on the wire did melt and there was some soot that needed to be sanded off. If you're ever stuck in a situation where your iron won't do the job or is not available at all, try this method. — *73, Walter Schoenbach, WB8FEC, 7263 Sandy Beach Dr, Waterford, MI, 48329,* **wb8fec@att.net**

Oversized Punches

Need to punch some holes? Greenlee punches are one size, of course, but what if you need a hole that doesn't match any of the Greenlee punches? Conduit punches may fit the bill. They easily punch through the ⅛ inch aluminum sheet used for 19 inch rack panels and also the steel and aluminum used in chassis construction, but their hole sizes are somewhat larger than the Greenlee equivalent. Note that conduit punches use a ¾ inch draw bolt. See **Table 1** for a list of the various conduit punch sizes. [An alternative to punches worth considering are stepdrills. — Ed.] — *73, Edward Barbacow, K3ZCY, 149 Geary Dr, Connellsville, PA 15425,* **k3zcy@arrl.net**

Wire Joining Techniques

If you have ever needed to solder tinsel wire, then you know it's not an easy task. Tinsel wire is formed of spiral conductors made of foil (like copper) wrapped around fibers. These conductors are extremely flimsy and sometimes melt when soldering. You are likely to find tinsel in self-coiled cords (like microphone and telephone handset cords), modular telephone set cords, headphone cords, wires with cloth insulation, switchboard cords, and some musical instrument cords.

The secret to soldering these wires is to wrap the flimsy conductor with a stiffer conductor before soldering. The best wire for this splinting job is 24 AWG solid bare copper.

Start with a 10 inch piece of solid wire and strip about 6 inches. Carefully strip about an inch of the tinsel wire. Next, wrap the solid wire in a tight coil around the tinsel wire (see **Figure 20**). You will find that the bare copper coil supports and stiffens the tinsel so the coil looks a little like a segment of heavier solid copper when it is finished. When you have about ¼ inch of coil in place, you can

Figure 19 — Here are three different levels of HSC on an aluminum bar. Number 1 is a big blob that will inhibit heat transfer. Number 2 is a big smear (sometimes looks like ridges), which is better but still too much for good heat transfer. Number 3 shows a correct application where the HSC is almost invisible. [Phil Karras, KE3FL, photo]

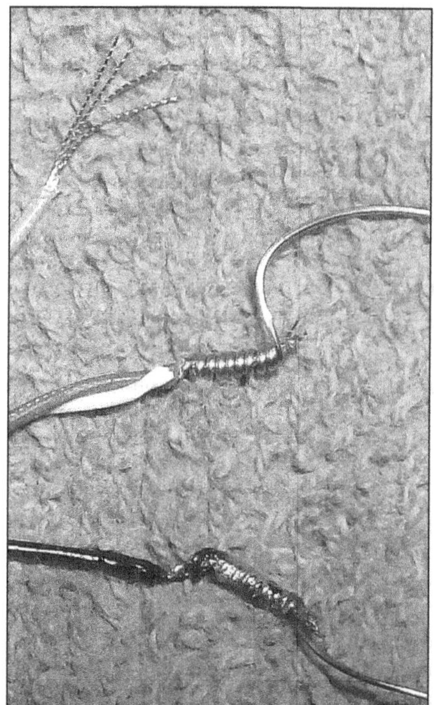

Figure 20 — Coil solid copper wire around the tinsel (red lead) then trim and tin the coil (black lead). The remaining solid wire pigtail can be trimmed off or used to make connections. [Frank Ingle, KG4CQK, photo]

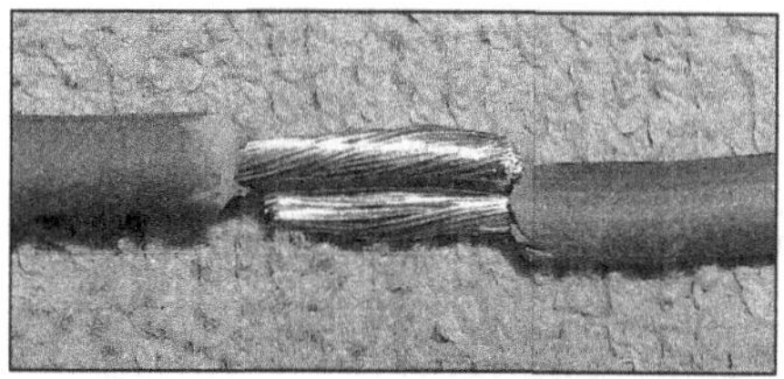

Figure 21 — To prepare a parallel splice, strip the ends of the wire and position them alongside each other. [Frank Ingle, KG4CQK, photo]

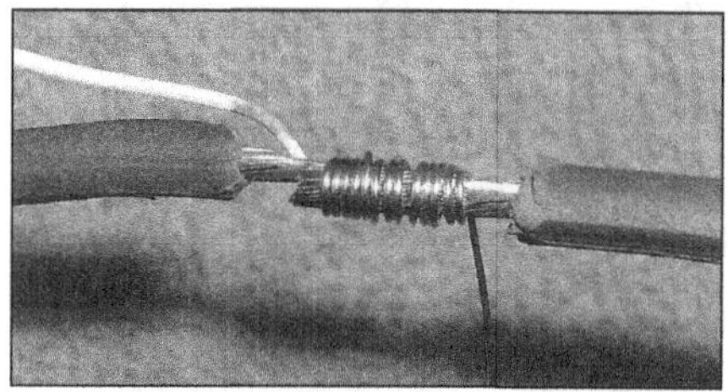

Figure 22 — Wrap both exposed wires with #24 AWG solid copper, then solder. [Frank Ingle, KG4CQK, photo]

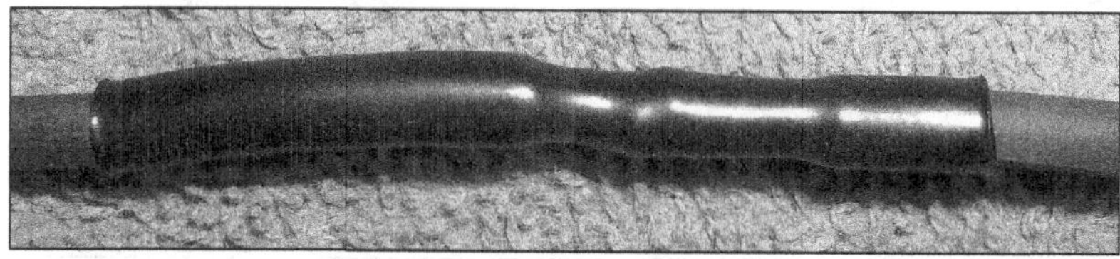

Figure 23 — Finish the splice with a piece of heat-shrink tubing for a neat and sturdy splice. [Frank Ingle, KG4CQK, photo]

trim the ends of the solid wire, leaving only the coil. When you tin the coil, it will bond with the tinsel and will become stiffer. Now you can solder the coil to a solder terminal, secure it with a screw terminal or even add a crimp on connector.

Sometimes I find I need to solder a stranded conductor to a terminal that is too small or too full to fit all the strands. Using the same wrapping technique, I can wrap the stranded wire for ¼ inch and then leave a short pigtail of 24 AWG wire (see **Figure 20**). This little pigtail will fit though all but the tiniest holes and can be soldered. Unless you are talking about high currents, the reduction in diameter for ⅛ inch will not be an issue.

In none of these cases will the wrapping provide a very strong mechanical connection. The boundary between the end of the solder and the start of the insulation will be a bit weak and you may need to improvise a mechanical support above the solder joint. A small zip-tie around the tinsel insulation and a nearby anchor will often work.

Another application for this technique is in parallel splices. This is similar in concept to the Western Union splice, but much neater. Expose about a half inch of conductor on two wires to be joined, then lie the clean, exposed parts next to each other (see **Figure 21**) and join them with the 24 AWG solid wire (see **Figure 22**). Tin the coil and you have a neat joint that is stronger than the original wire. When finished, cover the joint with a piece of heat-shrink tubing for a nearly invisible splice (see **Figure 23**). — *73, Frank Ingle, KG4CQK, 2580 Park St, Jacksonville, FL 32204,* **kg4cqk@comcast.net**

SMT Soldering Hints

In the January 2011 issue of *QST*, Jim Koehler, VE5FP, described a method of reflow soldering surface mount components.[1] After reading his article, I purchased a toaster oven and an inexpensive VOM that included a temperature probe, and tried his technique with good results. I found that the hardest part was applying the solder paste to the surface mount pads, and the subsequent placement of the components on the board. I have now made four boards, and have developed tech-

niques that have simplified the processes for applying paste and positioning components.

Which Size?

Surface mount capacitors and resistors come in various sizes — all considerably smaller than their leaded counterparts. I try to use larger sizes such as 0805 and 1206. [0805 = 0.08 × 0.05 inches — *Ed.*] For resistors, I generally use the 0805 size. On my last project, I installed two 18 pF capacitors, size 0402. These were very small and were difficult to work with. Also, I recommend that you buy a few more components than you need because it is quite easy to lose them.

I use small outline integrated circuits (SOIC) when available (see **Figure 24**). These units are about half the size of the standard DIP package and have a lead pitch of 0.05 inches. Some devices are only available in smaller packages; I try to use the largest package available with side leads so I can see the final solder bond.

SOICs with symmetrical leads have Pin 1 marked with a dot or a line. When using one with nonsymmetrical leads, refer to the data sheet to determine Pin 1's location.

Paste Application

Now that you have your parts and, presumably, a board, it's time to apply the solder paste to the pads. To simplify this process I use a small syringe filled with "no-clean" flux solder paste, Digikey (**www.digikey. com**) part number SMD291AX-ND, combined with a KDS22TN25-ND 0.019 inch diameter nozzle.

I squeeze out just barely enough paste to cover the pad (see **Figure 25**). Don't use a big dab of paste, and keep the nozzle clean. If you end up with too much paste on a pad or paste between two adjoining pads, I have found that dental picks are effective at removing excess solder paste as well as for cleaning the area between pads. A large sewing needle would probably work too.

Figure 24 — This photo shows a selection of passive components and small outline ICs on a PC board after reflow soldering. [Bill Kaune, W7IEQ, photo]

---

[1] Koehler, VE5FP, "Reflow Soldering for the Radio Amateur," *QST*, Jan 2011, pp 32-35.

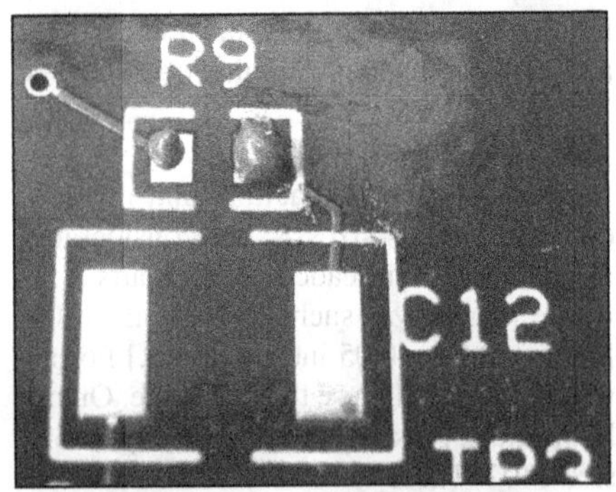

Figure 25 — These are pads for an 0805 resistor. The pad on the left has the correct amount of solder paste, while the pad on right has too much. [Bill Kaune, W7IEQ, photo]

Figure 26 — I built this "bridge" to help place SMT components on my project PC boards.

I was able to obtain used picks from my dentist for free, so you might try asking yours.

Magnifiers

I purchased an inexpensive 10 power jeweler's loupe from a surplus house (Marlin P. Jones, **www.mpja.com**) that I use to examine each pad after I have applied paste. I have friends who have purchased microscopes for this kind of work.

Component Placement

With the paste applied, I use tweezers to pick up each component and place it on the board. To steady my hands I made a "bridge" from a piece of aluminum stock, about 2 inches wide by ⅛ inch thick and 14 inches long with ½ inch thick blocks of wood at each end (see **Figure 26**). I place this fixture over the board to form a bridge over the work area. I rest my hand on this bridge as I place each component on the board. With this technique I can place components more accurately. If they are out of position, I use the dental tool to nudge them back.

A significant problem I had when starting to work on SMT boards was inadvertently disturbing a component already on the board while installing another. I once placed half a dozen capacitors with great effort and then, after installing an IC, I found that they all had been moved from their pads and two were completely missing. I now install components working from left to right and from the back to the front of the board. I always keep my hand, and the support bridge, over areas of the board that have not yet been populated.

Nudging SMT components tends to smear the solder paste on the pad. At first I was concerned about this, but I discovered that during the heating process, the melted paste is drawn back onto the pads. In the four boards that I have made I have only had one solder bridge and that was because I had too much paste on the pad.

Make a List, Check it Twice

I prepare a list of components to be installed, in the order of installation, and then check off each as I install it. When capacitors and/or resistors surround an IC, I usually install the IC first and follow with the surrounding components. Using these techniques has mostly eliminated the inadvertent disturbing of components already on the board.

Heat Things Up

When I have all the components placed, I carefully carry it over to my toaster oven and process it following Jim Koehler's procedure. I have found that the reflow soldering process seems to correct slight part misalignments and draws stray solder back onto the pad.

After the board has cooled, I inspect it with my jeweler's loupe and fix any problems, which have proven to be remarkably infrequent. Finally, I install the components with wire leads. It takes patience and some trial and error, but with some effort you can become quite good at making surface-mount circuit boards. — *73, Bill Kaune, W7IEQ, 160 Cedarview Dr, Port Townsend, WA 98368, w7ieq@arrl.net*

Vacuum Variable Capacitor Substitute

Antenna tuners and power amplifiers need vacuum variable capacitors that will withstand high voltages. The 811 amplifier (see **Figure 27**) I use for my 600 meter experimental transmitter (WD2XSH/6) needed vacuum capacitors, which cost about $150, but I needed a more affordable alternative. To solve my problem, I built two oil-filled variable capacitors.

Previously, I had used a stack of two capacitors. These worked until the frequency range of the future 630 meter ham band was announced. I realized I would need more capacitance to make my transmitter tune correctly, but there was not enough space in the cabinet to stack another capacitor. Because vacuum capacitors were beyond my ham radio parts budget, I had to find a better way.

It is possible to use readily available materials to build an inexpensive substitute for vacuum capacitors. The concept is based on the fact that an oil-filled variable capacitor has both more capacitance and a higher breakdown voltage rating than the same capacitor in air. The increase in capacitance

Figure 27 — Four bottles and some mineral oil get this amplifier humming on the 600 meter band. [Patrick Hamel, W5THT, photo]

is directly related to the dielectric constant of the oil used. The breakdown voltage rating increase is related to the dielectric strength of the oil.

The Oil

I built the capacitors using drugstore mineral oil, which does not have any toxic metals dissolved in it. However, mineral oil may have some dissolved water present. Usually this will not be present in large enough amounts to cause problems, unless the voltage on the capacitor is close to the breakdown potential.

Better oils do exist, but be leery of motor oils, which may have "recycled" content and other additives that could cause unwanted variations in the dielectric characteristics of the oil. Draining old oil-filled filter caps may sound affordable, but they may be high in PCB content.

One of the best choices — although more expensive than mineral oil — is vacuum pump oil. It is readily available from refrigeration supply houses. This oil is water-free and very pure.

Case Construction

To hold the oil, I needed a case and I decided to use plastic, which is inexpensive, readily available, and is easy to drill and thread, and also easy to glue with the correct acrylic cement.

A local hardware company was able to make the case components. The thickest acrylic plastic they had was ¼ inch, so I prepared a drawing of the case and had the hardware company cut the pieces. At home, I drilled and tapped all the holes prior to assembling the case.

I assembled the case using Craftics Thickened Cement #33 (**craftics.net**). I recommend that you ask the salespeople at the store for their cement suggestions. When gluing the pieces, be sure not to move any section while the glue is setting up. This will cause air bubbles in the glue and ultimately these bubbles will cause oil leaks. Note: model airplane glue will stick the pieces together, but one bump will break them apart, and the joints leak.

By placing the stator at the bottom of the case it is only necessary to fill it ¾ full. This allows expansion room for the oil, which will heat up when operating. However, this does not prevent diurnal pumping, an effect caused by changing air pressure that "pumps" humid air into sealed enclosures. Preventing diurnal pumping requires a bladder or diaphragm that will compensate for pressure changes while keeping outside air separate from inside air.

To solve this problem, I returned to the hardware store where I found a drawer full of lamp parts. Using a ⅜ inch OD nipple,

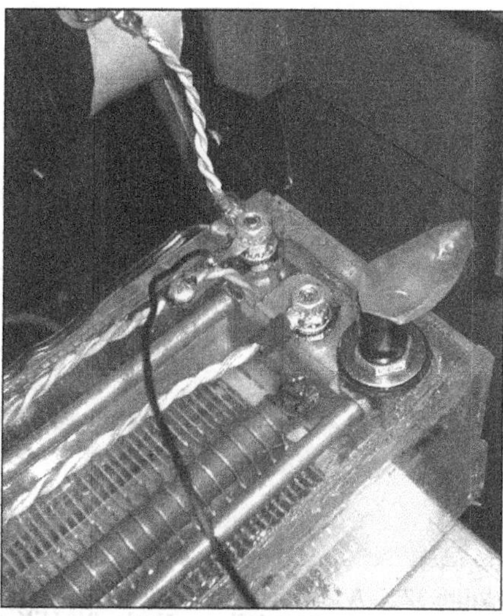

Figure 28 — **This makeshift diaphragm keeps the pressure equalized between the inside and outside of the case without allowing outside humidity to contaminate the capacitor. [Patrick Hamel, W5THT, photo]**

Figure 29 — **The Hexseal prevents the oil from leaking out while still allowing the capacitor's rotor to be adjusted. [Patrick Hamel, W5THT, photo]**

rubber washers, Form-a-Gasket and locknuts, I made a vent over the fill-and-drain hole to which I attached a finger from a vinyl exam glove (see **Figure 28**). A year or so of use should prove how well (or poorly) it works. All the electrical connections come out of the top through threaded holes with Form-A-Gasket to seal them.

A Sealing Solution

The next problem was the capacitor rotor shaft, which must extend outside the case without leaking. I tried several solutions and eventually found the answer — an N9030X1/4 Hexseal (**www.apmhexseal.com**). The seal allows the shaft to come through the case at or below the liquid level (see **Figure 29**). Since the capacitors did not have bushings, I used a panel bushing from an old potentiometer. I pushed the bushing through the hole in the side of the case from the inside and mounted it with rubber washers and #2 Form-A-Gasket on the threads. I then fed the capacitor's shaft through the bushing, and threaded on the Hexseal to

hold it all together. I did not try to thread the plastic for the ⅜ inch OD bushing.

Testing Results

I have no way to safely test the arc-over voltage of a capacitor or its Q, but I did some interesting measurements on the two identical capacitors I have rebuilt. I have the Autek and MFJ analyzers. They both measure the reactance and can calculate the capacitance of a load. Here are the readings I obtained:

Using the Autek, capacitor #1 measured a maximum capacitance of 210 pF when dry, and 510 pF in oil.

Capacitor #2 was measured with both instruments. When dry the Autek measured a maximum of 259 pF and the MFJ, 253 pF. When in oil the Autek measured a minimum of 64 pF and a maximum of 551 pF. When in oil the MFJ measured a minimum of 55 pF and a maximum of 562 pF.

With common sense safety (lots of fan cooling for a tube transmitter) it is possible to build an enclosure to allow an available variable capacitor to serve in place of the commercial vacuum variable capacitor. — *73, Patrick Hamel, W5THT, 1157 E Old Pass Rd, Long Beach, MS 39560-5050,* **w5tht@arrl.net**

Tube Holder

To hold a container of glue (or a similar dispensing tube) and keep it from tipping over and spilling, simply attach two clothespins, side by side vertically, onto a small, vertical surface. Then, place the tube into the groove made by the handles of the two pins (see **Figure 30**). The tube will fit perfectly into the handle portion of the pins and will stand up straight. — *73, Theodore Turk, WB8ADA, 20171 Miller Ave, Euclid, OH 44119,* **hammytee@hotmail.com**

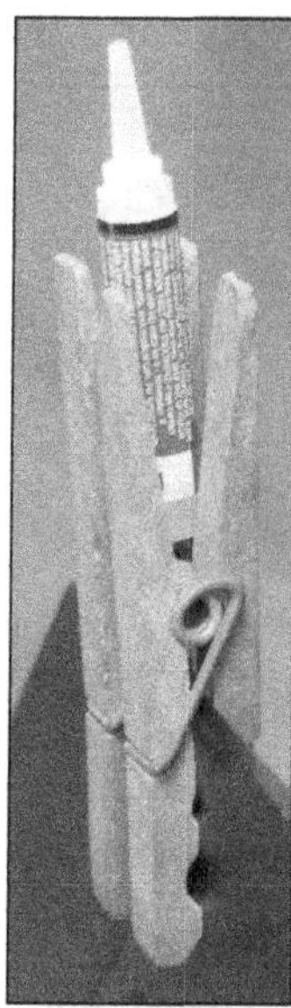

Figure 30 — **Clothespins make a good holder for tubes of glue, lubricant, or other chemicals, keeping them upright and accessible. [S. Sant Andrea, AG1YK, photo]**

Removing Wire Corrosion

I was faced with the task of resurrecting a Garmin handheld GPS-12XL, the alkaline batteries of which had burst and corroded. To make matters worse, corrosion had wicked into one of the power conductors. The damage could not be cleaned with heat and rosin, and the wires were fragile enough that scraping would only do more damage.

I recalled a similar situation at work. I consulted our materials and processes engineer, who suggested using an acetic acid solution. Acetic acid was not available at my local pharmacy, so I tried the next best thing — plain white cider vinegar, which is 5% acetic acid. The vinegar did an excellent job of dissolving the corrosion. Just set up a small container of vinegar, dip the end of the corroded wire in, and let it soak. It took about half an hour to eat away the corrosion. After the corrosion is dissolved, rinse the wires in running water or a solution of baking soda and water to neutralize any remaining acid.
— *73, Bob Chadwick, W1MTX, 1785 Teak Rd SE, Palm Bay, FL 32909,* **rchadwic@att.net**

Yaesu MH-34 Microphone

I have an older Yaesu FT-60 transceiver and the matching Yaesu MH-34 speaker/microphone. The microphone PTT switch became intermittent. I found that the small microswitch on the printed circuit board either gets dirty or corroded. Unfortunately, the local Yaesu repair facility does not sell the microswitch.

Here is how I cured the problem. First, remove the microswitch from the board. Note that the seal portion needs to be removed as the switch itself does not sit in the seal; rather the rubber seal is between the switch and the plastic lever that presses on it (see **Figure 31**). Next apply a few drops of low-water-content isopropyl alcohol directly to the button with a cotton swab while pushing

Figure 31 — The white rubber seal, visible at the upper right, mounts in front of the microswitch. Remove the seal to free the microswitch and then remove the switch to apply alcohol for cleaning. [John Powell, KF6EOJ, photo]

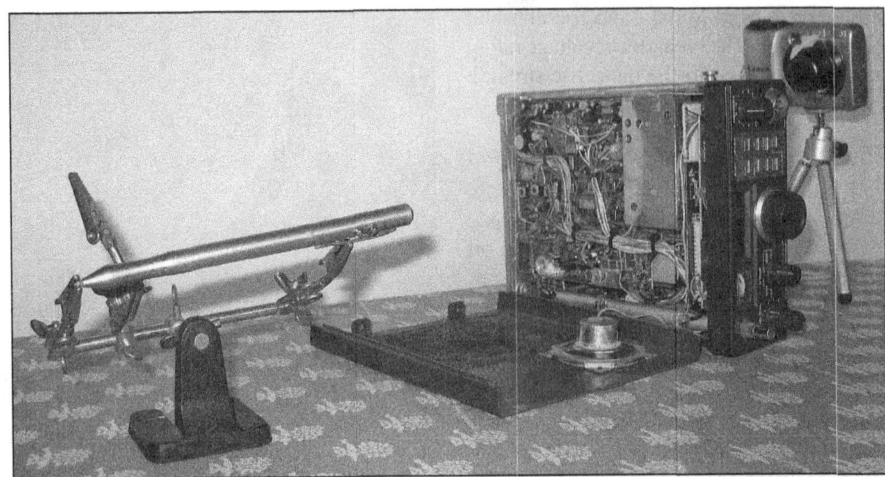

Figure 32 — A laser pointer and a small digital camera help to locate the printed circuit pads of a faulty component. [Luca Norio, IV3TEK, photo]

the button several times to help the alcohol penetrate. Finally use canned air to blow it dry and resolder the switch to the board.

Reassemble the unit, taking care to get the case seal back in place properly. I strongly suggest a magnifying glass for this project.
— *73, John Powell, KF6EOJ, 8325 Otto St, Downey, CA 90240,* **john.powell@csulb.edu**

Laser PCB Locator

My transceiver developed a problem and had to be repaired. It wasn't easy to find the faulty component but the real difficulty was locating the pads of the bad component on the printed circuit side of the board.

The manual includes a complete schematic with pictures of the component side of the board showing the key number of all transistors. It does not, however, include such information about the printed circuit side. The faulty board was very crowded, making it impossible to follow the tracks by looking through the board to the components side. After considering the problem for a while, I had the idea to use a laser pointer.

Safety First: avoid looking directly into the laser beam no matter what the power of the laser source.

I aimed the pointer's laser beam at a point close to a "reference" transistor on the component side that was near the faulty component (see **Figure 32**). This produced a bright red spot on the printed circuit side, which I recorded by using a digital camera (see **Figure 33**). This allowed me to easily pinpoint the solder points for the faulty part. — *73, Luca Norio, IV3TEK, via Umberto I 116, Maniago, PN 33085, Italy,* **iv3tek@arrl.net**

Build a Current Shunt Resistor

I needed to take current readings that exceeded the 10 A rating of my digital multimeter (DMM). The solution was to build a 0.01 Ω current shunt resistor capable of

Figure 33 — The red laser dot marks the spot. [Luca Norio, IV3TEK, photo]

handling up to 20 A of current flow.

A current shunt resistor is placed in series between the equipment load to be measured and its power source. A voltmeter is connected in parallel with the resistor and is used to measure the voltage drop across the resistor. Ohm's Law is used to determine the amount of current in amperes I for a given voltage E and resistance R using the formula $I = E / R$. For example, if you were to measure 0.2 V across the 0.01 Ω resistor then the corresponding current would be 0.2 / 0.01 = 20 A.

I decided to build a 0.01 Ω current shunt resistor because it produces a negligible 0.2 V drop in the supply voltage when presented with a 20 A load. Also, the current through the shunt is 100 times the voltage across the shunt; that is with a resistance of 0.01 Ω, each 0.01 V that appears across the shunt indicates a current flow of 1 A. So if 0.03 V appears across the shunt, by Ohm's Law, 3 A must be flowing through the shunt (0.03 / 0.01 = 3).

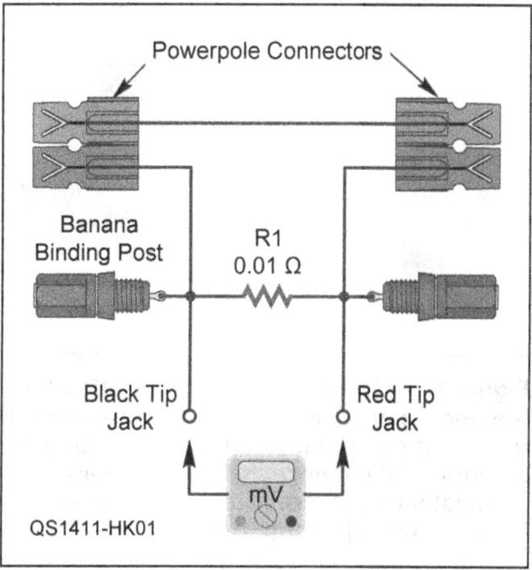

Figure 34 — Current shunt resistor schematic.

Construction

I made the current shunt resistor from a 6-foot piece of 12 AWG insulated solid copper wire (see **Figure 34**). The resistance of 12 AWG wire is 1.588 Ω per thousand feet or 0.001588 Ω per foot and an ampacity [The maximum safe current carrying capacity. — *Ed.*] rating of 25 A. I formed the wire into a multiturn coil having an inside diameter of 1¼ inches, which allows it to fit into a 5 × 2½ × 2 inch plastic project box (see **Figure 35**). Banana binding posts and Anderson Powerpoles are used for connecting the resistor (see **Figure 36**). Tip jacks are used for connecting the millivoltmeter or DMM test leads across the resistor.

Setup and Use

Attach the positive lead (B+) from your power source to the shunt binding post next to the red tip jack, and your load equipment supply lead (B+) to the other binding post. Your power source's negative lead (B–) is connected directly to the equipment load.

If your equipment has Anderson Powerpole connectors, then you can connect your power source to the connector pair near the red tip jack and your equipment load to the other pair. Observing meter polarity, insert the test leads of your millivoltmeter or DMM into the black (meter-negative) and red (meter-positive) tip jacks. Turn on your power source and load equipment. Observe the voltage reading on your meter, which will be in millivolts.

Calibration

Calibration of the shunt requires a calibrated ammeter, a voltmeter with a 10 MΩ or greater input impedance, and an adjustable load. I used a 100 W HF transceiver (in CW mode) terminated into a RF dummy load as my adjustable load; I changed the amperage level by adjusting the RF drive level as needed. I used my DMM for the calibrated ammeter and voltmeter. To calibrate the shunt you need to record several current and voltage readings as follows:

1. With the power off, connect the load to the power source, turn on the power, and set the load to draw a minimum current; turn the power off. Set the DMM for an ammeter scale and connect it in series with the load. Turn on the power and adjust the load until the ammeter reads 1 or 2 A. Note the current.

2. Turn the power off. Replace the ammeter with the shunt resistor. Set the DMM for a voltage scale and connect it to the tip jacks on the shunt.

3. Turn the power on. Note the voltage reading on the DMM.

4. Turn the power off and repeat the procedure, raising the current an ampere or two each time until you reach the limit of the ammeter.

Using Ohm's Law $R = E / I$, divide the measured voltage E by the measured current I to obtain the corresponding resistance value R for each current point you recorded. Find the average of all the resistance values R_{AVG}.

If your shunt has too much resistance, you can trim a calculated length off your wire. Two calculations are needed. First, determine the wire resistance per inch R_{INCH} by dividing R_{AVG} by the length of the wire. Next calculate the number of inches to be trimmed using the formula: $(R_{AVG} - 0.01) / R_{INCH}$. Cut about ½ inch less than your calculated trim length to compensate for the wire attached to the banana binding posts.

Final Comments

The current shunt resistor can also be used on ac power circuits. For example, you might have a portable generator and need to know how much current a load is drawing. This can be done by connecting the ac line (L, black) conductor to the two binding posts, with one side going to the power source and the other going to the load. The ac neutral (N, white) and protective ground (PG, bare copper or green) conductors would be connected directly between the ac power source and load. [When measuring ac, make sure your DMM is set to an appropriate ac scale. — *Ed.*] — *73, David Warner, W7SZS, 19658 Schaefer Dr, Oregon City, OR 97045,* **w7szs@arrl.net**

See That SMD

SMD parts are so tiny I have trouble seeing the connections to solder them to the circuit board. I have a headband magnifier that helps, but even that is not enough magnification for some tiny SMD parts. I have found that wearing drugstore reading glasses under the headband magnifier solves this problem. You get the magnification of the lenses in the headband magnifier, plus the magnification of the reading glasses. You can also select reading glasses of different powers to get more or less magnification to suit various jobs. — *73, Bob Sumption, W9RAS, 61250 Cass Rd, Diamond Lake, Cassopolis, MI 49031-9406,* **w9ras@arrl.net**.

ADI Microphone Repairs

The Premier Communications ADI seriers came with two types of mobile microphones.

If you have an older radio, you'll have the black, inverted teardrop-shaped hand microphone, similar to the one on the older ADI-146/446 radio. This microphone does not have a backlit keypad, but typically has no issues other than the usual loss of rubber

Figure 35 — The resistor is made from a 6-foot length of 12 AWG copper wire formed into a multiturn coil having an inside diameter of 1¼ inches. [David Warner, W7SZS, photo]

Figure 36 — The current shunt resistor mounted in a small plastic project box. [David Warner, W7SZS, photo]

on the coil cord just below the strain relief.

The newer microphone, with the white "popple-pad" backlit buttons is the real troublemaker. It too has the loss of insulation problem on the coil cord below the strain relief.

If the degradation of the coil cord's rubber hasn't gone too far, you can shorten either microphone's cord and do a splice. This repair requires cutting the wires above and below the funnel shaped strain relief. The wires within the strain relief can be removed by securing the strain relief in a workbench vise and carefully drilling out its center to create a new passageway. Take your time and use a drill bit smaller than the old coil cord's diameter. You'll rout out the old wires more easily that way and avoid drilling through the side of the strain relief.

Pass the shortened cord's new wires through the strain relief and do your eight splices, carefully matching the connections, insulating them, and squeezing them back into the housing. This is painstaking work, but much less expensive than trying to find a replacement coil cord. A bit of silicon RTV in the now wider strain relief channel, or some shrink tubing on its outside at the bottom will make the repair more rugged.

The newer style microphone also has a strange PTT switch return mechanism. It's not a spring, but a compressible cube of a rubber material that looks like a pencil eraser. I've found that replacing this with a cut segment of retractable ballpoint pen spring gives an improved PTT action. I've done the same upgrade on the Kenwood hand microphone on my TM-733A mobile transceiver. You may need to experiment with different lengths of spring to get an acceptable feel.

As a last resort, the ADI-147/247/447 eight-pin round microphone connectors are wired the same as those on Alinco (**www.alinco.com**) and Kenwood (**www.kenwood-usa.com**) radios. I've used an Alinco EMS-10 microphone with great success on my ADI-247 before figuring out how to do all of the above. — *73, Anthony Bogusz, W9MT, 10536 S Coyote Melon Loop, Vail, AZ 85641-2593,* **w9mt@arrl.net**

Powerpole Soldering Jig

I devised a simple jig to keep everything in proper alignment when soldering Anderson Powerpoles. I made the jig from a 2 × 4 cut down to about 1½ inches square and 6 inches long. Then I made a 22.5° cut on one end and attached a ¾ inch black binder clip as shown in **Figure 37**. The bend on the end of the Powerpole contact seats nicely on the rolled edge of the binder clip maintaining proper alignment. The rubber bands are used hold both electrical conductors in place. — *73, Dean Sheeley, W7DWS, 3957 Purple Sage, Prescott, AZ 86301,* **dean sheeley@gmail.com.**

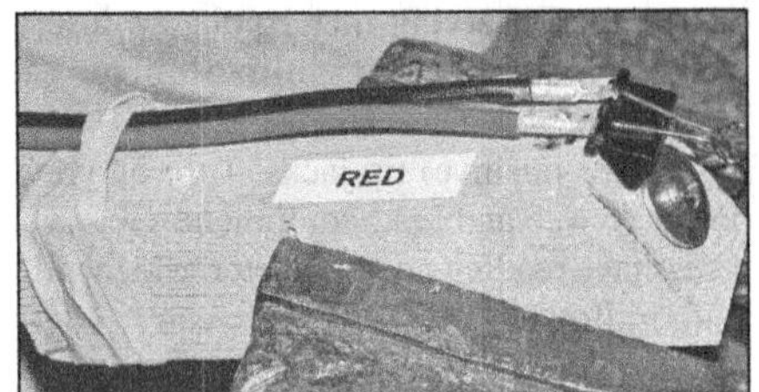

Figure 37 — This simple wooden jig simplifies the soldering of Powerpole connectors. [Dean Sheeley, KG7DKC, photo]

Mini TNC Case

I want to suggest an inexpensive, alternative case for those who have purchased the tiny Mobilinkd APRS terminal node controller (TNC) that was mentioned in the January 2014 issue of *QST*.[1] The Mobilinkd unit works quite well, but originally, there was no case option available from the manufacturer.

The manufacturer has made a case available for an additional $17, but I have found a less expensive — and waterproof — alternative. Walmart (in their camping and sporting goods department) sells a waterproof polycarbonate case (20-285489-08) made for storing credit cards during outdoor activities. The case accommodates the Mobilinkd TNC very nicely. This case is extremely strong and, when closed, has sufficient space for the TNC with a right-angled audio cable plugged into it. You will need to drill a small hole on the side of the case for the cable, or just use the case the way it is. These cases come in several clear colors that light up brightly at night from the TNC's light-emitting diodes. — *73, Paul Miller, NØBII, 3515 Meade St, Denver, CO 80211,* **teachdude@hotmail.com**

TO-3 Top Hat

Have you ever touched the top of a TO-3 transistor when it is dissipating 100 W? I don't recommend it, because even though it is bolted to a big piece of aluminum, you may still burn your finger. That is, unless you are using one of my top hat heat sinks. Taking heat from both the top and bottom of the transistor greatly improves the cooling action, and at 100 W you need all the cooling you can get.

I used a piece of 0.01-inch thick copper sheet, copper being a better heat conductor than aluminum. I cut a strip 3 inches long and 1 or 2 inches high, and drilled a hole at each end for a 2-56 mounting screw (it's easier to drill the screw holes while the copper is still flat). Next, I wrapped the strip around a ¾-inch wooden dowel to form a circle to fit the

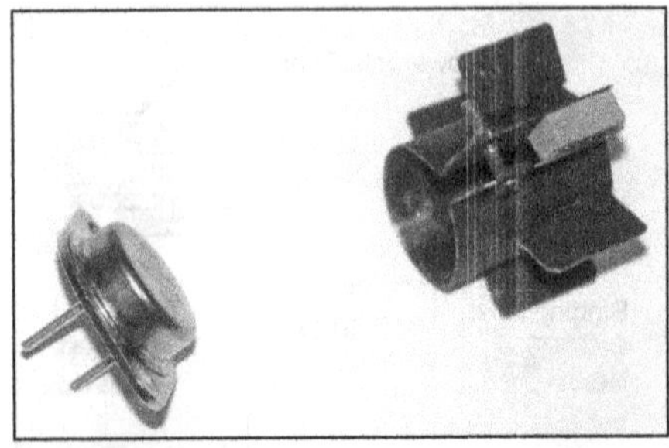

Figure 38 — The top hat heat sink added to the upper part of a TO-3 or similar packaged transistor can greatly improve the heat dissipation of the device. Note that the area of the heat sink that contacts the transistor is not painted. [George Carsner, WØPPF, photo]

TO-3, leaving two short ears bent up 45° for the mounting screw (see **Figure 38**).

You can add additional cooling area by soldering on ⅜-inch copper fins to the circumference of the strip. Paint the top hat black for best radiation, except where it contacts the TO-3 case. [It would also be a good idea to add a *thin* layer of thermal grease to improve heat transfer to the top hat. — *Ed.*] — *73, George Carsner, WØPPF, 411 Terrace Rd, Iowa City, IA 52245,* **w0ppf@mchsi.com**

Tampering with Anti-tamper Torx

I was trying to open the case of a transmitter, which was fastened in place with what I thought were Torx screws. Finding myself unable to remove the Torx screws, I took a close look and found that there was a pin in the center of each screw to prevent me from using my Torx screwdriver bit to remove these fasteners. I found out that some manufacturers assemble products using this security Torx to keep consumers from opening cases with "no user serviceable parts inside."

Needing to get inside, I took a trip to Harbor Freight and found a 33-piece Security Bit Set for under $10. It is item number 68459 (the Wiha 10-bit set is another choice) and it includes Torx, tri-wing, hex, and spanner pin bits; a ¼-inch drive; a magnetic bit holder, and a PVC case. [Note that some of the inexpensive Torx sets may have the pin installed off center. Inspect the pin placement before purchasing a set. — *Ed.*]

Save yourself some hefty service fees, have the fun of seeing what's inside, and remove those screws yourself. — *73, Bob Haynes, WB4AKA, 4316 Lithia Pinecrest Rd, Valrico, FL 33596,* **wb4aka@arrl.net**

A Straighter Cut

A project required cutting off a piece of light gauge sheet metal. From past experience, I knew that if I used tin shears, I would be certain to distort the metal. If I used a hacksaw, it would be impossible to make a

---

[1]Ford, WB8IMY, "Eclectic Technology," *QST*, Jan 2014, p 50.

Figure 39 — A second blade will not only help cut a straight line, but the added width helps when making a cut to fit two pieces of tubing together. [Hal Rogers, K8CMD, photo]

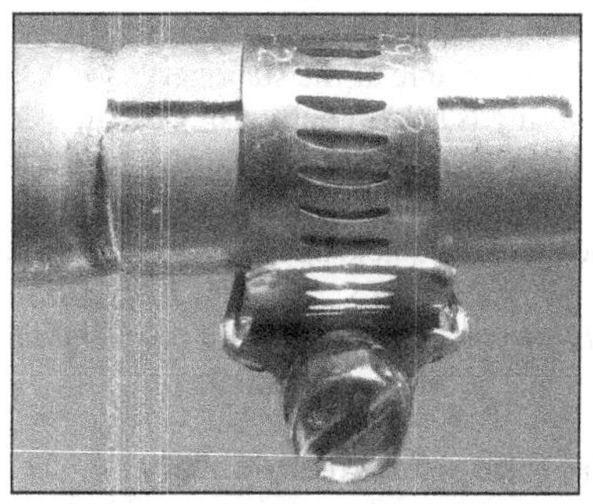

Figure 40 — A straight, wide cut and a hose clamp make fitting antenna elements easy. [Hal Rogers, K8CMD, photo]

straight cut. The blade always shifts, twists, and goes off course. I discovered a solution to this aggravating problem: stiffen the blade with a second blade (see **Figure 39**). Don't expect perfection, but with a more rigid cutting tool, you'll be amazed at how easy it is to cut a straight line. [If you don't have a second blade handy, try holding the single blade at a 45° angle to help maintain a straight cut. — *Ed.*]

Some tips: be sure both blades are new or in a good condition, have the same number of teeth per inch, and are running in the correct direction (there's usually an arrow printed on the blade that points forward). It's also good to use blades from the same batch. Some hacksaws have longer blade mounting pins, which makes installing the extra blade easier, so shop around. Also, the mounting pins are angled, so don't over-tension the blades, which could cause them to bow away from each other.

Using a dual blade offers another bonus. If, for example, you need to cut slits in the end of an aluminum tube to extend an antenna's element, the cut will not only be straight but it will be wide enough that you'll only require two slits — which you can make with one cut (see **Figures** 39 and **40**). Finally, deburr your cuts with a file, sandpaper, or a pocket knife. *—73, Hal Rogers, K8CMD, 7811 Dogwood Ln, Parma, OH 44130,* **k8cmd@arrl.net**

Cutting Epoxy-glass PCBs

I needed to cut some epoxy-glass printed circuit board (PCB) material. I have done this in the past by scoring the material and snapping it. This often resulted in splintered or rough edges. Also, the smallest size that can be cut using this method was limited. I tried a hacksaw, with terrible results.

Being a homeowner and a home improvement guy, I have a plethora of tools in my garage. I decided to see what I might have that could tackle this job. My tile saw, a small table saw used for cutting ceramic tile, has a diamond blade and is used with water when cutting ceramics. It also has guides to facilitate making straight cuts.

I decided to give it a try without water. I donned my safety glasses and dust mask and made several cuts on the PCB material. The results were very satisfactory. The edges were as smooth as or smoother than the "factory" edges of the PCB material. The blade is narrow, so the material waste is minimal.

I don't know how cutting this PCB material will affect the life of the blade, but the blades are inexpensive, so replacing it for the next tile job should not be a problem. — *73, Joel Bryant, WM4P, 1542 Indiana Ave, Palm Harbor, FL 34683,* **wm4p@arrl.net**

Quick Soldering Gun Tip

I was doing some wire antenna work when my 1955 vintage Weller (100 W/140 W) soldering gun tip broke. I've replaced the tip many times over the decades, but this time I couldn't find a replacement tip at the regular hardware stores. A large commercial supplier wanted $12 for a two-pack of tips, so I decided to try a homebrew solution. I snipped off a 10-inch scrap of #12 AWG residential ground wire and bent it into the shape of the broken gun tip. I pinched the tip end together to make it beefy and add mass. Finally, I made sure I inserted the nuts onto the new tip ends first, then made the right angle bends that go through the nut's side holes and snipped the ends before screwing the nuts into place. The new homebrew tip works just as well as the original and now I never have to worry about finding a replacement. — *73, Jim Wyckoff, K3BT, 8005 Frontier Dr, Severn, MD 21144-1619,* **jimwyckoff@comcast.net**

Powerpole Crimping Problem

Despite my best efforts at assembling Anderson Powerpoles using a quality crimping tool, occasionally a wire will pull out of a connector. I have found that this problem can be eliminated if the copper wire is first tinned with solder. The solder coating allows the Powerpole contact to "bite" into the soft coating, creating a tighter grip on the wire lead. — *73, Ken Slusher, N2DF, 156 Rose Rd, Woodbourne, NY 12788,* **n2df@arrl.net**

Aluminum Tubing Source

Recently, while laying out a satellite antenna (from the May 2010 issue of *QST*), I found that ¼-inch aluminum tubing was hard to find and rather expensive. I finally located a source that is both inexpensive and easy to find. I am using ¼-inch aluminum fuel line found in 10- and 20-foot coils at most auto parts stores and straight tubing (in 36-inch lengths) from my local model airplane hobby store. The coiled tubing is very easy to straighten. I rolled mine under a board on the tile floor after cutting to length. These seem to work great for my 2 meter and 70 centimeter antennas, but for bigger antennas I would use larger tubing. — *73, Terry Halladey, AF7W, 5940 Chapman St, Cocoa, FL 32927,* **af7w@arrl.net**

Lead-Free Solder

I just finished repairing a circuit board and what I found is worth passing on to others. A "questionable" solder joint caused the fault. When I started to repair the joint, I had problems with the solder not taking. After a few attempts, I realized that the circuit board used lead-free solder, a situation true for any circuit board made in the last 15 – 20 years.

A regular soldering iron will not completely melt lead-free solder and will not bond regular tin-lead solder to a pad that has lead-free solder on it. I invested in a temperature-controlled iron so that I could work on these newer lead-free circuit boards, and I had to raise the iron's temperature up to 560° F to make a good solder joint. [Find more information on lead-free solder in the *Handbook*.[1] — *73, Joseph Birsa, N3TTE, 321 Oak Rd, Pittsburgh, PA 15239,* **n3tte@arrl.net**

Quick Coax Checker

The following simple design for a coax checking device is an easy afternoon construction project. This checker finds intermittent connections, shorts, and opens at dc, but

[1] Available from your ARRL dealer, or from the ARRL Store. Telephone toll-free in the US 888-277-5289, or 860-594-0355, fax 860-594-0303; **www.arrl.org/shop**; **pubsales@arrl.org**.

will not find problems that only occur for RF signals. There are no switches or settings; just connect both ends of a cable to be tested.

My circuit idea is a Y pattern (see **Figure 41**). The incandescent lamp feeds current to the two branches, each with an LED and a resistor. The LEDs indicate continuity on their respective circuit, either through the center conductor or through the shield. The combined current through both LEDs isn't enough to light the lamp; however, if the center and shield are shorted, bypassing the LED resistors, the lamp will light.

If two green LEDs light, the cable is good. One LED means you have an open connection and indicates whether it's the center or the shield that is open. If the lamp lights instead of the LEDs, it means the center of the coax is shorted to the shield (see **Figure 42**).

The LEDs are typical stock parts; anything you have in your junk box will do, just make sure to adjust the value of the resistors to accommodate the LEDs you use. The incandescent lamp can be any low-voltage

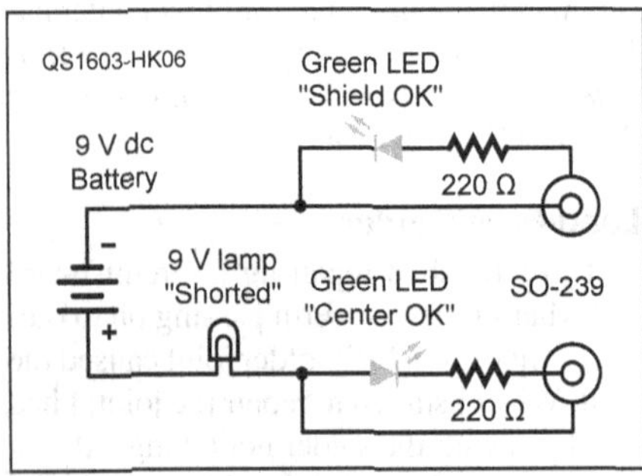

Figure 41 — **This simple checking circuit will test lengths of coax for shorts and/or broken shields and center conductors.**

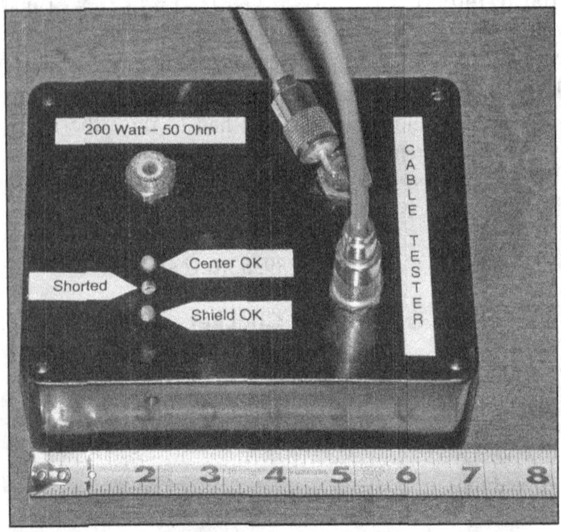

Figure 42 — **A simulated center to shield short causes the shorted lamp to illuminate amber. With a good cable connected to the checker, both the green LEDs will illuminate. Note that the green glow of the LEDs in this photo is light reflected from the camera flash; they are not actually illuminated. [Mark Phillips, N2RPZ, photo]**

bulb, such as used in a flashlight, as long as it lights up with a 9 V battery. Parts are not at all critical for this project; use whatever you have available. The components need to be mounted in a plastic box. As an added feature, I chose to include a dummy load resistor in the same box. The dummy load is separate from the rest of the tester circuit and is not included on the schematic. [Note: a dummy load resistor mounted in an unventilated box can easily overheat. Apply power in short bursts with ample cooling time between. — *Ed.*]

As a final test, with the cable ends connected, I pull and wiggle the cables looking for any flicker in the lights, indicating an intermittent failure. — *73, Mark Phillips, N5RPZ, 806 Clover Park Dr, Arlington, TX 76013,* **blueharley@gmail.com**

Lever Nuts

I am sure that many hams are faced with the problem of "semipermanent" wiring, particularly as it applies to the tangle of 12 V dc supply wires for all the accessories that accumulate in a shack. This jumble often requires connecting a multitude of wire sizes (How do you *neatly* connect #12 AWG solid to #22 AWG zip cord?) in dark and difficult corners.

I have found a product called the Lever-Nut [A trade name for a splicing connector made by Wago (**www.wago.us**). — *Ed.*], that is well suited to many applications around the shack. Intended as a replacement for wire nuts in the construction trade, these connectors can interconnect #12 AWG through #28 AWG, solid, stranded, or flexible stranded wire in any combination.

Better yet, they are reusable and do not damage the wire ends, so the wires may be reused without cutting back and restripping. They are also convenient for experimenting with wire antennas, although I do not know about their longevity in an outdoor environment. Lever-Nuts are rated to 600 V at 20 A, are finger safe, and are UL listed. — *73, Andrew Jarosik, K1APJ, 21 Taylor Dr, Portland, CT 06480-1211,* **k1apj@arrl.net**

Surface Saver

How many times have you soldered wire ends, terminals, or other components on the workbench and wound up burning the bench top, or worse, burning that expensive electrostatic discharge (ESD) conductive mat on the benchtop?

A simple piece of ceramic tile can be used to prevent this. Available at building supply stores, or left over from your last renovation, they come in various sizes and shapes from 4 – 12 inches.

I use a smooth 4¼-inch tile from my last bathroom renovation, which seems to work just fine; however, I think a tile that has some kind of irregular surface might keep parts from sliding. Another ham I know says he uses a plain concrete brick, which may be better if you are using a torch. — *73, Don Dorward, VA3DDN, 1363 Brands Crt, Pickering, Ontario, L1V2T2, Canada,* **va3ddn@arrl.net**

Project Box Covering

I like to reuse enclosures for my homebrew low-power equipment. If I find the enclosure has too many holes or cutouts from a previous project, I just cover it with vinyl covering (a good source is a cover from an A4 ring binder; you can choose your favorite color or texture at the stationery store). The covering is held in place using double-sided sticky tape (carpet tape) and also by the nuts holding switches, potentiometers, etc. It is inexpensive, quick, and gives a nice, finished appearance. — *73, Malcolm Joyce, EI8FH, 111 Mount Prospect Dr, Clontarf, Dublin 3, Ireland,* **maljoyce8@gmail.com**

Protoboard Battery Safety

I want to express my appreciation for Ward's, NØAX, "Hands-On Radio" series. Although I've used circuits like many of those in the series, it's good to review them in the simplest state to keep them fresh in your memory.

In Experiment #155, Ward recommends using battery packs, including D cells or gel cells for experimental split-voltage supplies. My concern is that, particularly if using protoboards, it is very easy to accidentally short one or both supplies. When a "wall wart" or the like is used, this will probably only result in a hot supply or wire. However, because of the low internal resistance of batteries, even a shorted pack of D cells can produce a great deal of current — and heat — very quickly. A gel cell of even small ampere/hour capacity can source tens and even hundreds of amperes that will have your protoboard on fire before you realize your mistake.

A former colleague tells the story of shorting a pair of 6 V 10 Ah lead-acid cells and having his protoboard in flames within seconds. Luckily, the fire was contained, but the burned spot is a sobering reminder.

The simple answer is to place a small fuse in the "hot" lead of each supply. A note to that effect in the article would be a good reminder to both experienced and new hams that energy sources, particularly ones that have very low internal resistance, can deliver large amounts of energy very quickly, and are not to be trifled with. — *73, Tom Fowle, WA6IVG, 1552 McCullen Ave, Eureka, CA 95503,* **wa6ivgtf@ fastmail.fm**

Silver Tape Inductor

Excellent low-loss inductors can be made from ¼-inch silver-plated copper tape, which is sold in craft and hobby stores. It has an adhesive back, and is used in making stained glass items. **Figure 43** shows such an inductor, which uses the tape wound on a plastic bottle form (a plastic form requires careful soldering; glass would afford better heat dissipation). I'm using it for an L-section tuner for my cage-umbrella vertical; running 500 W, there's no noticeable heating. [Note, careful testing should be done before using this type of inductor with high power levels. — *Ed.*] To keep the winding ends from eventually peeling off, secure each with a small piece of clear packaging tape. Finally, you might need some help when spooling out the tape. — *73, Jim Parkinson, W9JEF, PO Box 321, Tontitown, AR 72770-0321,* **w9jef@arrl.net**

Bench Helper

When working at the bench, I often find I can use an extra hand or two. I have one of the multi-arm bench helpers, but often find it cumbersome to use. In response to that, I devised this simple alligator clip component holder for soldering small parts. My holder only requires four parts:

- a short-legged desktop camera tripod, with the ¼-20 camera thread mount
- a ¼-20 nut

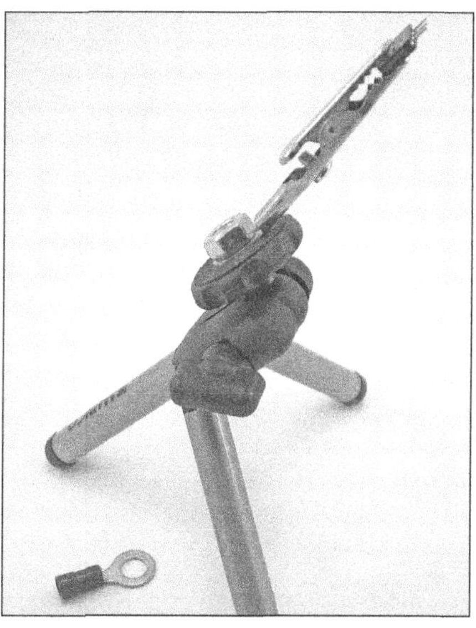

Figure 44 — A camera tabletop tripod and an alligator clip make a simple component holder, adding a third hand to your bench. [Don Dorward, VA3DDN, photo]

- a crimp-on ¼-inch ID ring terminal
- a good-quality alligator clip like the Mueller #BU-60

Figure 44 shows the finished holder with one clip. Just remove any plastic insulator from the ring terminal and crimp or solder it to the BU-60 clip barrel. Then secure it to the tripod using the ¼-20 nut. — *73, Don Dorward, VA3DDN, 1363 Brands Crt, Pickering, Ontario, Canada L1V2T2,* **va3ddn@arrl.net**

Intraoral Camera Magnifier

Looking at the markings on surface-mount devices (SMD) or trying to see into the back corner of a chassis can be difficult. The standard tools for such activities have been magnifying glasses and dental mirrors. Today's technology has brought us an improved solution — the intraoral dental camera. These units are small and lightweight (see **Figure 45**), plug into your computer's USB port, and are recognized by *Windows* as a USB camera device. Intraoral cameras have a focus range of about ¼ – 2 inches and include six blue-white LEDs surrounding the camera lens to provide light in dark areas (see **Figure 46**). A 4-megapixel camera can be found on the Internet for under $50. — *73, Scott Reaser, K6TAR,* **sreaser@verizon.net**

Figure 46 — This close-up of surface-mounted parts shows the image quality and magnifying capacity of an intraoral dental camera. [Scott Reaser, K6TAR, photo]

Figure 43 — Silver craft tape, wound on a plastic bottle, makes an effective inductor. Here it is being used for an antenna tuner. [Jim Parkinson, W9JEF, photo]

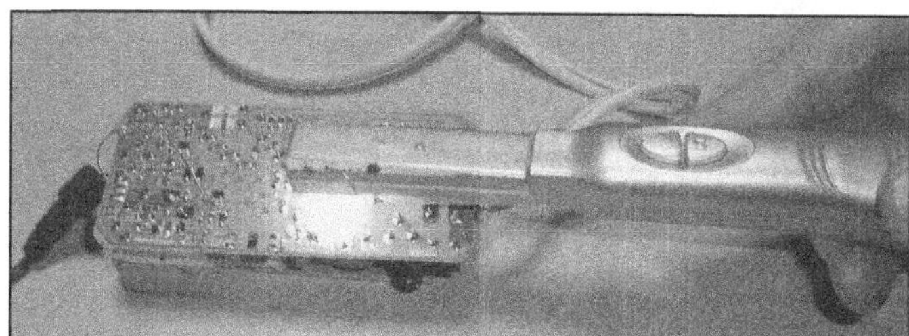

Figure 45 — An intraoral dental camera is a slender, pen-like device containing a 4-megapixel camera and its own lights. It can be used to magnify small components or view the dark interior of equipment cabinets. [Scott Reaser, K6TAR, photo]

Equipment Tips and Mods

T/R Delay for the Eagle

I love my Ten-Tec Eagle (**www.tentec. com**) transceiver, but soon found that in CW the amplifier relay output follows the keying even if a CW T/R delay is set. Therefore, the Eagle is not compatible with non-QSK (break-in) amplifiers.

This was a problem for my Ameritron AL-811 amplifier (**www.ameritron.com**), which cannot follow keying. Some correspondence with Ten-Tec confirmed that this is the way it operates and the issue cannot be addressed through a firmware update. My solution was to build an outboard delay relay circuit to provide semi-break-in operation of the amplifier.

I searched the *QST* archives for a solution and found a "Hints and Kinks" article by Charles Darrow, K8GZQ, which addressed the same issue with the Ten-Tec Century 21 transceiver.[1]

Upon reviewing his circuit, it became apparent to me that Darrow's arrangement would not work off the Eagle's relay output. His approach was to pick up the receive line voltage from the transceiver and use that to trigger an NE555 one-shot multivibrator to provide an amplifier relay output with an adjustable delay. The NE555 starts when the voltage on the trigger input (pin 2) goes low, so Darrow's use of the 12 V receive line voltage filled the bill (it goes low on transmit). The Eagle's relay output is a transistor switch to ground, however, so Darrow's circuit needed some tinkering, which resulted in the circuit shown in **Figure 1**.

It is like Darrow's circuit, except I trigger the NE555 by using the Eagle's amplifier switch to ground the trigger input, which is held at a positive voltage by the 47 kΩ resistor connected between it and the +12 V dc supply. The 47 kΩ resistor between the trigger input and the reset terminal (pin 4) of the NE555 is important, as it keeps the NE555 constantly resetting by applying voltage to the reset terminal during every

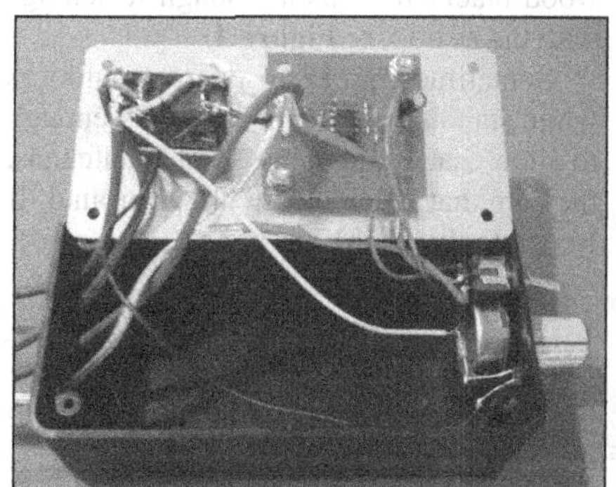

Figure 1 — This is the circuit for the adjustable T/R delay for the Ten-Tec Eagle.

Figure 2 — The Eagle T/R delay circuit requires only a few simple components that all fit neatly in a small project box. [Thomas Valenti, K3AJ, photo]

key-up period. Otherwise, the delay relay would open after the multivibrator reaches its time delay set point, then the circuit would reset and the relay would reclose on the next character sent.

Using the component values shown, the circuit will provide a delay from almost nothing to many seconds. Any small 12 V relay should work. I used a RadioShack double-pole, double-throw (DPDT) miniature PC relay, catalog number 275-0249, so I could provide two outputs (one for my amplifier and one for a VHF transverter). [Caution, the recommended output current for the NE555 is ≤ 200 mA. Total device dissipation must be < 600 mW. — *Ed.*] The diode (a 1N4005) across the relay terminals prevents the coil's reverse transient voltage from affecting the NE555. I constructed the circuit on a RadioShack general purpose IC board (catalog number 276-159) and housed the unit in a plastic project box (see **Figure 2**). I fastened the relay to the cover plate of the box with double-sided foam tape. Shielded conductors were used for the relay input and output lines; non-shielded, two conductor speaker wire was used to pick up operating voltage from my station is 12 V dc buss.

Now when using the Eagle with an amplifier, I can set the delay to minimum when operating SSB and dial in a suitable delay when operating CW. Be aware that if set for

[1]Darrow, K8GZQ, "Semi-Break-In for the Century 21," *QST*, Dec 1979, p 57.

a very small delay (little or no resistance on the potentiometer) the circuit will break into oscillation when keyed rapidly. I keep the setting high enough to avoid this, but you could install a suitable fixed resistor in series with the potentiometer to prevent oscillation. The Eagle has a 17 ms delay between the amplifier relay output and the generation of RF, which is plenty of time for the delay relay to pull in and switch the amplifier. I chose to incorporate an LED to indicate when the relay output is closed and a power switch to disable the delay relay when running the Eagle barefoot. — *73, Thomas Valenti, K3AJ, 17210 Foreston Rd, Upperco, MD, 21155-9410,* **tomv307@hotmail.com**

The Rigpod

My Yaesu FT-817ND is a great radio to take along anywhere — even backpacking. It has all the bands I like including AM/FM radio. One problem plagues me each time I set up in a new location: Where do I put this thing so I can conveniently operate it? There is never a rock flat enough or at the right height. DXing with the FT-817 hanging from the shoulder or resting on the belly is no fun. Even sitting flat on a table, the operator must scrunch down to see the screen and controls. Yes, there are after-market accessories to prop the radio up at some pre-determined angle for better viewing but there is a better way. You can make this mount from items you probably already have and it will work on all surfaces, at all reasonable heights and angles.

I call my creation the rigpod. The items needed are a tabletop camera tripod (I tried a lighter duty tripod designed for digital cameras, but the radio's weight wouldn't allow the tripod to keep an angled position), a block of pine (end piece from a 1 inch board) and a ¼ inch × 20 threaded fastener.

I cut the block to 5 × 7 inches, about the size of the FT-817. The fastener is called a T-nut. It has four barbs designed to hold

Figure 4 — The completed rigpod ready to keep your radio on the level no matter how crooked the terrain. [Phillip Grant, N1YPS, photo]

fast in wood (see **Figure 3**). First, you must drill a pilot hole in the center of the block deep enough to accommodate the length of the threaded portion. I used a ¼ inch drill the size of the stud found on most camera tripods. I do not recommend hammering it into the wood. Put the assembly in a bench vise and squeeze it in. Even though I love the look of natural wood, I spray painted the wood black. It looks as though it belongs with the radio (see **Figure 4**).

Sometimes, I like to operate outdoors while standing. My tripod extends enough to allow that and if using flexible antennas, they are that much higher above ground. If I operate the FT-817 at too steep an angle,

I wrap the supplied shoulder strap around the radio and block to hold it there. An elastic band the size the post office uses will also work.

Enjoy your FT-817 and don't worry about it sliding off a rock and into the pricker bush. — *73, Phillip Grant, N1YPS, 119 Hoe Shop Rd, Bernardston, MA 01337-9638,* **phill112643@verizon.net**

KX3 Mounting Bracket

The KX3 was designed as an ultra-portable transceiver. However, many hams purchased it as a base station or a backup base station. The transceiver normally sits at a low angle on the desktop, which many hams find inconvenient. Several manufacturers offer Plexiglas and wooden stands, which are great if you use the accessory paddles. I don't use the paddles and, because space is at a premium, I wanted the KX3 up off of the desktop.

My solution was to design and make a simple, yet sturdy, under-shelf mounting bracket. The bracket consists of three pieces of wood cut from a 1 × 6 inch pine board (see **Figure 5**). Two of the pieces are cut to a size of 3 × 5 inches and then layered to form the horizontal board that attaches to the shelf. This layering lowers the KX3 enough to clear the shelf. One edge of the lower board in the sandwich is beveled to angle the KX3 for proper viewing of the display. The angle will vary depending on the specifics of your setup, so some trial and error is needed here.

The third piece is 3½ × 8 inches and forms the transceiver back support. It is mounted to the beveled edge of the lower board once the appropriate angle for comfortable viewing and operation has been established. With the mounting board centered against the shelf boards, the mounting board will extend about 1½ inches on either side of the

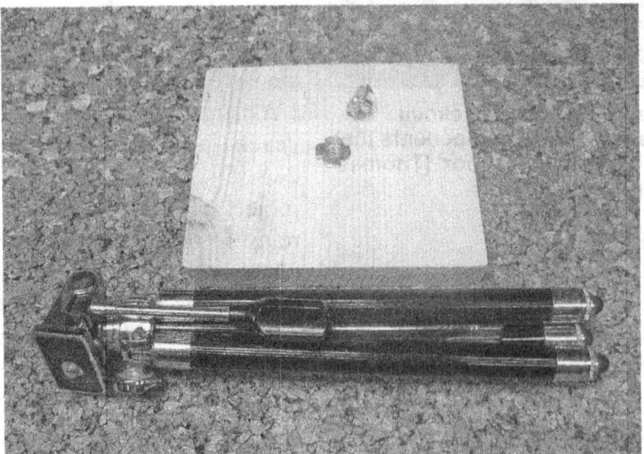

Figure 3 — The T-nut is mounted to the board to provide an attachment point for the table tripod. [Phillip Grant, N1YPS, photo]

Figure 5 — Three pieces of pine board will turn the KX3 into a space-saving base station. [Mike Lichtman, KF6KXG, photo]

Figure 6 — The KX3, mounted to an under-shelf bracket, makes an excellent backup transceiver or primary rig for a "space challenged" shack. [Mike Lichtman, KF6KXG, photo]

shelf board. The extended back legs of the KX3 go over these extended sides and hold it to the shelf by gravity. [The use of some mounting putty or notches in the mounting board for the legs should be considered for a more solid installation. — *Ed.*]

I used short drywall screws to join the boards and also to fasten them to the underside of the shelf (see **Figure 6**). Finally, an application of flat black paint will help the mount blend in with the KX3. — *73, Mike Lichtman, KF6KXG, 1717 El Rito Ave, Glendale, CA 91208,* **kf6kxg@arrl.net**

SignaLink USB Modification

The SignaLink USB (**www.tigertronics. com**) is a handy digital interface for your rig and it is an excellent choice for operating most digital modes. If you like to chase the very weakest signals this modification may be of interest. Please note that this modification will void your SignaLink USB warranty, if it's still in effect.

The SignaLink USB has an internal noise source that can potentially be vexing when working very low-level digital signals. The noise is a significant low-frequency broad-band noise between 10 Hz and 700 Hz. To measure this noise, I used a free software tool called *Spectran* from I2PHD (**www.weaksignals.com**) that turns your PC sound card (or SignaLink USB) into a spectrum analyzer (technically an FFT analyzer).

It turns out that the bias voltage supply (set at 2.5 V) for the four ac-coupled inverting op amp circuits in the SignaLink USB is directly driven from the relatively noisy external USB power source with only some capacitive decoupling. Essentially, any noise from the USB power source that is not removed by the low-pass filter feeds directly into the audio stream.

One way to reduce the noise on the op amp's bias voltage is to use a voltage regulator. This is the solution I chose because there is already an on-board low-noise source (a Micron MIC5205BM5) supplying 3.7 V for the comparator reference in the PIC microcontroller and also for the Audio USB IC A/D supply.

Referring to **Figures 7** and **8**, remove the 1 kΩ resistor in series with the 5 V USB supply line. Next, add a jumper with an inline 470 Ω resistor (to return the bias to the 2.5 V required by the op amps) between the Micron regulator output and the inboard solder pad of the removed 1 kΩ resistor. The outboard resistor pad should remain empty. The easiest place to access the low-noise 3.7 V source is at JP4, although you can solder directly to the Micron IC if your skills are good enough. You can either use an inline resistor or an SMD resistor mounted to the inboard pad where the 1 kΩ resistor used to reside.

More information on SignaLink USB modifications can be found at Peter Frenning's, OZ1PIF, website (**www.frenning.dk/ oz1pif.htm**). — *73, Mat Breton, AB8VJ, 35229 Rosslyn St, Westland, MI 48185,* **ab8vj@arrl.net**

Icom Memory Playback Keypad

Many Icom radios have a four-memory digital voice recorder built in. The memories are accessed by using the front panel "F" keys to trigger the individual memories. Unfortunately, the procedure of reaching across your keyboard, paddle, and operating desk a few hundred times during a contest

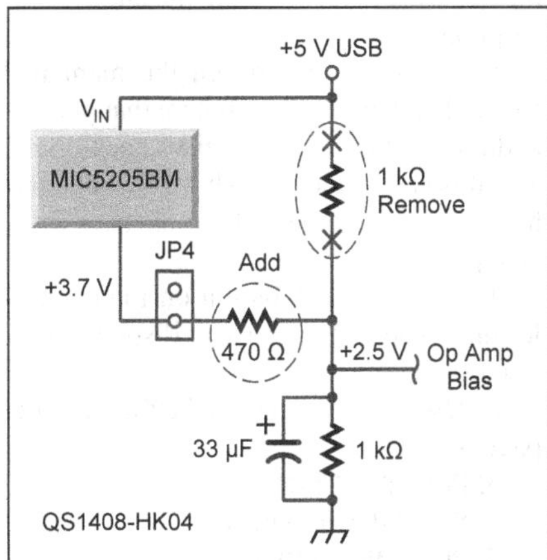

Figure 7 — This schematic shows the SignaLink USB bias supply before (right) and after modification.

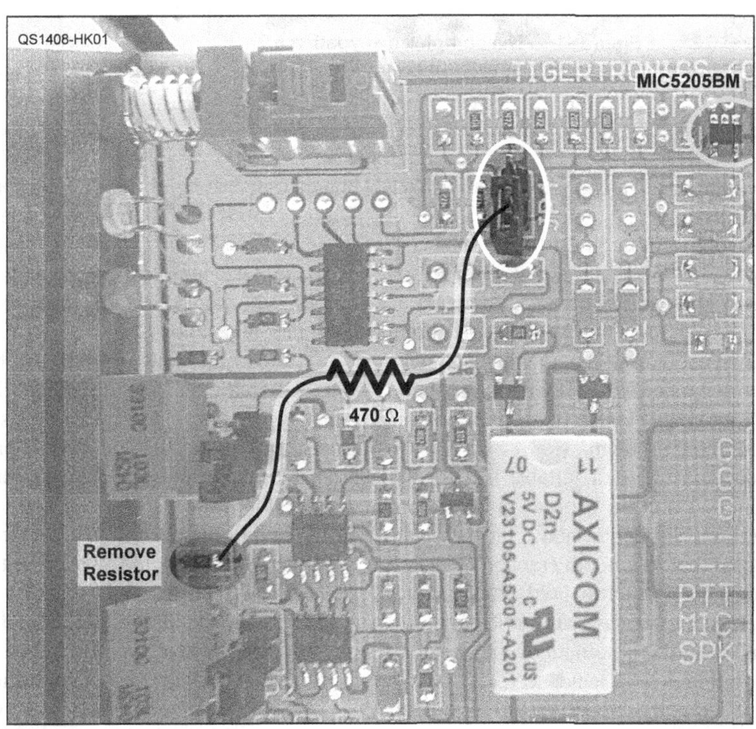

Figure 8 — Connecting a jumper and resistor between the JP4 connector and the inboard pad of the removed resistor applies regulated bias voltage to the op amps. [Mat Breton, AB8VJ, photo]

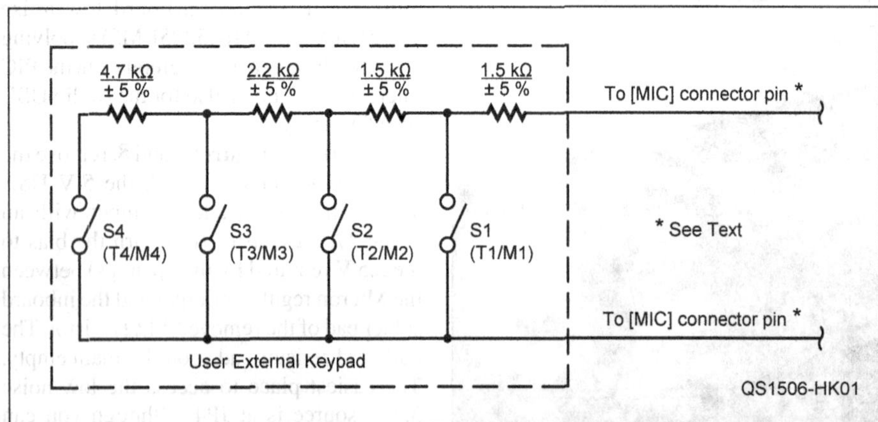

Figure 9 — This is the Icom schematic for the external keypad. The four switches each trigger one of the four digital recorder memories, freeing you from having to punch the radio's function keys.

Figure 10 — This is the upper shell with switches, resistors, and the new cable in place. The red No 1 button is on the left. [Lee Jennings, ZL2AL, photo]

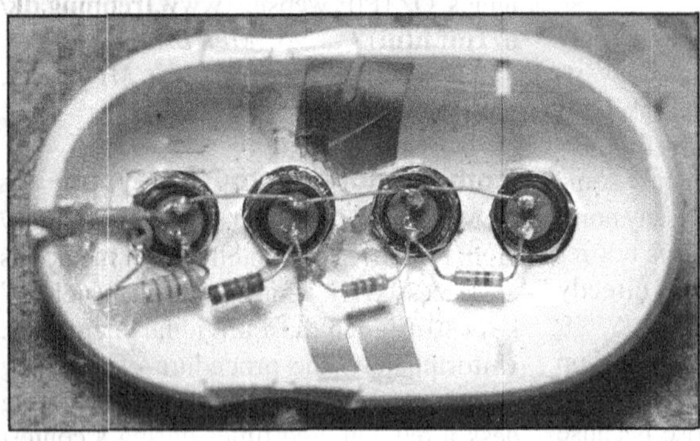

Figure 11 — The completed keypad, ready to use. [Lee Jennings, ZL2AL, photo]

socket to connect the keypad to the radio. Pin 3 is the resistor network and Pin 7 is the ground for my Icom radio, but check your manual to verify the connections for your radio. Finally, go into your Icom radio's menu system (again, check your manual for a detailed description of the menu options) and activate the keypad feature. A second menu item must also be set to allow the keypad to operate the four memories on both CW and SSB modes.

The keypad makes contest operation a pleasure. Accessing memories from it is so much easier than pushing buttons on the front panel. The cost of the keypad is nearly nothing if you can find an old mouse, and it is a great project for a rainy afternoon as it can be built in a few hours. — *73, Lee Jennings, ZL2AL, 87 Auckland Rd, Greenmeadows, Napier 4112, New Zealand,* **leezl2al@gmail. com**

Using Heil Headsets on an Icom-7100

I recently acquired a new Icom-7100 transceiver. The touchscreen is a pleasure to work with, and it has more bells and whistles than I will ever use. I have owned transceivers in the Icom 706 series and found them to be excellent mobile and base stations. About 5 years ago, I bought the Heil HS-706 headset, which has the eight-pin modular plug to fit the 706 series. Pin 3 of the plug connects to the audio of the transceiver, so you can hear the audio in the earphone of the headset.

When I plugged the HS-706 into my new transceiver, however, I couldn't hear anything in the earphone. The speaker was still on, but I prefer to mute it so it does not disturb my spouse. (We live in an RV, so listening with headphones is sometimes a must.) Several e-mails to Bob Heil suggested that Icom has not included the audio output in the modular plug on the remote unit. He suggested I make up an adapter where I could break out the audio and ground from the headset cable and route the audio to the earphone jack.

But in reading through the manual, I found that there is a menu item that puts the audio output of the transceiver back on Pin 3. If this is a problem you have encountered, here is how you can get the audio back in the headphone:

1. Press the **SET** button on the controller unit. The **SET** menu will display on the touchscreen.

2. Use the down arrow key to scroll to page 3.

3. Press **FUNCTION**.

4. Scroll down to page 7.

5. Press **MIC AF OUT**.

6. Press **ON**.

7. Press the back arrow several times until

to press the keys becomes tedious and rather uncomfortable.

In looking through my Icom manual, I discovered a schematic (see **Figure 9**) for an external keypad. The circuit consists of four resistors, four single-pole-single-throw (SPST) pushbutton switches, and a short two-wire connection into your microphone plug. This simple circuit became the answer to my fatigue problem.

I needed to find a suitable box for the switches. Looking around my shack, I noticed an old Apple mouse on the shelf. I realized that its smooth ergonomic design would be comfortable to hold through a long contest. Any small case will work, of course, just make sure it is something that will feel comfortable in your hand.

Use a small screwdriver to lever open the mouse. I discarded the circuit board and USB cable, leaving only the shell and base. I

bought four easy-to-push SPST pushbuttons. The electronics parts suppliers have many types to choose from. I bought a red button to be memory number 1 and to serve as an index for the whole keypad. I made the other three the same color.

Start the holes with a small drill bit and slowly increase the size as required. Mount the switches and solder the four resistors in place (see **Figure 10**). Connect a cable with a 3.5 millimeter plug on the end. Any old audio cable with the plug attached cut to length will do. I secured the cable in place with some hot-melt glue and mated the upper and lower shell halves together again. A little instant-bond glue will hold them in place if the tabs don't align as they should. The finished keypad is shown in **Figure 11**.

The next thing you must do is connect two small leads from inside your microphone plug to a female 3.5 millimeter line-mount

you are back to the main dial showing the frequency. (See the "Set Mode" section in the IC-7100 manual for more information on this procedure.)

Now the audio will be routed to Pin 3, which will enable you to use the headset and control the volume with the main volume control. The speaker in the control unit will still be active. To shut it off, just plug in a standard ⅛-inch phone plug with open connections. [There appears to be a **SPEAKER OUT** menu item that will mute the speaker. — *Ed.*]

If you want to use the standard HM-198 microphone that comes with the radio, just unplug the Heil headset, plug in the HM-198, and everything operates as normal. — *73, Art Horovitch, VE3AIH, 7-841 Sydney St, Suite # 233, Cornwall, ON K6H 7L2, Canada,* **suzanandart1@gmail.com**

Magnetic Fan Control

I wanted to add a fan to my repeater that would operate during transmit. I didn't want to tap the transmit control wire in the radio so I used a magnetic reed switch to sense the higher current in the power supply wire that occurs during transmit. You can use this same method to add a fan to any transceiver without digging into the radio to find the push-to-talk control line.

The reed switch I used has a current rating of ¼ A, which enables it to control the fan directly without any other parts in the circuit. I used the Littelfuse MITI-3V1 6-12.5 reed switch positioned inside three turns of 12 AWG wire to sense the repeater's 8 A transmit current (see **Figure 12**). More turns will trigger the fan at a lower current level, fewer turns for a higher current level.

My power supply wire is stranded and would not stay tightly coiled, so I made a coil by taking solid insulated wire (THHN house wire) and wrapping it around a Phillips screwdriver shaft. I then cut the power supply wire and inserted my solid wire coil in series. While you can solder to the reed switch wire leads, I chose to use some single wire sockets to connect the switch once I had affixed it to the center of the sense winding with RTV silicone.

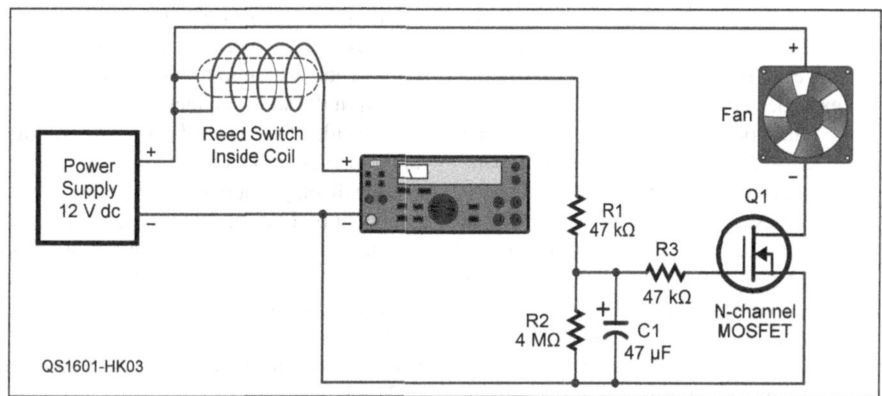

Figure 13 — Schematic of the reed fan switch. The basic circuit just places the switch in series with the positive fan lead. Adding a transistor can increase the current carrying capacity or be used for a timing circuit to allow the fan to operate a few minutes after transmit current drops.

If you want to use a higher current fan or just want it to operate for a few minutes after the transmit current drops, an FET and a couple of parts (as shown in **Figure 13**) is all you need. I used an RFP50N05 MOSFET from my parts collection; however, its 50 A rating is much greater than necessary. Note, R3 is optional. Use it only if you have a problem with ringing. — *73, David D'Alfonso, K6DVD, Rural Route 1, Gaviota, CA 93117,* **k6dvd@arrl.net**

The KX3 Grows Legs

I enjoy owning one of the low-power rigs from Elecraft, the KX3 HF – 6 meter transceiver. However, it is not my only rig; I also have a K3/KPA500/KAT500 system sitting on my desk. Needing more "real estate" at the operating position, I built a desk riser in my workshop, giving me several shelves to set my equipment on. The KX3 wound up on the top, along with the new arrival, the PX3 panadapter. Unfortunately, when mounted 10 inches above the desk, the KX3 and PX3 are difficult to read. Flipping out the KX3 and PX3's built-in legs helps, but it's still difficult to see the LCD panels.

There are options available from third party vendors, but I wanted something that would be portable and lightweight for camping, inexpensive, and small enough for my go-kit. Looking around my workshop, I saw a short length of white Schedule 40 PVC water pipe. The inside diameter looked about right and after some experimenting, I came up with this inexpensive solution.

Using a hacksaw, I cut the PVC into four pieces, each 2½ inches long. The inside diameter of Schedule 40 is about 0.610 inches. [Schedule 40 PVC pipe is available in a variety of inside diameters. The ½-inch size pipe has an inside diameter in the 0.6-inch range. Also, Schedule 40 pipe is available in black and other colors. — *Ed.*] The KX3/PX3 legs (with their rubber feet) are 0.55 inches wide. With those dimensions, the PVC extenders fit loosely over the KX3/PX3 legs. To make a tight fit, I wrapped a 5½ inch length of black electrical tape around the KX3/PX3 legs, which is just enough to make a snug fit on the PVC extenders. The extenders are easy to put on and snug enough to remain on when I tilt the KX3/PX3 back.

If you decide to try a set of these leg extenders, you may want to experiment with different lengths for the PVC. The correct angle will depend on how high the KX3/PX3 sits off your desk and how tall you sit in your operating chair. Mine worked out just right at 2½ inches.

The white PVC extenders clashed with the black KX3 and PX3, so, once I had trimmed the PVC so the viewing angle was correct, I removed the extenders and applied

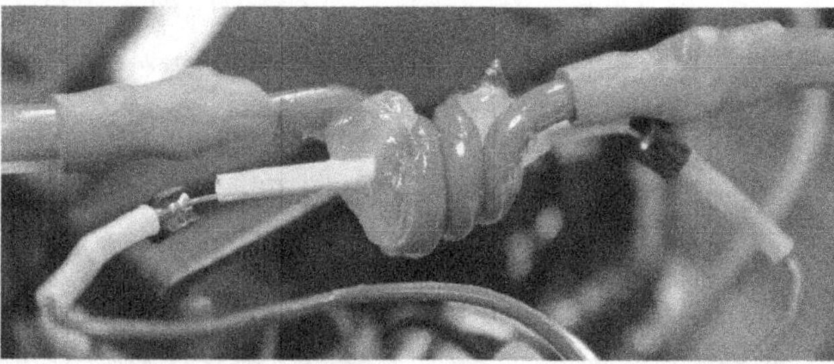

Figure 12 — Solid wire coil with magnetic reed switch held in place with RTV silicone. [David D'Alfonso, K6DVD, photo]

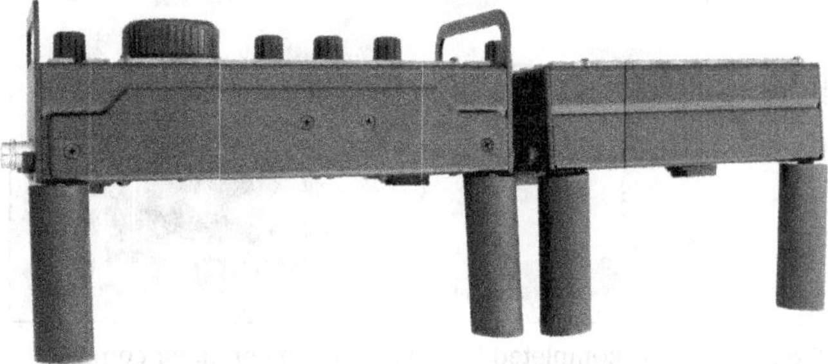

Figure 14 — Homemade legs can be used to adjust the viewing angle for a KX3/PX3 to accommodate your specific operating position. [James Bennett, W6JHB, photo]

a coat of flat black paint, allowing them ample time to dry before putting them back on the legs. You can see in **Figure 14** how the extenders look on the back of the two units. One note about these extenders: the stock KX3/PX3 legs are rubber and keep the two units from sliding around a smooth desk surface. These PVC extenders are smooth, so your equipment may not stay put when you push a front panel button. In my case, this isn't a problem, as I covered my desk riser's surface with black rubber shelf liner material, which prevents such slippage. You may need to find a similar solution. — *73, James Bennett, W6JHB, 224 American River Canyon Dr, Folsom, CA 95630,* **w6jhb@arrl.net**

A Cabinet for the DSO138 Digital Oscilloscope

The January 2016 *QST* DIY issue featured a kit digital oscilloscope offered by JYE Tech Ltd.[1] It looked like a fun project to build, and proved to be just that. It occurred to me that the finished project board assembly could be mounted in a case to complement a digital voltmeter.

First, I needed to find a box. My research determined that the Hammond 1591XXTS-BK is an almost perfect fit with dimensions of 4.8 × 3.2 × 2.2 inches. The DSO138 fits into the lid, needing only a minor amount of filing with a jewelers round file on the lid and box holes to line these up with the DSO138 mount holes. Don't file the PC board; it uses plated through holes.

My original plan was to have two layers in the box, with the scope on top and a battery pack below. After seeing the available

[1]Danzer, N1II, "JYE Tech Ltd DSO138 DIY Oscilloscope Kit," *QST*, Jan 2016, pp 62 – 64.

box depth, I devised a better plan. I used three layers, with a lower level to store the scope probe, and a bottom hatch for access.

You fit the slide and tactile switch bodies to the inside of the box lid and make cutouts for them. A central area is cut out of the lid to pass the display mount jacks and taller board components. Two capacitors, the power supply filter inductor, and the signal BNC jack are remounted on the underside of the main board because these parts are outside the outline of the display board. The lid work and primary components are shown in **Figure 15**.

The Trigger LED is repositioned outward to be clear of the display board overhang. The calibration point loop is replaced with a Keystone test point (#5011). It is positioned so that the point base is flush with the box surface.

The cover for the display unit is made from hard paperboard stock. It is formed as a box so that the sides mate with the main box when assembled. I used black masking tape to cover the completed part. This creates a textured finish that looks like ABS plastic. When making the cover, be sure to cut the edges straight, as they are very visible.

I used four Eagle 3 millimeter Mini Post snap-in standoffs in the display board mount holes to support the cover box. I cut off the top insertion snaps of these standoffs, which gave me four mounting pads that are even with the display screen surface. Dabs of instant-bond glue secure these pads to the cover box.

I used six AA cells to power the unit. These mount in an Eagle holder (Mouser #12BH361-GR). Supplied wiring is relocated to the sides of the battery holder. I mounted the holder on a shelf made from a small piece of 0.016 aluminum stock with

two 6-32 flathead machine screws. This shelf in turn sits ¾ inch up from the bottom of the box, and is held by small rivets or 4-40 screws.

I moved the power connection under the main board and removed the coaxial power jack. I could not locate a mate to the supplied two-pin power header and replaced it with a TE 2-644486-2 header and a TE 644563-2 plug. The header is side mounted with instant-bond glue. Mounted power supply parts are shown in **Figure 16**. You can see how the parts described so far come together in **Figure 17**.

My added attraction is a bottom compartment for storing the test probe. **Figure 18** shows how I cut a hinged door into the box. The hinges are nylon parts normally for model airplane use, which I mounted with instant-bond glue. I used a scrap of Micarta stock for the latch. The latch pivot and striker pin are ³⁄₃₂-inch rivets that are easily press-fit into the box by light finger pressure. I secured these parts with more instant-bond glue, but 4-40 machine screws will work. You can see how the box is used to store the probe in Figure 18. I have a box liner that is retained by the nuts that hold the battery pack to the intermediate level shelf. The liner was part of a frozen dinner box. The plastic feet just came out of my junk box.

The buttons on the supplied tactile switches do not rise above the box cover. I fixed this issue by adding Mountain Switch round black switch caps (Mouser #101-0200-EV). I put these in place with the box and board assembled as shown in Figure 17. First, place a minute amount of instant-bond glue on the switch button. Then position the switch cap through the mounting hole. Center the cap and push down for a few seconds until the

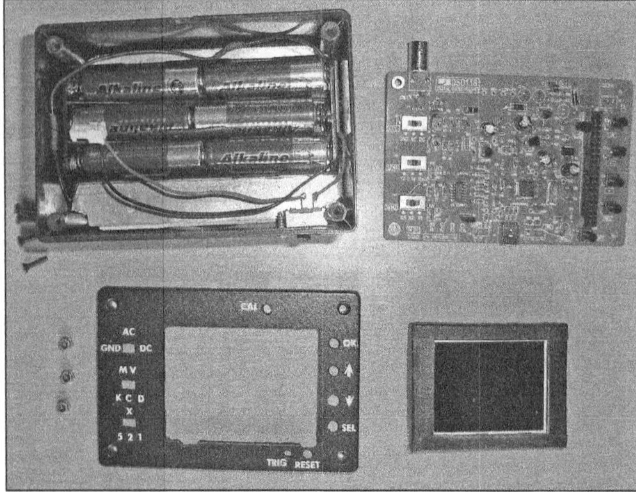

Figure 15 — The completed front panel and other major components of the scope housing project. [Scott Reaser, K6TAR, photo]

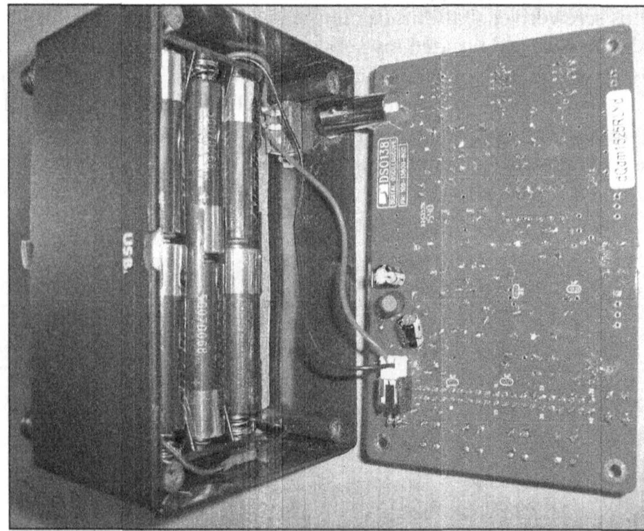

Figure 16 — The battery pack wiring arrangement is shown together with the three capacitors and BNC connector that were relocated to the bottom of the board. [Scott Reaser, K6TAR, photo]

Figure 17 — The box assembled with the modified lid in place and the TFT LCD display ready for mounting. [Scott Reaser, K6TAR, photo]

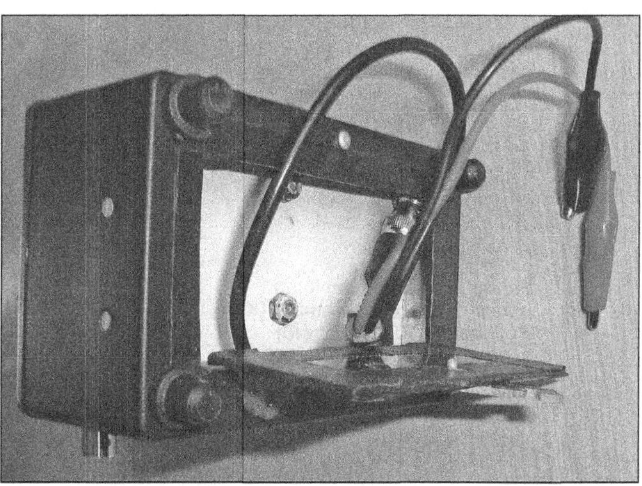

Figure 18 — The "trap door" on the bottom provides access to the probe storage area. [Scott Reaser, K6TAR, photo]

glue sets. This process helps eliminate small misalignments.

The slide switches do extend above the box surface, however, the handles are small and hard on the fingertips. I found a nice fix. The diagonal handle dimension fits a 6-32 machine screw. I made handle caps from 6-32 elastic stop nuts I had available. I carefully started these on to the slide switch handles. You do not want to force things, but you can cut two threads into the plastic. The installed handle caps, and the box in operation, are shown in **Figure 19**.

I did the lettering with wet transfer decals. I prefer this method because you can nudge the script around to adjust alignment and spacing. I used two light spray coats of clear, semi-gloss, water-based urethane over the lettering, to protect it.

The original article notes that you need to exercise patience. That applies to building a case for the scope. Once done, you will have a nice, quality instrument packaged for on-the-go troubleshooting. — *73, Scott Reaser, K6TAR, 1121 Villa View Dr, Pacific Palisades, CA 90272,* **sreaser@verizon.net**

TS-440S Power Connection Upgrade

My TS-440S transceiver had occasional problems with poor contact between the Astron power cable and the Molex-type power receptacle on the transceiver. Even when working, the S meter's pilot light would dim when transmitting. I made several attempts to tighten the contacts by inserting an ice pick into the cable end, but these attempts were only marginally successful.

I decided to install Anderson Powerpole connecters on the cable and on the TS-440S. I was able to make a neat and secure mount without any cutting or drilling on the transceiver. **Figure 20** shows all the hardware needed for the project (except the Powerpole connectors). The two screws on the right are the original screws removed from the transceiver.

The mounting bracket is attached to the radio by these two screws. I found a bracket in my junk box that already had slots. Fortunately, one of the mounting screws aligned perfectly with one of the slots so it was only necessary to lay out and drill the bracket for the other mounting screw and for the 4-40 screw that secures the Powerpole block to the bracket. I had to run a number 32 drill bit (~ ⅛ inch) through the center hole in the connector block to provide clearance for the 4-40 × 1-inch-long mounting bolt.

Figure 21 shows the Powerpole connectors assembled and mounted to the bracket. I highly recommend using a tool specifi-

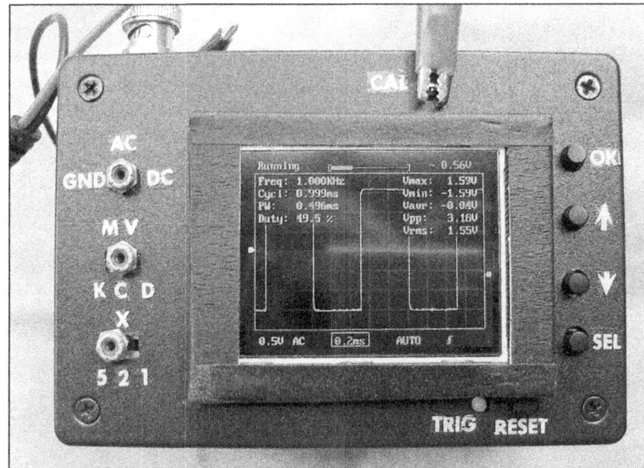

Figure 19 — The completed oscilloscope in its new case ready to help troubleshoot the next problem. [Scott Reaser, K6TAR, photo]

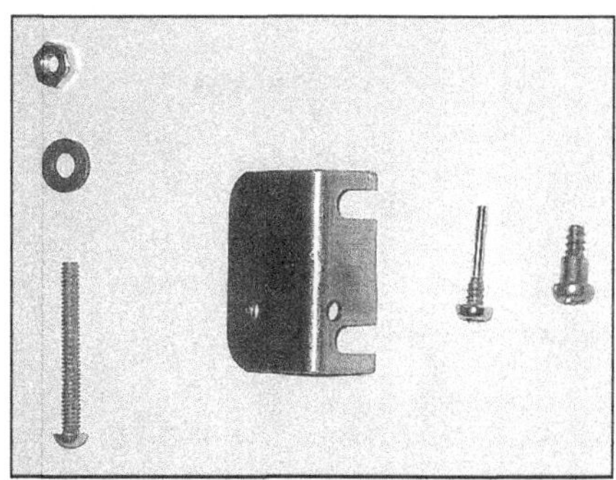

Figure 20 — The parts needed for the mod. The hardware on the left is for mounting the Powerpole connectors to the bracket. The hardware on the right are the mounting screws of the original connector. [Ken Hollenbeck, K9CAX, photo]

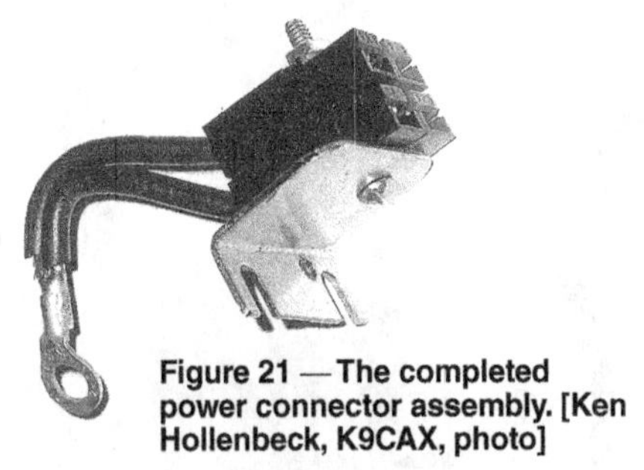

Figure 21 — The completed power connector assembly. [Ken Hollenbeck, K9CAX, photo]

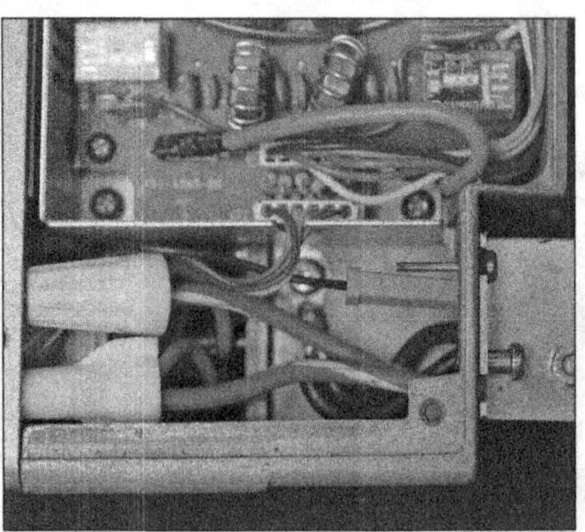

Figure 22 — The Powerpole assembly mounted in the TS-440 chassis. [Joanne Miglore, photo]

Figure 23 — The completed installation with the Powerpole connectors attached to the rear of the transceiver. [Joanne Miglore, photo]

cally designed for crimping the Powerpole connections. The new pre-wired ground connections have also been installed. Because the positive lead connections in the transceiver are a little difficult to reach, I decided to cut the two red wires off near the original connector and splice on the new lead to the Powerpole. These connections were twisted together, soldered, and capped with yellow wire nuts, as shown in **Figure 22**. Note that in Figure 22, a partial view of the reconnected negative lead is shown and the elongated grey ground connector has not been attached to the screw extension.

Figure 23 shows the completed installation. With this improved power connection, there is no dimming of the pilot light during transmission. This upgrade is definitely worth the small investment in time and material. — *73, Ken Hollenbeck, K9CAX, 2433 Cherrywood Ln, Sister Bay, WI 54234,* **marken52@charter.net**

Interference (RFI and EMI), Operating, and Software and Computers

Interference (RFI and EMI)

Switching Supply Receiver Noise

Twice in the past 5 years, switching mode power supply noise has suddenly appeared in the receivers at N4GG. This type of noise is easy to identify once you know how it sounds — it tends to peak every 20 kHz or so and drifts in frequency. It is most noticeable on 160 and 80 meters.

Care has gone into the shack design at N4GG and one key element is to use linear power supplies wherever possible, particularly for the 13.8 V dc "house power" that runs to many accessories. Everything is properly grounded and most coax in the shack is high quality and double shielded. Where was this new noise coming from and why did it start suddenly?

I have two switching supplies in the shack — I decided they would not be worth the trouble to replace with linear supplies. One is for my Dell notebook computer. The other powers an LCD monitor. Both supplies lie on the floor together with myriad runs of coax. While I operate, my feet tend to rearrange the spacing between these items. Both of my sudden jumps in receive noise were due to one of the supplies winding up on top of a coax cable that was carrying receive signals. In both cases, the quick fix was to move the switching supply a few inches away from the coax. That's all it takes to eliminate the noise. The final fix was to hang those little switchers up off the floor.

What's happening? Electromagnetic radiation causes the switch mode noise we are most familiar with — both conducted and radiated. Within an inch or two of a switching supply, there is also a magnetic field emitted from the power transformer and the internal wiring. Coax is a good "shielded cable" for electromagnetic fields; it is not a good shielded cable for magnetic fields. The magnetic field from a switching power supply induces unequal currents in the shield and the center conductor of coax, which appears as a signal at the receiver.

This phenomenon is easy to demonstrate. Hook any receiver to a dummy load through a short run of coax and rest a small switching supply on top of the coax. Hear the noise? Notice how the noise increases when the supply's load is on and decreases or even disappears with the load off? Under load, the power supply's internal currents are higher — hence the radiated magnetic field is higher.

Also, it's not just computer products that have small switching supplies. SteppIR antennas are one of many examples of amateur-market products that come with their own switching power supplies. So keep those switchers away from your coax. — *73, Hal Kennedy, N4GG, 5033 Winding Hills Ln, Woodstock, GA 30189-2566,* **n4gg@arrl.net**

Low Voltage Lighting RFI

Most of us who use the HF bands have experienced RFI problems from solid-state power supplies, even in our own houses. One of the most common examples is the router. Increasingly popular are the low-voltage lighting systems, which have been appearing in remodels and new construction. They may be single lamps or a bar with four or six lamps. Most often, these are powered by a solid-state supply, contained in the fixture for convenience. This is a rapidly increasing and major source of RFI. It's characterized by repeating noise peaks at regular intervals.

Since many states are regulating against the use of incandescent lighting, homeowners are attracted to these devices as a decorative choice, such as the colorful pendants seen in light stores. They are not energy saving devices. A regular magnetic transformer, instead of the solid-state type that typically radiates RF noise, could power these lamps. Most manufacturers offer this option but the nonradiating magnetic type is too large to contain within the fixture.

Here in California, kitchens must now be designed with 51% or more energy saving lamps — fluorescent or LED. Many opt for the low-voltage lighting for the other 49%.

I had a neighbor one block away who installed two bars of six low-voltage lamps with a solid state power supply. They were made by George Kovacs Co. I was paralyzed with noise on 40 meters. The neighbor was not about to go to any expense to replace the transformer. I finally asked an electrician to place a two section Corcom EMI filter (**cor.com**) in the back box of each bank of lights. They barely fit but eliminated the problem completely.

If you know someone who is contemplating a remodel, ask them to use a magnetic transformer in all low-voltage lighting systems. — *73, Jim McCook, W6YA, 1029 Passiflora Pl, Encinitas, CA 92024-2308,* **w6ya@cox.net**

HDMI RFI

I have local cable service for my TV and Internet. The TV's digital video recorder (DVR) box started to lock up for no apparent reason. Nothing could be done except to unplug it and restart the unit.

Well, my first and only thought was a failure within the Motorola DVR unit. I simply called my provider who swapped it for a new unit. All went well for a week or so then the problem returned. I made another call and got another unit but the lock-up reoccurred. By this time I questioned whether or not it was the Motorola DVR.

One day after it happened again, it dawned on me that maybe I was causing the problem when I was on the air. To prove out the theory, I fired up the rig one band at a time and, sure enough, the DVR locked up every time I used 80 meters. The other bands seemed to have no adverse effect.

RFI was the obvious culprit. I attached a split-core ferrite choke to the ac line cord to no effect. How about the remote speaker cables? Nothing. My next approach was to disconnect the HDMI cable between the box and TV. That was it! I tested it again with the amplifier on at 1 kW and there was still no problem. I wound the HDMI cable onto the split-core chokes, fired up 80 meters again and there was no lockup.

I wanted a more permanent fix so I searched eBay and found HDMI cables with ferrite chokes in each end at a reasonable price. I ordered two HDMI cables, which solved the problem.

Another plus was finding that USB cables with ferrite chokes installed are available. That took care of my computer interface problems. There are several sources offering these cables, which can be found by doing a quick Internet search. — *73, Dave Metzger, K8GVK, 895 Haslett Rd, Williamston, MI 48895-9313,* **k8gvk@arrl.net**

RFI from Electronic Lighting Ballast

I use F96 T12 lamps in my shack that required the older Advance Transformer Company# SM-2E75-S-1-TP light fixture ballasts. I decided to replace the old ballasts with the newer Philips Advance Model ICN-2P60-SC electronic ballasts. When I turned on the lights I was confronted with an S9+ signal on several 2 meter radios, which appeared to cover the entire band. The "electronic ballasts" also interfered with X10 modules used to control other lighting in the house.

I was able to eliminate the interference by installing a single Fair-Rite (**www.fair-rite.com**), type 43 split core material ferrite snap-on filter (p/n 0443164151) on the power cable to the lamps. — *73, Dwight Holtzen, N3ARU, 795 N McComas St, Wichita, KS 67203-4832,* **n3aru@cox.net**

RFI by Remote

Radio frequency interference (RFI) can severely limit your ability to receive weak signals. Sometimes this interference can originate in your own home. I experienced several RFI issues that originated from my own electronic gadgets.

My first RFI problem started when I began listening to the local Sunday HF ARES® net. I noticed interference in the form of a ticking sound. I was unable to locate the source but accidently discovered it when I disconnected a nearby TV. The TV had a remote control, which means that its remote sensor is always on, even when the TV is off. I placed ferrite beads on the power cord, but they had no effect, so I installed a switch in the TV's power line; turning the TV completely off solved the ticking RFI problem.

A second episode occurred when an S-9 noise appeared on my receiver. At first, I suspected my switching power supply, but changing to battery power didn't help. I realized that the only recent change in the house was the installation of a digital to analog TV converter. Unplugging the converter (again, turning it off didn't help because it also had remote control) dropped the noise from S-9 to S-6. I switched to the battery and the noise dropped to S-5.

The next RFI problem I ran into was somewhat different. One day I noticed strong interference (strong enough to open the squelch) on the frequency of a local repeater. This time, rather than broadband noise, the interference sounded like a high speed digital transmission. I checked my mobile VHF rig but heard no interference. A bit stymied, I just turned up the squelch. One day I was participating in my local ARES net. I turned off my HD digital radio receiver, which had been on in the background, and the interference disappeared. The RFI was present as long as the HD radio was on, regardless of whether I was using AM, FM, or CD.

My last RFI adventure occurred when my cable Internet connection, which uses a wireless router, started mysteriously disappearing. After several days I noticed that this would occur whenever the remote phone was being used. After each call ended, I had to "repair" the Internet connection. Unfortunately, the only solution I have found for this problem is to use the wired telephone when on the Internet.

In looking for RFI around the house keep in mind that remotely controlled equipment, such as TVs, can cause interference even when turned off. Recheck some of the possible culprits that did not seem to be a problem after the major cause of interference is eliminated. — *73, Andrzej Przedpelski, KØABP, 7260 Terrace Pl, Boulder, CO 80303-4638,* **k0abp@arrl.net**

Light Emitting Noisemaker

I am interested in LED lighting. The technology has grown by leaps and bounds, with today's lights requiring half — or less — of the power of an incandescent lamp to generate an equivalent number of lumens.

I wanted to replace the lights in my radio room and adjacent areas with LED lighting. The radio room was previously equipped with two dual-bulb fluorescent fixtures. Each bulb was rated at 40 W and it is not unusual for these lights to stay on 12 or more hours a day. Another room, about 20 feet away, has a single-bulb fixture that is on all the time and there is also a two-bulb 24-inch fixture in the area just outside the radio room.

I purchased five 48-inch LED replacement tubes for the radio room and the single tube fixture (see **Figure 1**). I also purchased two 24-inch tubes for the smaller fixture.

While the LED tubes are smaller in diameter than the original T12 bulbs, the conversion to LED from fluorescent tubes is not a simple plug and play. It is necessary to remove the fluorescent ballast completely from the circuit. It is not a problem to leave it physically in the fixture, but the connecting wiring to the ac input and the "tombstones" (the fluorescent tube sockets) must be removed or the LED tube will be damaged. Once the ballast is removed, simply run an ac lead to each tombstone to complete the installation.

[Safety First! AC line voltages can be *lethal*. Before modifying any light fixture as described here, turn off the main circuit breaker supplying power to it and use a voltmeter to verify that the power is off. If you are unsure about working with high-voltage electrical wiring, seek help from an electrician. — *Ed.*]

Once I made the conversion in each fixture I was fairly pleased with the result. The light was whiter than the fluorescent tubes, was dispersed in a more downward direction than before, and seemed brighter.

Shortly after the install, however, I turned on my Icom 756 Pro II and discovered a tremendous noise level and regularly spaced spikes on the rig's spectrum display. The S meter was reading 10 dB over S-9.

When I turned off the LED lights, the noise level returned to a more normal S-2. Even the LED fixture in the room 20 feet away caused unacceptable levels of noise on my receiver. The affected bands were 20 and 15 meters and, to a lesser extent, 10 meters.

I do not know the manufacturer of the lights. I purchased them through eBay and am in the process of returning them to the seller with an explanation of the problem. It is interesting to note that the smaller 24-inch LED tube is not causing interference. It was obtained through a different source and is manufactured by Exetik Systems, Cambridge, Massachusetts.

Figure 1 — The product label for one of the noisy lamps. [Andy Corbin, W4KDN, photo]

If you are considering converting fluorescent fixtures to use LED replacements, I would advise you to shop around carefully and buy lamps with the option to return them should they prove incompatible with your radio equipment. [You might also like to read Mike Gruber's, W1MG, article on light bulb RFI in the October 2013 issue of *QST. — Ed.*[1]] — *73, Andy Corbin, W4KDN, 1210 Mountain View Rd, Vinton, VA 24179,* **nitespark@cox.net**

Cooling a Hot RFI Problem

I replaced my 25-year-old gas furnace with a new Lennox furnace that had a 96% efficiency rating. This furnace has a "modulated" (variable speed) dc fan. When the new furnace was powered up, I immediately noticed a significant variable buzzing noise on all HF bands; turning the furnace off stopped the noise.

I called the installation company. I found it very helpful to be able to show the service technician the noise signal on my radio's thin-film-transistor (TFT) display. The service company contacted Lennox, who immediately sent out a filter to place on the dc input to the fan. However, installing the filter didn't solve the RFI issue. I purchased and added some more filters, grounded the furnace with copper braid, and added shielded cable from the furnace to the thermostat. None of these solutions were effective. [When grounding an appliance's metal cabinet, make sure all the parts of the cabinet are *electrically* connected. As with cars, many appliance metal cases are coated with a non-conductive finish electrically isolating the various parts. This nullifies the cabinet's RF suppression ability. — *Ed.*]

Finally, at my suggestion, the installation company crafted a sheet metal enclosure for the motor, which was installed and properly grounded. This solved the problem.

I hope this provides a shortcut solution for anyone facing a similar problem. (It also helps to have superb installation company that stuck with me throughout the entire issue, charging it off as an installation cost at no additional cost to me). — *73, Tom Traughber, WØZX, 8500 Montgomery Ct, Eden Prairie, Minnesota 55347,* **w0zx@arrl.net**

Mobile HF Power RFI

After installing my FT-857D transceiver in my 2002 Ford F-150, I was not surprised by the significant ignition noise that was present. I went through the regular checks, determined it was radiated and not power cord related, and postponed my "fix" efforts for a later date. Just by chance, I drove over a rough stretch of road — and the radio quit! Upon investigation, I discovered that the crimp on the + battery lead, which looked good on visual inspection, was actually loose inside the connector. When I replaced it with a properly crimped connector, the ignition noise was almost completely gone. Apparently, this marginal connection was somehow rectifying the ignition noise and actually radiating it. The lesson here is that if your mobile is experiencing RFI, remember to check your power connections along with other RFI checks. [The "Hands-On Radio" column in this issue has a discussion of troubleshooting techniques. — *Ed.*] — *73, Mike Zonnefeld, WØLTL, 2701 N Camino Valle Verde, Tucson, AZ 85715,* **mikezonn@aol.com**

Old Enough to Vote

For over 10 years, I have tried to find a solution for a seemingly insurmountable problem. When running high power (1500 W) my residential alarm would generate a panic alarm to the alarm central office that, in turn, would summon the police. As a retired police officer, it is always nice to see my colleagues at the front door, however, a bogus alarm is a waste of police resources and could potentially delay a response to a real emergency call.

To solve the problem I tried grounding the alarm panel and cabinet using 1½-inch copper braid welded to an outside ground rod that was then bonded to my existing ground system. I placed ferrite beads on every sensor cable connected to the panel and several snap-on ferrites around the panel cable bundle and the few sensor cables I could access. I also placed snap-on ferrites on both ends of all transmission lines to the antenna. The shield on the antenna coax feed line was grounded at the base of the tower. I even tried common mode coax filters and commercial ac and dc line filters to cover power to the panel. Nothing worked.

I stumbled on the solution when I was preparing coax cables for my HF station. I noticed an old RG-8 patch cable that connected an HF amplifier to an antenna switch. Because I generally use double-shielded coax (LMR-600), I decided to replace the RG-8 with the other style cable. I then tuned the transmitter on various bands to identify any antenna matching anomalies. The antenna SWR on most bands was nearly flat where, in the past, there had been matching issues. The biggest news — no false alarms! Apparently, the alarm issue stemmed from radiated RF coming from the RG-8. It is still hard for me to believe my 50-year-old RG-8 was causing the problem. The lesson: if your coax is old enough to vote, it's old enough to be retired.

If you suffer from a recurring RFI issue, first check your station ground and the associated coax's age and connections. — *73, Tom Traughber, WØZX, 8500 Montgomery Ct, Eden Prairie, MN 55347-1403,* **w0zx@arrl.net**

Operating

Bug Screen

Bugs on Field Day? If you are bothered by various flying creatures buzzing your head and messing up your log, here's a way to save your temper for the electronic kinds. Just hang your light bulb about 3 feet above the operating position and suspend an old screen window (or door) horizontally about 6 inches below the lamp. The creatures will get around the edges of the screen and do their buzzing around the bulb. Then after they burn up on the hot bulb, they land on the screen, not on your papers. In the morning you just empty the screen, which is much better than wiping off the log every 10 minutes. — *73, Doyle Strandlund, W9NJD, 2849 N 035 W, Huntington, IN 46750-4012*

Lightweight Surplus Headset

Looking for a lightweight headset with great audio you can wear all day long? Try to find an old Dictaphone headset (see **Figure 2**) on the surplus market. They are still available and definitely worth the search. — *73, Joe Morse, AD4W, 317 Westlawn Rd, Columbia, SC 29210-5622,* **ad4w@sc.rr.com**

Hearing Loss Help

Like all people, hams can lose much of their hearing as they age and with it the comprehension of high-speed code diminishes. I have gone through 40 years of progressive hearing loss that is presently bordering on a loss of over 90 dB. I have tried various headphones, hearing aids and combinations of the two.

Luckily I came upon an arrangement that works well even for my level of hearing loss. The hearing aid is completely eliminated and only a set of special phones plus an amplifier

Figure 2 — Dictaphone headsets are lightweight and have excellent sound quality. [Joe Morse, AD4W, photo]

---

[1] Gruber, W1MG, "Light Bulbs and RFI — A Closer Look," *QST*, Oct 2013, pp 42 – 45.

are used. The phones are a dynamic type made by Sennheiser (**www.sennheiserusa. com**). Widely used by audiologists for testing hearing loss and in locations of high ambient noise, they are capable of being driven to very high levels without distortion (113 dB/1 V RMS). They have the highest gain allowed by present US standards if operated at a maximum of 0.5 W input power.

The headphone jack on my rig requires amplification to drive the phones to their maximum limit. The MFJ Model 616 (**www. mfjenterprises.com**) speech intelligibility unit I already use is capable of putting out about 2 W per channel. If the headphone jack on the MFJ is used, a pair of attenuators in series with the jack to reduce the volume must be bypassed, this is a simple procedure explained in the MFJ manual. Any other amplifier could be substituted but be careful not to exceed 0.5 W into the phones. Such levels could damage the phones and possibly accelerate long-term hearing loss. [As hearing loss is a medical issue, we strongly advise the reader to speak with their hearing specialist before adopting this method. — *Ed.*]

The 616 has the advantage of shaping the frequency response to suit the individual's hearing response. I purchased a new pair of Model HD 280 Sennheisers on the Internet for under $80 shipped, although they can also be purchased from Sennheiser America for about $100. They have an impedance of 64 Ω. This arrangement should help comprehension for all but the most advanced hearing loss cases. — *73, Fred Ryan, W3NJZ, PO Box 406, New Alexandria, PA 15670-0406,* **fredmerkr@comcast.net**

Advantages of Binaural Reception

How many times have you been working a DX station, only to have it fade into the noise at a critical time during the contact? I have found that many times, the fade is not a drop in the strength of the signal, but rather a change in its polarization. For example, if one is receiving with a vertical antenna, the incoming signal will sound the loudest when it is vertically polarized. But if the signal path changes and the polarization shifts to horizontal, the signal strength will drop dramatically. In such a situation, it is useful to receive with both vertical and horizontal antennas.

I use a Yaesu FT-5000MP with both vertical and horizontal antennas. Yaesu discusses polarization fading on page 44 of the *FTDX-5000 Series Operating Manual.* They also suggest a remedy, which I have found to be useful. When operating 20 meters, for example, I select a half-wave vertical for VFO A receive and transmit. When I encounter fading while operating simplex, I select my 17 meter horizontal dipole for VFO B receive. Then I select VFO A for the left headphone or

speaker and VFO B for the right headphone or speaker, then balance the two channels for equal noise volumes. Many times, when the signal fades for VFO A, I can still copy it with VFO B, and vice versa. As the signal path and polarization slowly change, there is a spatial perception of the signal moving from side to side in front of me. I use this stereo receive configuration often for making contacts that would otherwise be impossible with a single antenna.

There can be additional benefits to using both horizontal and vertical antennas in a stereo receive configuration. First, when there are other stations calling on top of the station I am trying to receive, their signals typically come in via different paths, and therefore different polarizations. The spatial perception is that these interfering signals are spread out in front of me, from side to side, and not all on top of the signal I want to hear. This effect can sometimes make interference significantly less of a problem. The desired signal and interfering signals are separated in my mind, because they appear to be coming from different points in the imaginary space in front of me.

The second benefit can be a perceived improvement in the signal-to-noise ratio. If you think of the noise as a large number of interfering signals, they appear to be smeared uniformly across the imaginary space in one's mind. Each component of the noise appears to be coming from a different point in space, whereas the desired signal appears to be coming from a single point. This perception helps the brain separate the desired signal from the noise, creating an apparent improvement in signal-to-noise ratio.

A third benefit from stereo reception can be an *actual* improvement in the signal-to-noise ratio. When the signal strength is the same for both the vertical and horizontal antennas, for example, for a circularly polarized signal or a 45° linearly polarized signal, the combined audio signal from the two antennas is double that from each antenna, because the two signals are coherent. Random skywave noise, however, tends to be incoherent and uncorrelated for vertical and horizontal polarizations. The result of combining the coherent and incoherent signals results in an improved signal-to-noise ratio.

Note that there are cases where the signal-to-noise ratio will be degraded by switching from monaural to stereo reception. If the signal can be heard only on the VFO A channel and not the VFO B channel, then VFO B is only adding more noise. Even in this case, depending on how the received signal is perceived, signal-to-noise ratio might not be degraded and might still appear to improve.

Stereo receiving with orthogonal antennas is unfortunately limited to simplex operation, and doesn't always improve the copy of a signal. But I have used it on many occasions

and found it to be a very helpful trick. [If you find binaural reception interesting, you might want to read about a binaural receiver built by Rick, KK7B, describedin the March 1999 issue of *QST.* — *Ed.*[1]] — *73, Arlen "Skip" Young, K6KZM, 250 Scripps Ct, Palo Alto, CA 94306,* **k6kzm@arrl.net**

Software and Computers

APRSdroid

We all know that personal computers have played an increasing role in the ham shack. PCs have now given way to smartphones, which I believe are going to be seen increasingly as the choice for operations in the field.

APRS

APRS is the Automatic Packet Reporting System founded by Bob Bruninga, WB4APR (**www.aprs.org**), which is used to transmit GPS location data and text information over radio. One of the most low-cost ways to implement APRS involves using a VHF handheld transceiver, an APRS tracker, a GPS unit, battery, and some cables. The *APRSdroid* app simplifies these requirements.

APRSdroid

APRSdroid is a $4.95 app that installs on any Android smartphone or tablet running *Android 1.5* or later. Installed on an Android smartphone, *APRSdroid* is a complete implementation of APRS for sending and receiving tactical information, including messages. *APRSdroid* includes not only the map software; it also serves as an APRS tracker. Using the smartphone's GPS, display, and keyboard, it becomes a complete APRS implementation, lacking only the radio.

There is one hitch. Because earphone/microphone jacks for smartphones don't have a push-to-talk connection, a radio with voice-operated transmitting (VOX) capability is needed. While external VOX devices exist, I learned that the Baofeng UV-3R and -5R handheld transceivers have VOX capability. This VOX capability allows the APRS data stream to trigger transmit automatically.

Connecting to the Radio

APRSdroid has an "AFSK via Speaker/ Mic" feature (in **PREFERENCES**) that will send and receive data over the smartphone's speaker and microphone or with a cable, over the smartphone's four-conductor earphone/microphone jack. I happen to have the first version Samsung Galaxy Tab Wi-Fi, a 7-inch tablet running *Android 2.2.1* and the Baofeng UV-5R (see **Figure 3**).

[1]Campbell, KK7B, "A Binaural I-Q Receiver," *QST,* Mar 1999, pp 44-28.

Figure 3

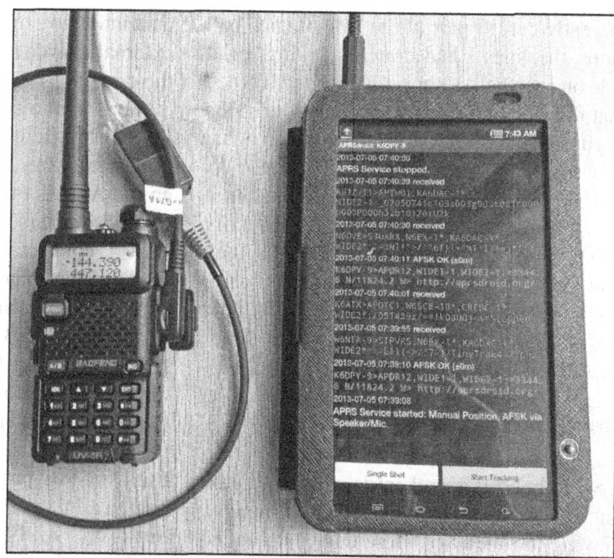

Figure 3 — The Baofeng UV-5R radio and Samsung Galaxy Tab Wi-Fi connected by cable with *APRSdroid* in action. [Daniel Yang, K6DPY, photo]

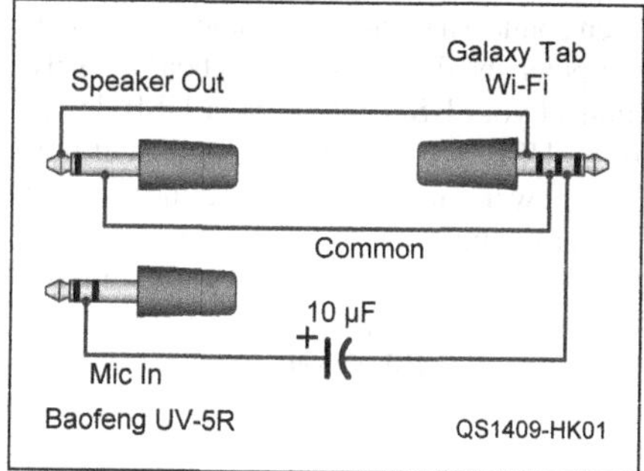

Figure 4 — Schematic of the UV-5R to Galaxy interface cable.

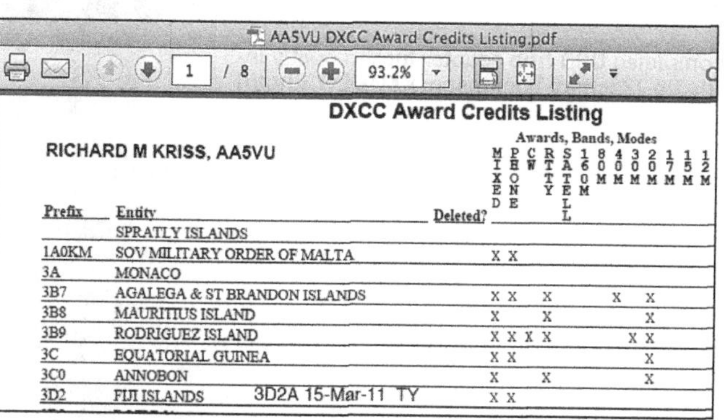

Figure 5— The completed interface cable.

Audio Cable

The pinouts for the UV-5R are the same as the pinouts for some Kenwood handhelds. Note that the UV-3R has different pinouts.

I have a cable for my Kenwood TH-G71A that was suitable for the UV-5R. Because it terminated in an RJ45 connector, I decided to use that format for the cable to the Galaxy Tab (see **Figures 4** and **5**). Note that only three of the eight wires of the RJ-45 connector are used. The volume controls on the smartphone and the UV-5R are used to match the output levels, so the diagram for the cable turned out to be simple.

The trick to making this work is in matching the output levels to the microphone input levels on both the UV-5R and the Galaxy. These are both low-level inputs. The output on the Galaxy is a low-level output (matching the low-level microphone input on the UV-5R), *but* the output on the UV-5R is a *high*-level output. Normally this would require a matching circuit, but because both output levels are controlled by the volume controls on the devices, in the interest of keeping costs low, I decided to forgo the matching circuit. This makes it a bit tricky to set up the levels.

So with only an inexpensive handheld transceiver and my smartphone, I have the ability to use APRS to broadcast my GPS position and short text messages to others via either ham radio or the Internet. In an emergency, such flexibility can be valuable. — *73, Daniel Yang, K6DPY, PO Box 2812, Palos Verdes Peninsula, CA 90274,* **k6dpy@ arrl.net**.

DX Quick Check

Like many others when I hear, copy or see a cluster spot for an unusual prefix my first reaction is "Do I need that call for my DXCC?"

I load my logging program and search for the prefix or call. This is fine, but it is a slow and time consuming process. Thanks to the ARRL DXCC Desk, I now have a fast method for checking whether I need a certain DX prefix.

With each new or update application the DXCC Desk returns the QSL cards with a paper copy of your DXCC Award Credit Slip and a DXCC Award Credits Listing matrix. The matrix is a really cool and useful tool for quickly checking a DX prefix.

Before we started using computers in the ham shack, many hams would keep a copy of the matrix on their desk to do a quick scan and check off new entities worked. Now most of us have very complicated logging programs that take time and keystrokes to do the same thing.

To speed things up, scan the hard copy of the matrix and save it as a PDF file on your computer. Place it on the desktop or in the dock for quick access. With the matrix in PDF format you can use the free *Apple OS X Preview* or *Adobe Reader* (for Mac or PC) applications to view the matrix for a quick visual check of your contact status for a DX prefix (see **Figure 6**). Both programs are very fast and will let you scroll or search by prefix or entity. I also have a freeware Mac utility that will let me annotate the PDF file to show items I have worked since my last update with the DXCC Desk. If there is no "X" in the matrix box, I need to work that prefix.

I keep a shortcut to the PDF matrix on my desktop (or Dock on the Mac) and one mouse click pops up a detailed listing of my DXCC credits. Having a current DXCC Award Credits Listing matrix is one of the benefits of doing an annual DXCC update. I maintain my personal logging program, but the DXCC Award Credits Listing is a very useful tool for quick checks to see if I really need a particular DX station. — *73, Richard Kriss, AA5VU, 904 Dartmoor Dr, Austin, TX 78746-5163,* **aa5vu@arrl.net**

DXCC Award Credits Listing

RICHARD M KRISS, AA5VU

Awards, Bands, Modes

Prefix	Entity	Deleted?	MIXED	PHONE	CW	RTTY	SATELLITE	160M	80M	40M	30M	17M	15M	12M
	SPRATLY ISLANDS													
1A0KM	SOV MILITARY ORDER OF MALTA		X	X										
3A	MONACO													
3B7	AGALEGA & ST BRANDON ISLANDS		X	X	X			X		X				
3B8	MAURITIUS ISLAND		X		X					X				
3B9	RODRIGUEZ ISLAND		X	X	X	X			X	X				
3C	EQUATORIAL GUINEA		X	X						X				
3C0	ANNOBON		X		X					X				
3D2	FIJI ISLANDS 3D2A 15-Mar-11 TY		X	X										

Figure 6 — Once scanned and saved to your computer, the DXCC Award Credits Listing matrix becomes a quick and easy to use tool to help you decide if a DX station belongs in your log. [Richard Kriss, AA5VU, photo]

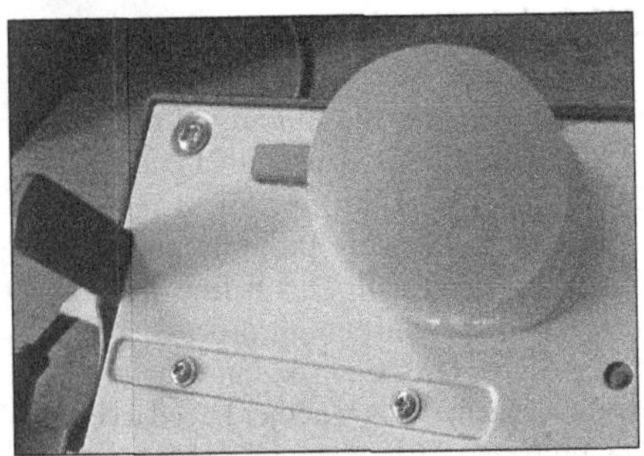

Figure 7 — A close-up of one of the balls glued to the underside of the keyboard near the remaining extended leg. [Randal Schulze, KDØHKD, photo]

Simple, Effective Keyboard Repair

While setting up one of our stations for Field Day, a computer keyboard was knocked off the table, breaking a leg.

After Field Day, I brought the keyboard home. Considering the problem, I remembered a display of "Super Ball" toys at a local store. They were about 1½ inches in diameter and were selling for about $1 each. I purchased two.

At home, I grabbed a roll of tape, with an inner hole of about 1 inch in diameter. I used this as my template to draw a circle on each of the balls. I then used a handheld rotary tool to neatly cut off part of the side of each ball along the lines I had drawn, creating a flat side on each ball. Using silicone glue, I glued that flat side of each ball to the bottom of the keyboard near the back edge, about an equal distance apart (see **Figure 7**).

The result is that the keyboard sits at a comfortable angle, and the balls actually work better than the flimsy legs, as they provide a non-slip footing for the keyboard. — *73, Randal Schulze, KDØHKD, 500 East 105th Ter, Kansas City, MO 64131-4333,* **rschulze@everestkc.net**

Homebrew USB Hub

If you are like me, you tend to overload a USB hub. USB sound cards, wireless LAN, and Bluetooth adapters are all power-hungry. Having too many of these devices plugged into a USB hub can cause flakiness and electrical noise. While powered USB hubs are widely available to solve this problem, non-powered hubs are much cheaper, and you probably already have one or two available. Using some spare parts, a soldering iron, a phone charger, and a little wire, you can create your own circuit to turn that extra USB hub into a powered hub.

In a standard USB cable, there are four wires — D+, D−, +5 V, and GND. D+ and D− refer to a differential system used to transfer data. This device simply isolates power and data into two separate connectors, so that an external power source can be inserted into the circuit.

For this project, we will need two micro USB B breakout boards, a standard USB A port, some wire, and a small enclosure (I used an old Altoids tin). The concept behind this is very simple (see **Figure 8**). From the USB A port, connect the +5 V dc (red wire) and GND to the micro USB B power breakout board. Then, connect the D+, D−, and GND to the micro USB B data breakout board. (If you are using an Altoids tin, you can use the tin as GND.) There is no need to worry about the ID pin (shown as n.c. [not connected] in Figure 8) on either breakout board; it's not used for this project. Once you have connected all the wires, assemble it within your enclosure (see **Figure 9**).

To use the hub, connect a standard micro USB phone charger (verify the charger's output voltage) to the power breakout board. Then connect a micro USB cable between the computer and the data breakout board. Lastly, connect your USB hub or other USB device to the USB A port. The USB hub will now have power supplied by the phone charger, allowing the use of more power-hungry peripherals. — *73, Zachary Thompson, KM4BLG, 12 Emerson Poore Dr, Candler, NC 28715,* **km4blg@arrl.net**

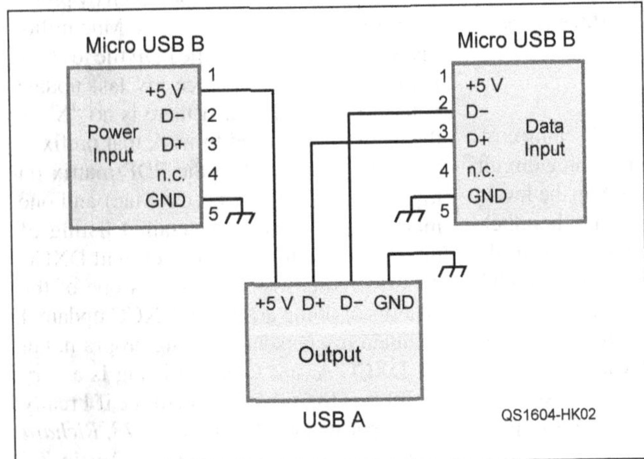

Figure 8 — Schematic of the USB hub power adapter.

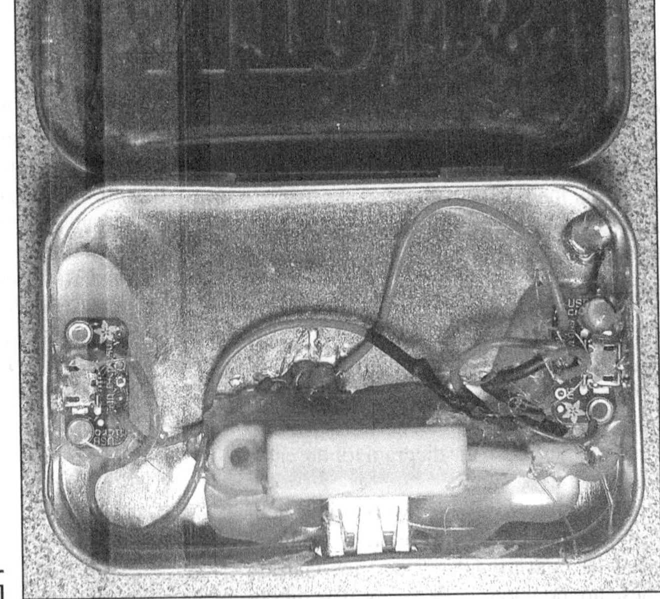

Figure 9 — The completed USB hub adapter tucked neatly inside an Altoids tin. [Zachary Thompson, KM4BLG, photo]

Heathkit Drive Belts

My favorite part of the ham radio hobby is restoring vintage (tube) gear. I especially enjoy bringing 1960s and '70s Swan and Heathkit gear "back from the dead." I found an HW-101 transceiver at a hamfest that needed a lot of restoration. It "spoke" to me, so I took the plunge after haggling an agreeable purchase price.

A very common problem with the HW-100/101 and SB-100/101/102 is the dry rot that causes the rubber drive belts for the driver/preselector variable capacitors to split and eventually break. Then, of course, they fall off. This HW-101's belt set was no different.

Several sources on eBay regularly sell replacement belt sets for about $4, plus shipping. Not bad, but there's an easier, more convenient and less expensive source. Take a trip to the store of your helpful "hardware man" and rummage through his collection of O-rings.

For 59 cents apiece, I found suitable replacements for my HW-101. They are 2½ inch OD, 1¹⁵⁄₁₆ inch ID and ³⁄₃₂ inch thick. This makes them a bit thinner than the ⅛ inch originals, but they work just as well. For about a buck and a quarter, including sales tax, you're back in business.

By the way, ever since building my first Heathkit GR-64 receiver in 1970, I learned early how much I *hated* stringing spring-tensioned dial cords. Well, my newly acquired HW-101 had no dial cord strung for its plate-loading capacitor. So, another O-ring to the rescue.

Looking deeper into the hardware store's collection, I found a suitably larger diameter O-ring, ³⁄₁₆ inch thick, for another $1.15. I was able to coax this third belt around the variable capacitor's lower and upper drive wheels by slightly stretching it, totally avoiding the need for a replacement dial cord.

All three new belts are shown in **Figure 1.** All work just great and will be equally cheap and easy to replace again in another 10 or 15 years. — *73, Tony Bogusz, W9MT, 10536 S Coyote Melon Loop, Vail, AZ 85641,* **w9mt@arrl.net**

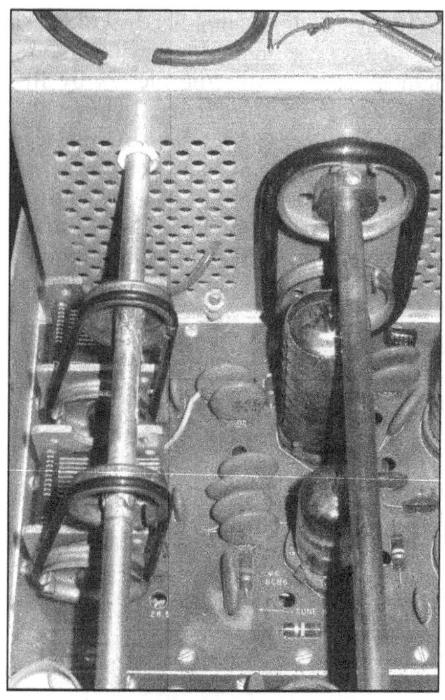

Figure 1 — The photo shows how inexpensive O-rings can be used as an alternative to the more expensive drive belts when restoring Heathkit transceivers.

Coil Core Removal

Slug-tuned coils often suffer from a major deficiency that becomes apparent over time. After years of being set in place, the core becomes frozen in the phenolic-paper tube. When applying pressure, the alignment tool snaps the core. When this happens the core is cracked and any attempt to turn it will fail. When sharp tools are put into play, the form is usually damaged, and replacements are no longer available. To remove single cores without damaging them, find a long screw that will pass through the core and check that its head will pass through the coil form. A screw with a fillister head works best, but others may be used if filed or ground to clear. Be sure the screwdriver slot is retained.

Next, find a couple pieces of metal tubing that will clear both the screw threads and the coil form. Place one piece of tubing over the screw, insert the screw through the core, place

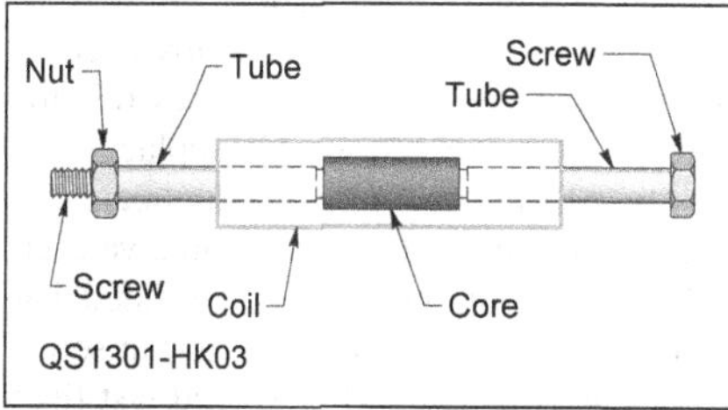

Figure 2 — A long screw and a couple of pieces of tubing can help get that stuck core free from its coil.

the second tube on the screw and tighten a nut to grip the core by its ends (see **Figure 2**). With a screwdriver, turn the screw-slug assembly to remove it from the coil. Be sure to replace the slug with the same type of material. A word of wisdom — be careful that you don't lock two slugs together in the center of an IF transformer — those are hard to remove. If there are two cores, remove the free one before applying the fixture.

Using end pressure to hold a slug together makes it seem feasible to apply the same principle with small, slotted cores by mounting a piece of drill rod in the chuck of a lathe and one in the tool holder. Then clamp the core between the two rods and remove it by turning the core.

When you wind cores on phenolic tubes, too much pressure can reduce the tube diameter, making the core bind. Heavy wire should be wound like a spring, placed on the form and secured with adhesive. If a core is too loose in a coil form, a bit of lacing cord may be placed lengthwise in the form before screwing in the core. Rubber bands are sometimes used, but they deteriorate rather quickly. — *73, Doyle Strandlund, W9NJD, 2849 N 035 W, Huntington, IN 46750*

LED Upgrade for L-1500 #47 Lamps

There are still a great many pieces of equipment out there using the old #47 lamp or one of its cousins. My Ameritron AL-1500 is one of them.

I discovered a nice white LED at RadioShack (part number 276-0024) that I have been using to replace some of the old incandescent bulbs in my equipment. A similar part is available from Mouser (part number 749-R20WHT-F-0160). Mouser's price is much lower per unit, but for a small quantity the RadioShack price is competitive when you consider the shipping cost and delay.

If you have tried to use a more conventional bullet shaped LED to illuminate a dial or meter the result is typically a bright spot of light on part of the face with the rest dark. This is due to the narrow viewing angle of most LEDs. These new LEDs emit a white light with a wide viewing angle of about 160°. This makes them an excellent choice for back or side lighting meters.

Unlike the incandescent lamps used in my Ameritron, LEDs need a series resistor to limit the current flow. My replacement uses two LEDs per meter to duplicate the original two lamps. I was going to give each LED its own limiting resistor but decided that it would be easier to put two LEDs in series with a single resistor. This worked just fine; I couldn't see any difference in brightness when using separate limiting resistors or with two LEDs in series with a single resistor.

The LED has a recommended forward current (I_f) of 20 mA and a forward voltage drop (V_f) of 3.5 V. To compute the limiting resistor value (R_L), simply subtract the V_f from the supply voltage (V) and then divide by the forward current. I used two LEDs in series so the LED V_f needed to be doubled. The general formula for this is:

$$(V - (2 \times V_f)) / I_f = R_D$$

This is the minimum value of the resistor. You might want to experiment with the resistor value to adjust the illumination level. I found a 560 Ω resistor worked in my AL-1500.

Warning: This is a high voltage amplifier. Even though we are working in the low voltage section, this amplifier can kill! Be sure to unplug the amplifier and discharge the high voltage before doing anything else. [Refer to the AL-1500 manual and the ARRL Handbook for Radio Communications for more information on high voltage safety. — Ed.]

In my AL-1500 amplifier (**www.ameritron.com**) the incandescent bulbs are held in place above the meters by little 3AG fuse clips. **Figure 3** shows the mounting clips holding the LED replacements. I fashioned a small mounting from a piece of prototyping board by cutting it to fit between the clips. This is about 2.5 inches.

The spacing between the pins on the LEDs is 0.3 inches so I cut the board 0.4 inches wide. I then broke away the glass on the old bulbs, leaving only the metal base. [Wrap the bulbs in a piece of cloth when breaking the glass to prevent injury from flying chips. — *Ed.*] I cut a slot in each end of the board large enough to hold the lamp base and then glued them in place.

I mounted two LEDs on the board, one near each end with the series resistor in the center (see **Figure 4**). I connected the wire from ground to the cathode of one LED and the supply voltage to the anode of the other LED. The assembly then simply snaps in place over the meters. I repeated the process for the other meter and — presto — once again I had illuminated meters. Additionally, the power for the grid current meter light comes from the high voltage delay relay circuit. The grid meter LEDs illuminating indicate that the high voltage has been applied and the amplifier is ready to operate.

I think the color and intensity of the LEDs look better than the original lamps and I am expecting a much longer life from the LEDs. Since all of the Ameritron amplifiers in this series are constructed essentially the same way, I am sure this modification will work for any of these amplifiers. These LEDs can also be easily adapted to replace the lamps in almost any piece of equipment. *Note*: if your equipment uses low voltage ac to power the lamps you will need to include a rectifying diode as the LEDs have a very low reverse breakdown voltage and cannot handle ac. — *73, Alfred T. Yerger II, K2ATY, 1312 Union Ave, Newburgh, NY 12550-8907,* **k2aty@arrl.net**

Cardboard Wire Separator

I needed to replace a failed power transformer. I wanted to bench test the replacement before installing it. To keep the numerous leads from touching one another, I cut a 2 inch wide strip of corrugated cardboard so as to expose the individual cells. Then I slid the wires into the cardboard cells (but not far enough for them to contact each other) to keep them separated. To prevent wire slippage or movement I taped the wires where they entered and exited the cardboard separator. In addition, you can make notes or measurements on the cardboard. — *73, Roy Lehner, WA2SON, 5464 Oakwood Dr, North Tonawanda, NY 14120,* **wa2son@toast.net.**

Fixing Ceramic Capacitors

I discovered this by accident while replacing a badly drifting temperature-compensating capacitor in one of my vintage VFOs. After removing the part and giving it the heat-gun test with my capacitor tester, it proved to be quite stable. Heating the leads apparently fixed it. I put it back into the circuit where it has been working since.

Heat each lead with a soldering iron until the body is too hot to touch. That seems to fix the problem. I've done this only on the old *dog-bone* style ceramic capacitors, all in the range of 30-100 pF. I don't know if it will work on any other types or values. [While this technique may work, the capacitor is suspect and should be replaced. — *Ed.*] — *73, Tom Webb, W4YOK, 3533 Teakwood Ln, Plano, TX 75075-1783,* **sam9lives1@verizon.net**

Figure 3 — Here are the LED boards mounted above the meters. The pilot lamp bases are used to secure the board to the old lamp clips. [Alfred Yerger II, K2ATY, photo]

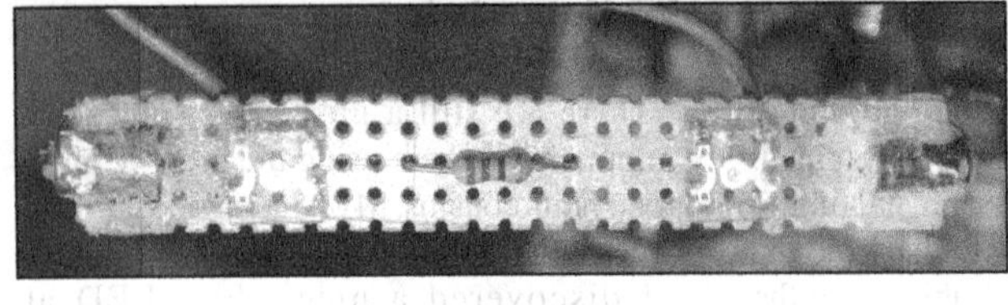

Figure 4 — This is the completed LED assembly. Note the pilot lamp bases glued to the board ends for mounting and the single resistor wired in series with the two LEDs. [Alfred Yerger II, K2ATY, photo]

Johnson Dial Cover Restoration

A couple of years ago, I restored three Johnson Viking Valiant transmitters. In the course of making them "pretty," I needed the rubber gasket for the VFO dial cover, as the old one was dried and cracked. I discovered the perfect product in the large-scale model aircraft section of a hobby store. Fourmost Products' "Cockpit Coaming; large #120" (**www.fourmost.com/index.cfm?fuseaction=detail&id=88321&product=42**) is made to be cockpit trim on large model aircraft, but it fits the VFO/dial cover used on the Ranger, Valiant, Navigator, Pacemaker and others. Be sure to cut it a tiny bit long to compensate for the corners. — *73, Tom Dailey, WØEAJ, 270 S Lafayette St, Denver, CO 80209,* **daileyservices@qwest.net**

Repairing the IC-27/37/47 Squelch

Back in 2009, I submitted a hint about replacing the defective and obsolete volume control pot in an Icom IC-37A transceiver with an outboard unit.[1] A ham friend of mine had Icom service replace his bad volume control in the 1990s. Recently, when his squelch control failed, he asked me for help repairing it.

The squelch control circuit is a "different animal" from the volume control. Where the volume control is a single 10 kΩ control, the squelch circuit uses two 10 kΩ pots ganged together. One 10 kΩ ganged pot section controls IC1's squelch gate circuitry. The set point of this pot is critical. The second 10 kΩ ganged pot section simply sets a balance level between the squelch gate circuit's two buffer amplifier stages in IC2. I found this setting is *not* critical and used that information to replace the second pot with a fixed resistance.

Begin the modification by finding the J3 and J4 connectors on the bottom board of your radio (see **Figure 5**). After unplugging them, use a small pair of diagonal cutters to clip all five wires from the original squelch control's daughter board. These are the yellow and brown wires going to J3 and the blue, green and black wires going to J4.

To find the correct set point for proper minimal squelch I connected two 10 kΩ trimpots in place of the originals. After some experimentation, I found that 7.6 kΩ was the minimum setting (for both pots) for good squelch action. Using Ohm's Law for parallel resistances, I determined that a parallel combination of 100 kΩ and 8.2 kΩ ¼ W resistors can replace the pot for IC2. I put these parallel resistors (R_p) inside of some shrink tubing.

Connect everything as shown in **Figure 6**. The R_p is connected to the green and blue wires of J4. The parallel resistors inside the

shrink tubing fit nicely tucked behind the front panel of the radio. J4's black wire connects to the ground end of the new squelch pot, P. J3's brown wire connects to the wiper and its yellow wire to P's high resistance end.

If you want to stay with a trimpot for the new squelch control, a dab of silicone RTV will secure it to the top circuit board near the reset switch and scanning select switch bank under the snap cover (see **Figure 7**). If you want use an external knob to set the squelch,

Figure 5 — A bottom view of the transceiver showing the locations of the J3 and J4 connectors. [Tony Bogusz, W9MT, photo]

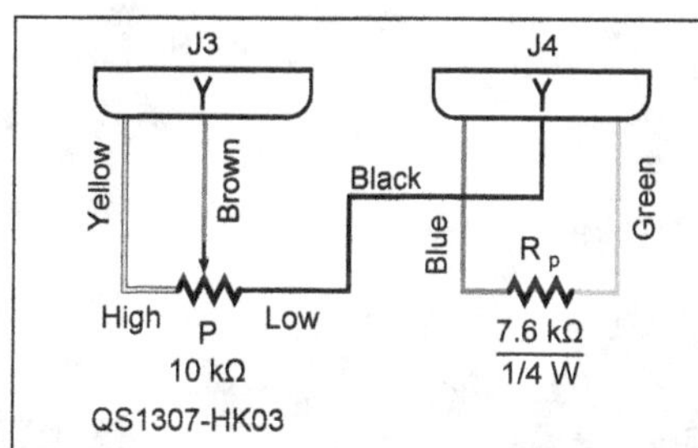

Figure 6 — The schematic diagram of the squelch modification.

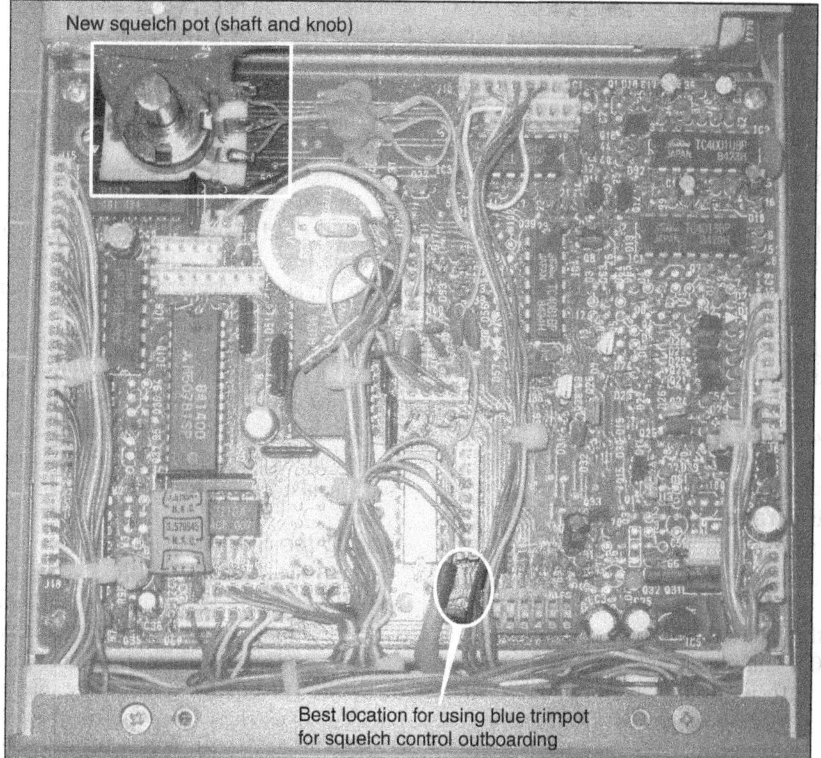

Figure 7 — A top view of the transceiver showing the mounting location for the squelch pot. The shaft extends through the case and a knob on the outside permits adjustment of the squelch level. [Tony Bogusz, W9MT, photo]

[1]Bogusz, W9MT, "Help Your Icom IC-27/37/47 Regain Its Voice," *QST*, July 2009, p 56.

you can use a pot with shaft mounted on double stick tape with more RTV to secure it. The best location for this is where the optional voice synthesizer would normally mount. Be sure you don't short the metal body of the control to the legs of the IC just to the front of this location. Careful positioning of the double stick foam tape will prevent a short circuit. You will need to splice an extension from the black, brown and yellow wires to go the extra distance. Insulate and secure these connections well.

Squelch operation employing the single pot and fixed resistance is both smooth and reliable. No degradation of receive sensitivity or squelch action was noticed when using the fixed resistance for the IC2 buffer-follower amplifier circuit.

If you already did the volume control modification from my 2009 hint, you can probably fit two controls in the voice synthesizer area, but it's going to be crowded. It may be better to save that area for the more often used volume control replacement and use the reset switch location for a less often used squelch trimpot. In any case, your radio should now be good for another 25 years of faithful service. — *73, Tony Bogusz, W9MT, 10536 S Coyote Melon Loop, Vail, AZ 85641-2593,* **w9mt@arrl.net**

Storing Crystals

If you're a vintage radio enthusiast like me, you probably have a large collection of crystals in various sizes piled on your operating desk. Organizing and storing them for easy access can be a problem. One solution I've found, especially for the larger FT-243 cases, is a large pill holder with separate compartments intended to hold 7 days of medicine. The larger FT-243 cases fit perfectly into the individual compartments. Each compartment pops open with a touch and closes securely to keep out dust and moisture. The pill holder I used is sold by pharmacies under the trade name "Push Button Pill Reminder 7-Day 2XL." Smaller pill holders are also readily available that can be used to store crystals in the HC6U holders typically used in Drake and Collins equipment. — *73, Robert Logan, NZ5A, 8712 Lone Tree Dr, Manor, TX 78653,* **bob.logan47@yahoo.com**

Battery Replacement in the Yaesu DVS-2

The Yaesu DVS-2 plug-in Digital Voice Keyer accessory has a button cell that's tab welded and soldered onto one of the internal PC boards. In mine it was a 2032 lithium cell. Its purpose is to retain the information in memory after shutdown or loss of power. The factory cell, and subsequent cells, should last about 10 years after installation.

Figure 8 — This is the component side of the cell board. At the center right is the connector that mates to the upper board and at the lower left is the original battery. [John Grout, KT4AD, photo]

While some information is available on the Internet, I couldn't find a procedure on how to get the DVS-2 open to replace the cell. Here's a detailed approach that works.

You'll need a small soldering iron, appropriate flux, solder wick, good eyes or a magnifier, shrink sleeve and the usual ham tools for detailed work.

A note of caution: Do not short the button cell, even for a second, or it can be ruined. When installing the new cell, remember it's hot, so don't let it, or the leads, touch any of the surrounding circuitry or you may damage components.

Making the Swap

You'll need several small, flat-bladed, screwdrivers in order to separate the unit. There are three hold points for the halves of the case. One is the Phillips screw near the cable. The other two points are internal plastic snap catches. One is across from the screw at the cable end on the side, below the "foot;" the other is in the center at the bottom. With the screw removed, carefully pry open and press in on the catch at the bottom. Then, separate the catch opposite the screw position hole.

Carefully separate the two small PC boards via the hard mounted connectors (see **Figure 8**). Note: The factory cell at the lower left has a yellow isolation ring to prevent shorts between + and – terminals.

Flip the PC board over and find the holes where the cell's tabs are soldered to the board. You'll have to carefully use solder wick (or solder sucker) to remove the solder, taking care not to overheat the copper lands on the board. Don't be in a hurry. If you don't reuse the welded tabs on the cell, you may want to cut one tab to ease access to the dead cell for removal. See Figure 8 to help reference the points for solder removal.

After removal of the original cell, you'll note the factory tape used to hold the battery in place. I didn't utilize tape in my replacement method.

On the solder side of the board, locate

Figure 9 — After the old cell is removed and the holes cleaned, the board should look like this. The battery + and – connections are marked on the board. [John Grout, KT4AD, photo]

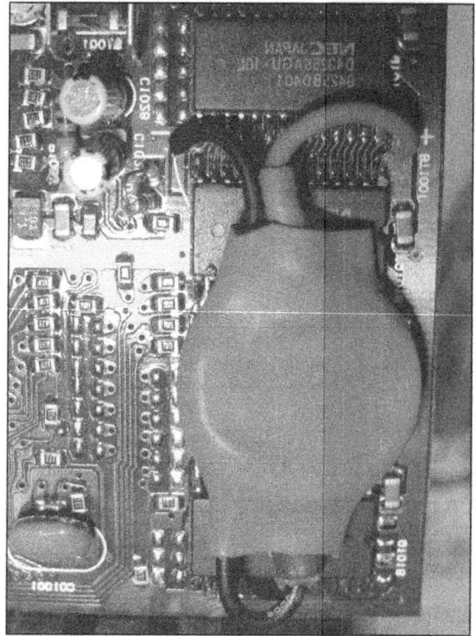

Figure 10 — The new cell, double insulated and installed on the DVR-2 board. [John Grout, KT4AD, photo]

where the cell's tabs were unsoldered. This will help in the wicking and removal process. Clean the area and ensure the holes are open to accept the new cell's leads later. You will install a replacement cell that is wired and encapsulated in a shrink sleeve. There is plenty of room.

After the holes are clean, select red and black pieces of wire that will fit in the holes. Since the current draw is minimal, a stiff, small gauge wire is adequate.

Figure 9 is a view of the area, ready to accept the new assembly. A nearby Batteries Plus store spot welded tabs on a new cell. [Both Batteries America and Mouser sell 2032 cells with solder tabs. — *Ed.*] Use some-

thing to isolate the tabs so they can't short to the opposite side of the cell or anything else before installation. Solder on your wires and add shrink sleeve to the contacts. Then shrink sleeve the whole assembly so it's isolated from the board. Strip the wires enough to clear the holes on the board's solder side and temporarily sleeve the exposed ends to prevent any erroneous contact with the PC board.

I used a wooden clothespin to hold the cell in place while soldering on the reverse side of the board, though you could use tape. It's critical to keep the exposed lead ends from shorting or touching any contacts during soldering. Check the cell voltage one last time before final installation to ensure it is still good. When complete (see **Figure 10**), secure the new cell in place if need be (I found the wire stiffness adequate to hold it) and reassemble the unit.

Before connecting the DVR-2 to your rig, push the unit reset button through the unit side access hole. It should now work for another 10 years of contesting fun. As a side note, there have been a lot of Internet comments about the excess audio drive when using this unit. Keep the microphone gain and drive down to minimum when recording and speak in a normal voice — don't shout. A lot depends upon the microphone or contest headset being used. — *73, John Grout, KT4AD, 15798 Fawn Pl, Montclair, VA 22025-1427,* **kt4ad@aol.com**

Equipment Drying Tip

A local ham brought me a water-damaged Heathkit for repair. I packed it with rice and some silica packets, and left it alone overnight. [Silica gel packets that have been in the open air for a while can be recharged by placing them in a 250°F oven for 5 hours. See **www.ehow.com/how_5035187_recharge-silica-gel.html** for more information. — *Ed.*] The next day it was as dry as a bone, but still harbored an unpleasant odor as a result of having been left in the ham's basement to dry out. Once again, I packed it in a box, this time using a copious amount of dryer sheets inside the radio and the box. The next day, I had in my hands a lavender scented Heathkit to return to my friend. He took it home, applied power, and found that it worked perfectly. — *73, Alin Steglinski, KC9YSN, 19 N Ashland Ave, Palatine, IL 60074-5401,* **kc9ysn@arrl.net**

LED Meter Light Upgrade

When one of the incandescent meter lamps inside my Ameritron AL-811 amplifier burned out, I decided to replace them both with homemade LED bulbs. [For another lamp replacement solution, see Charles',

[1]Rankin, WA2HMM, "Making the Switch to LEDs," *QST*, Dec 2013, pp 33 – 34.

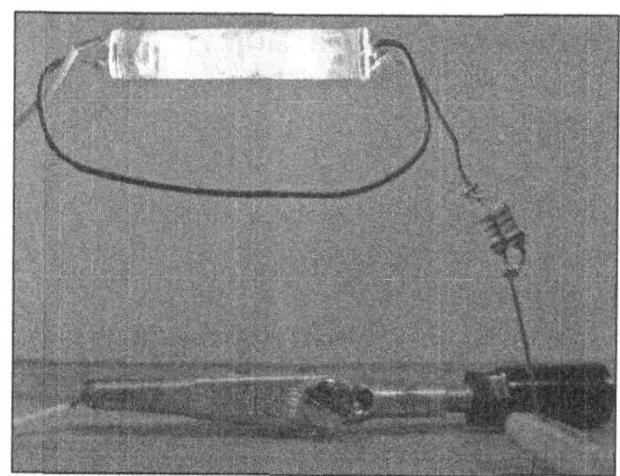

Figure 11 — A couple of LEDs, some plastic chips, and a piece of tubing make a replacement for small incandescent lamps. [George Martin, N2HO, photo]

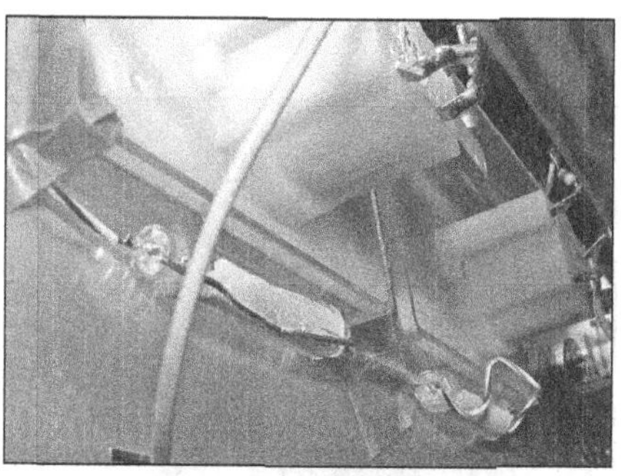

Figure 12 — The new LED bulbs are easily installed with just a few pieces of tape. [George Martin, N2HO, photo]

WA2HMM, article in the December issue.[1]] I took a clear plastic tube wide enough for the LEDs — about 1 inch long. Then I took a small piece of clear plastic cover from a broken CD case and crushed it using cutters into small pieces of random shapes and sizes (about $\frac{1}{32} - \frac{1}{16}$ inch). I then filled the tube with the small pieces of crushed plastic to diffuse the light. Finally, I inserted amber colored LEDs, which I sanded lightly to improve the diffusion, into both sides, securing them with silicone RTV (see **Figure 11**).

I opened up the amplifier, being careful to observe all high voltage safety protocols. I didn't remove the incandescent lamps and meters from the amplifier or disassemble anything to install the lights. Because their cases are translucent, I placed my homemade bulbs underneath each meter (see **Figure 12**). I used tape to hold them in place. [RTV might be a longer-lasting solution. — *Ed.*] All the LEDs (four total in two tubes) were connected in series with a 330 Ω 0.5 W limiting resistor directly to the original lamp's 14.8V dc power source. The project took about 30 minutes to complete and resulted in beautifully lighted meters! — *73, George Martin, N2HO, 2881 Evergreen Ave, Oceanside, NY 11572,* **n2ho@ yahoo.com**

Transceiver Conversion for the R-390A

The Collins R-390A is one of the best receivers of the tube era. I enjoyed band surfing with it and began to wonder if it could be used in conjunction with a transmitter for integrated operation.

After some research, I discovered that the R-390 can be modified to provide a standard 5 – 5.5 MHz VFO output to any transceiver from the 1970s, thereby allowing the VFO in the R-390A to control an SSB transceiver. I implemented this modification on my homebrew 20 meter transceiver, which follows a similar frequency plan to that of a 1970s SSB transceiver.

Transmit/receive operation with an R-390A can be achieved through bypassing the VFO connection from the VFO sub-chassis to the RF sub-chassis (see **Figure 13**). Fortunately, the VFO connections are coaxial and compatible with modern Mini-Circuits (**www.minicircuits.com**) or equivalent 50 Ω components.

Referring to **Figure 14**, the VFO stage (V701) covers 2.455 – 3.455 MHz with an output of 23 dBm. This signal is divided by using a 3 dB power splitter (SPLTR1), where half of the VFO signal is fed back to the RF subchassis and the other half is fed to the LO port of an external mixer (MXR1). The RF port of MXR1 is fed by an 8.455 MHz external oscillator, providing high-side mixing of the VFO. The IF output is amplified to 10 dBm by AMP1 and filtered by a 6 MHz low-pass filter FL1, passing only the shifted VFO output from 5 – 6 MHz. This signal is fed to the external VFO input of any transceiver (see **Figure 15**). The parts list for this modification is in **Table 1**.

In addition to the frequency shifted VFO signal, a method of muting the R-390A on transmit must be implemented by tying the PTT line of your transceiver to a relay, where this relay (SW1) pulls down the break-in terminal on the R-390A when PTT is high. Finally, a T/R relay (SW2) must be used to switch your antenna be-

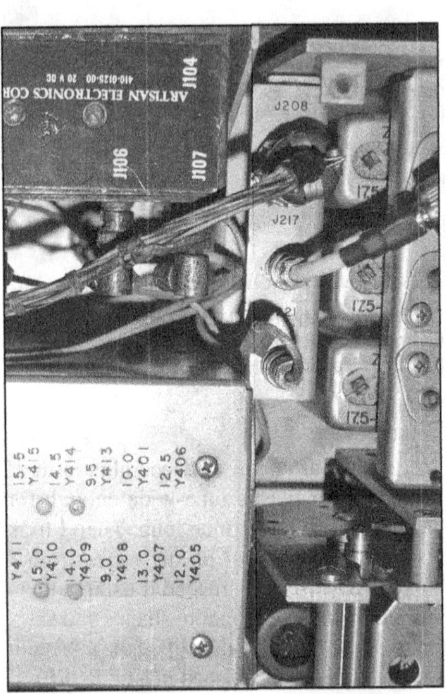

Figure 13 — This connector (J217) is the input to the third mixer. One output of SPLTR1 is connected here to operate the R-390A. [Gregory Charvat, N8ZRY, photo]

Table 1
R-390A Modification Parts List

AMP1	Wide band amplifier, gain = 10 dB
FL1	6 MHz, fifth-order Chebyschev low-pass filter
MXR1	Mixer, Mini-Circuits ZAD-3
OSC1	HP 3325A synthesized signal generator, DDS, crystal oscillator, or other source
SPLTR1	Mini-Circuits splitter ZMSC 2-1W
SW1	General purpose relay
SW2	High isolation relay, either a two-pole relay to ground the R-390A antenna port during transmission or a high-end coaxial relay

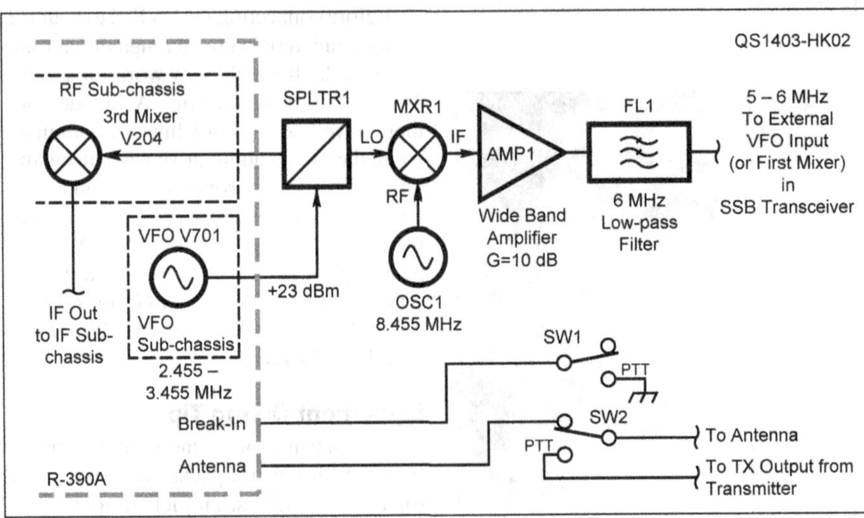

Figure 14 — This block diagram shows the R-390A modification to use its VFO to control an external transmitter.

Figure 15 — The external components are shown here. At the left is the Mini-Circuits ZAD-3 mixer, the small blue cube at the upper center is the splitter. The large aluminum case in the center is the amplifier and the blue case at the bottom is the 6 MHz low-pass filter. [Gregory Charvat, N8ZRY, photo]

tween your transmitter and your R-390A. Similar to SW1, SW2 must be actuated when PTT is high. I have a number of related videos posted at **www.glcharvat. com/Dr._Gregory_L._Charvat_Projects/ R390A.html** if you would like more information on this modification and the restoration of the R-390A.

This technique can be applied to almost any vintage receiver and transmitter. Imagine the fun you can have participating in boat-anchor nets or chasing DX using your R-390A. — *73, Gregory Charvat, N8ZRY, 630 Old Clinton Rd, Westbrook, CT 06498,* **charvatg@gmail.com**

A Part-Cleaning Jig

Old electronics and the miscellaneous tubs at hamfests are good sources of parts that can be recycled, but those parts often need to be cleaned up before we can reuse them. In particular, the terminals of switches, pots, tube sockets, connectors, terminal strips, and similar components must be cleared of old solder and scraps of wire. The usual technique uses flux, desoldering braid, or a solder sucker to remove solder, and often needle-nose pliers to remove twists of wire. This technique, however, calls for more hands than humans have; one to hold the part, another to hold the soldering iron, and a third to hold the desoldering braid or pliers.

A simple jig (see **Figure 16**) makes cleaning those parts' terminals far easier. A small panel is attached to a scrap of wood, which acts as a base. Holes in the panel accommodate the parts to be cleaned and the base is large enough that you can put your wrist on it to keep it from moving. To clean a part's terminals, mount it on the panel with the terminals facing the base and go to work. This is more stable than "third hand" devices that use alligator clips and you can save your workbench by soldering on the base.

The materials are non-critical. For the unit in the picture, I used some Masonite and a piece of ¾ inch plywood that I found in the garage. After preparing the holes, I fastened the panel to the base with screws and some wood glue. You can put rubber feet or a non-skid material on the base, to help hold the jig in place, if you like. For the panel, avoid plastic, which might melt, and metal, which might act as a heat sink. — *73, Bryant Julstrom, KC0ZNG, 1945 30th St S, Saint Cloud, MN 56301,* **kc0zng@arrl.net**

Cleaning Rotary Switch Contacts

My old Heath V-7A vacuum tube voltmeter (VTVM) had stopped working due to intermittent rotary switches. I fixed it by applying Noalox Anti-Oxidant Compound (**www.idealindustries.com**) to the contacts with a toothpick. Noalox is a blue, thick semiliquid suspension of zinc particles that will remove corrosion from the switch contacts. It's available at home improvement centers and electrical/commercial supply companies.

To clean the switch, I first removed the front and rear pieces of the V-7A VTVM cabinet to give me access to the two front panel switches. I removed the nozzle on the Noalox bottle and inserted a long screwdriver into it. I stirred the contents for a few minutes, as the suspension tends to separate over time.

I reattached the nozzle, which allows a measured amount of the Noalox to be dispensed from the bottle, and put about a half teaspoon of the Noalox into a clean saucer.

Using a toothpick, I applied the Noalox to the contacts, turning the shaft to make sure all of the parts had a thin coating. When I finished, I rotated both switch shafts several times to make sure the coverage was complete.

Note: When using Noalox on switches in an RF circuit, apply it sparingly. The Noalox runs and can detune a circuit. — *73, Thomas Webb, W4YOK, 3533 Teakwood Ln, Plano, TX 75075,* **sam9lives1@verizon.net**

[1]Measures, AG6K, "Parasitics Revisited" — Parts 1 and 2, *QST*, Sep 1990, pp 15 – 18, and Oct 1990, pp 32 – 35.

Plating Nichrome Wire

I was given a Heathkit SB-200 amplifier that was mostly stock. After doing online research, I decided to do most of the modifications described by Robert Norgard, KL7FM (**www.westlawn.net**). One of the modifications, originally described by Rich Measures, involves replacing the anode chokes with low Quality Factor (Q) parasitic suppressors made of Nichrome resistance wire.[1]

It is very hard to solder this wire. I tried acid core solder and various fluxes, but nothing worked. I contacted Caswell Inc (**www. caswellplating.com**), a company that specializes in small-scale plating systems, to find a way to put a solderable interface metal on the wire. They suggested their brush copper plating kit, which I purchased. [Another option would be to attach crimp-on connectors to the Nichrome wire ends. — *Ed.*]

The results of my first test were disappointing. The copper layer was very dark, suggesting copper oxide rather than pure metal, and it proved to be unsolderable. The instructions suggested using the plating solution for immersion plating rather than brush plating. I decided to give that a try, using the supplied 4.5V power adapter. It worked, sort of — the resulting copper layer was very soft, and more like mud than metal. After more research, I tried using a lower voltage for the process from two nickel-metal-hydride (NiMH) cells. This resulted in a perfect layer of pure metallic copper plated onto the wires (see **Figure 17**).

Sand the leads with fine sandpaper to make sure they are clean, and avoid touching them

Figure 17 — Using copper electroplating techniques; you can tame Nichrome wire and give it an easily solderable copper coating. [Bennett Wilson, AF2RF, photo]

Figure 16 — A simple aid for cleaning the terminals on recycled parts. [Bryant Julstrom, KC0ZNG, photo]

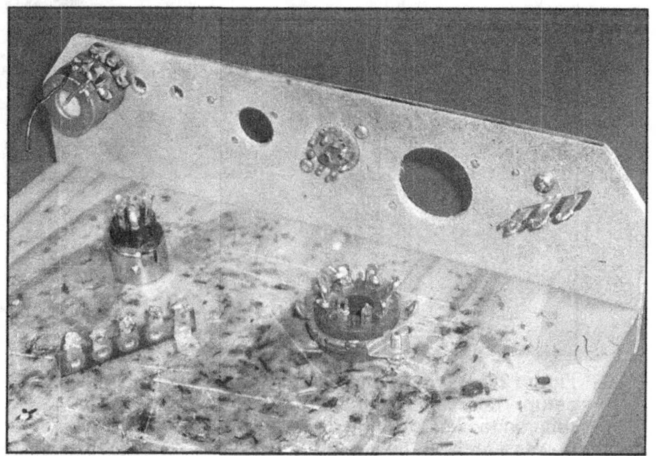

afterward. My anode is a piece of #12 AWG wire and the cathode is the inductor. I let the leads remain in the plating solution about 30 – 45 minutes to build up a thick layer of copper on the wires. Plate *only* the ends of the leads that will be soldered to the anode resistors — not the entire choke, which would defeat the whole purpose of this process. You can purchase just the plating solution if all you plan to do is plating of this type. — *73, Bennett Wilson, AF2RF, PO Box 210321, Brooklyn, NY 11221-0321,* **adastra_2001@yahoo.com**

HW-100 Dial Indicator Repair

A number of years ago I obtained a Heathkit HW-100 transceiver at a price I couldn't pass up (free). It had many problems, but the most significant cosmetic flaw was the original dial cover. It had become detached from the front panel (it was only attached with double-stick tape) and had been rubbing against the dial, causing part of the vertical index line to be scratched off.

Since the front panel had three extra holes from a receiver incremental tuning modification that was never finished, I decided to replace the dial cover with one that is bolted in.

The material I used was a 4 × 2 inch piece of Plexiglas cover sheet for a welder's helmet, but any piece of hobby shop Plexiglas will work. Instead of using paint, I decided to scribe the index line using an old-fashioned can piercing tool of the kind used to open condensed milk. In place of this tool, a narrow screwdriver, wood chisel, or a pair of shears may be used. Whatever tool you use, the end must be flattened to make a cut that is wide at the bottom. If you use a tool with a knife edge, the plastic will crack.

To scribe the plastic, place it on a non-marring surface (I used a washcloth, but several layers of paper towels should also work) and place the guide bar on top so the plastic is held very tightly. As always, practice this on scrap material first.

If you use a chisel, lay the flat side against the guide bar and keep it tight along the side of the bar. The corner of the scribe must be sharp at a right angle to the direction of pull. Set up the work so you can pull the tool to scribe the line. Pulling the tool will give you much better control of the process. Whatever tool is used, the bottom of the scribed line should be flat across the width of the line to make it easy to see.

Make several light passes, each of which should sound a little like hook-and-loop fastener strips being separated. The strokes must be slow and deliberate or the plastic will melt leaving a ridge on either side of the cut that is hard to remove. Check every few passes until you are satisfied with the line. When you are satisfied, carefully clean along the scribed line with your fingernail and mount the cover with the line facing the dial (see **Figure 18**). — *73, Paul Willsie, KC2REN, PO Box 228, Randolph, NY 14772,* **kc2ren@arrl.net**

Thoriated-Tungsten Filament Rejuvenation

There are some older tubes that use thoriated-tungsten filaments. The tungsten generates the heat and the thorium emits the electrons. Thorium is hard to get, so recent tubes have the thorium atoms as a coating on the outside of the filament, while the older tubes have thorium mixed throughout the filament. The older ones are the ones that can be rejuvenated.

As the tube works, the thorium near the outside of the wire is depleted, causing the electron emission level to drop. At that point, the tube must be rejuvenated or replaced.

The big-picture procedure is to run the filament wire above its rated voltage to allow migration of the thorium atoms to the wire surface and then run the filament at its rated voltage to settle it.

First, some cautions — it is important not to jar the filaments when hot. To prevent this, I attached the sockets to a board that I then placed in a vise, as shown in **Figure 19**. It is

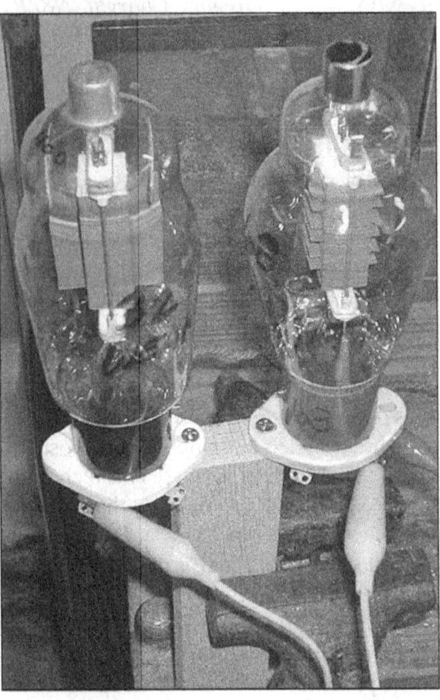

Figure 19 — A close-up of the tube support. When being rejuvenated, the tubes must be kept upright and not subject to any vibration during the heating process. [Patrick Hamel, W5THT, photo]

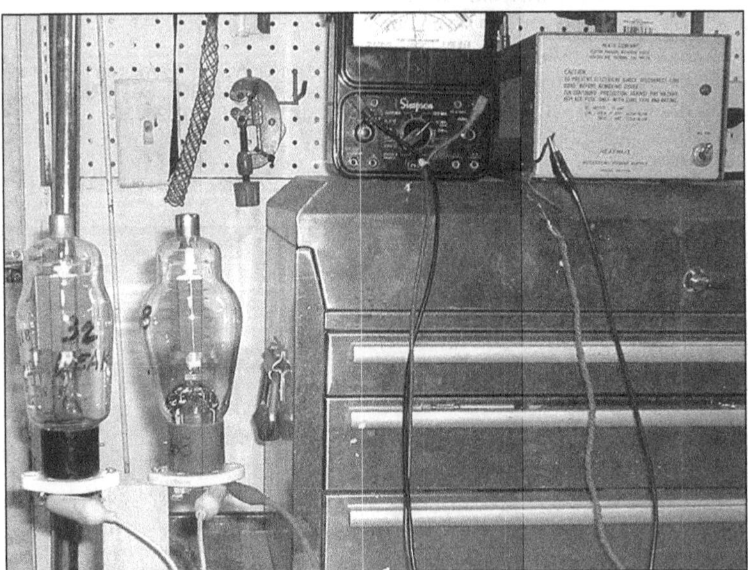

Figure 20 — Older tubes like these 811s can be rejuvenated using this simple setup. The weak 811 on the right is having its emission level boosted. The tube on the left is acting as a ballast for the filament voltage supplied by the power supply at the upper right. [Patrick Hamel, W5THT, photo]

Figure 18 — The finished dial cover bolted to the HW-100 front panel. [Paul Willsie, KC2REN, photo]

also important not to disturb the dimensions between the filament and grid, so make sure to keep the tube vertical during the heating process.

I had some 811 tubes that needed rejuvenation. The 811 uses a 6 V filament. I hooked two 6 V tubes in series across a 12 V, 6 A power supply. This arrangement uses one tube as a ballast for the one being rejuvenated (see **Figure 20**). Other tubes will require you to calculate the proper ballast resistance to use in series with the tube or you can use a variable-voltage power supply.

Mount the tube securely in a vertical orientation, set a timer for 30 seconds, and turn on the voltage, making sure that both tubes light. Short out the ballast tube or resistor with a jumper wire for 30 seconds, then run the pair at rated voltage (12 V total) for 30 minutes. Finally, allow the tubes to cool down for 5 minutes after shutting off the power and before moving and retesting the tubes.

You don't need a lot of fancy equipment to rejuvenate old tubes, just a tube checker, sockets, power supply, and jumper leads. I have successfully raised the emission level of old 811s from midscale yellow well into the green range using this technique (as measured on my old Heathkit emission tester). New tubes have always failed to rejuvenate.

One more tip — if an old tube from the shelf shows up as "gassy," 8 – 10 hours run-ning with just filaments at normal voltage in this type of fixture may reheat the getter material and save the tube from the recycling bin. [The "getter" is a material placed inside the tube to absorb extraneous gas, much as a dehumidifying packet in a food package absorbs humidity. — *Ed.*] — *73, Patrick Hamel, W5THT, 1157 E Old Pass Rd, Long Beach, MS 39560,* **w5tht@arrl.net**

Repairing IF Transformers

I have repaired many radios over the years and one of the areas that has always been a problem is the tuning of IF transformers. These IF "cans" contain coils wound on a tube of insulating, non-magnetic material lined with a coating of wax. A ferrite core, or "slug," is screwed into the tube to allow for the adjustment of the coil's inductance. Over time, the wax lining can harden or break down completely, making it difficult or impossible to adjust the transformer.

In the days of vacuum tubes, these coils were large, and repair technicians would simply insert a small rubber band down the inside of the tube and then screw in the slug. The rubber band tightened the slug in the tube, and provided a gripping surface for its threads. This worked fine at the time, but today's transformers are too small for that technique.

I tried a number of ways to solve the prob-

Figure 21 — A thin strip of Teflon pipe tape in this Kenwood TS-940S IF transformer allows the worn slug to be adjusted. [Ben Fisher, K9BF, photo]

lem. One method is to remove the slug, apply a small amount of RTV silicone sealant to the inside of the tube with a cotton-tipped swab, wait a day for it to harden, and then reinsert the slug. This worked, but it is messy and you have to wait 24 hours to continue the project.

These limitations led me to develop a better method, which is to insert narrow strips of Teflon tape — the type used by plumbers for sealing pipe threads. I use a small pair of scissors to cut a narrow strip of the Teflon tape, which I then insert into the tube. (see **Figure 21**). Sometimes it takes more than one to get the job done, but it works great. — *73, Ben Fisher, K9BF,* **k9bf@arrl.net**

Troubleshooting and Test Gear

Boat Anchor Noise

There had been a steady S9+ noise level at my station on all HF bands. The problem was eventually located after much snooping. I had purchased a Viking Valiant transmitter and an NC-300 receiver several months earlier with the hope of getting the old boat anchors on the air and trying 40 meter AM. Both units were off but were plugged into the 120 V line. When the Valiant was unplugged the noise disappeared! Television interference was a big deal in the '50s when these radios were in their heyday, so line cord bypass capacitors were common. The Viking has several bypass capacitors on the 120 V line *before* the power switch. One or more was leaking and creating quite a lot of noise, even with the unit off. So for any of you boat anchor fans with high noise levels, this might be the cause. — *73, Dick Dabney, K6BZZ, 42561 W Jail House Rock Ct, Maricopa, AZ 85138,* **k6bzz@arrl.net**

Wax Finds Waning Transistor

Here is a tip I think anyone who works on transistorized RF amplifiers will appreciate. This is a quick test to help determine the condition of RF output transistors. I'm sure most of us who have serviced amplifiers will relate to this scenario. You have an amplifier and suspect the RF output transistors may have suffered damage because you observed low RF output or a high quiescent current.

Quiescent current is the current measured when the amplifier is powered-up and keyed with no RF input. Basically an unmodulated SSB signal or an AM or FM signal with the drive (exciter) turned all the way down. Just because an amplifier is producing good RF output, doesn't mean the transistors are working properly. They can be partially damaged and still produce output, yet will draw excessive current at idle when keyed up. This is easily detected when you can't bias the amplifier properly or it gets hot faster than usual. Abnormally high quiescent current usually indicates an emitter-collector short on RF transistors. Base-emitter or base-collector shorts are often obvious as they tend to burn-up resistors in the drive circuitry that feeds the RF output transistor's base.

The normally accepted way to check collector current is to lift the collector lead from each output transistor and measure the current with a dc ammeter. This can be a very tedious process. Most RF power transistors use large foil-type component leads. These leads require a lot of heat to melt the solder so the lead can be lifted from the circuit board. Not only does this require the skillful use of a very large soldering iron, you also run the risk of damaging a good transistor and the circuit board it is mounted to.

I used to stick my finger down into the amplifier and physically feel the temperature of each transistor. This very low-tech method sometimes worked but seldom gave me much more than a burned finger and lots of questionable transistors.

Let me set the scene here. I was working on a Metron MA1000B linear amplifier whose output was down a bit and had high quiescent current. The Metron is a 12 V HF amplifier that draws about 85 A on voice peaks. It was manufactured in the late '80s with a rated output of 600 W. Its quiescent idling current should be less than 2 A.

I knew the amplifier had at least one damaged transistor and maybe more. I normally like to replace all the transistors with a matched set when one goes bad. This is the best way to repair any amplifier that uses multiple output transistors or tubes, in my estimation. The problem was the Metron uses eight rather expensive RF output transistors (around $35 each). Not only that, the amplifier still worked producing over 500 W on 75 meters. Not bad, but that quiescent idling current was very high (about 20 A) and would climb even higher as the transistors heated up, approaching a runaway condition.

I thought to myself, there has to be a better way to troubleshoot all these transistors rather than going through the pain of lifting each collector lead or feeling the transistors for excessive heat. Sure if you're lucky enough to have access to a data-logger that would work, although still a time consuming setup. Then I thought maybe I'll go buy one of those temp-gun digital thermometers. After thinking about this approach through, I realized it would also be time-consuming and have a potential for error.

To accurately measure all of the individual transistors with a temp-gun would involve moving components aside that blocked direct access to the top of the transistor. Not only that, I could only check one transistor at a time with a temp-gun or risk causing further damage due to the long key-down time required to check all eight RF output transistors. Another temp-gun method would

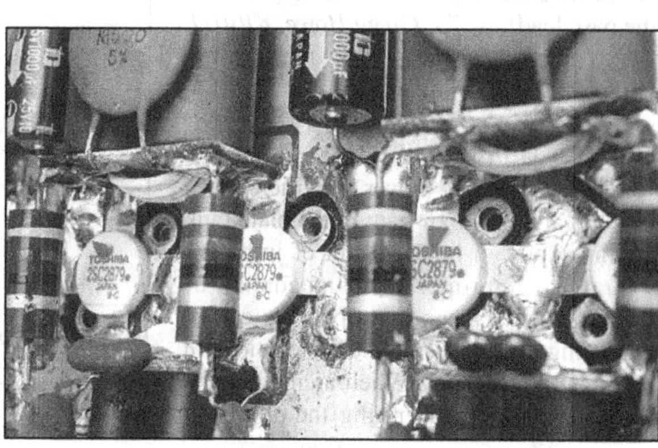

Figure 1 — To begin the test, place small chips of candle wax on each of the transistors to be tested. Here the wax chips are visible at the 12 o'clock position. [Gregg Howe, KI6IUJ, photo]

Figure 2 — The transistor on the right is getting hot enough to melt the wax while the one on the left is remaining cool. [Gregg Howe, KI6IUJ, photo]

require an exact key-down time for each transistor under test, with a rest period allowing things to cool before the next transistor could be tested. Also very time consuming.

After pondering the situation for a while I had an idea. What if I shaved off some candle wax and placed a shaving on each transistor? This would allow a quick temperature comparison of all eight transistors at the same time under the exact same conditions. I used red candle wax to aid the visual effect (see **Figure 1**).

Bingo! It worked. The wax melted on the bad transistor pulling excessive quiescent current (see **Figure 2**). I replaced the bad transistor, the quiescent current dropped to within specification and the amplifier stayed cool. Although the quiescent current was now correct, the power output was still down a bit. On to part two of the wax test.

In this next test, we're now looking for a weak or blown transistor that is not producing an acceptable level of RF output power. This test is done with RF drive into the amplifier. In most amplifiers, transistor or tube, the transistors that get hot or the tubes that glow red are the ones producing power, while the ones that run cool are suspect. So once again, with wax shavings sitting on top of the transistors, I applied RF power, this time with the amplifier keyed up. I expected the wax to melt on the good transistors indicating that, yes they are getting hot producing power. The cold ones (no melting) are weak or blown (assuming they're receiving drive at the base lead). — *73, Gregg Howe, KI6IUJ, 27221 Westridge Ln, Laguna Hills, CA 92653, s9radio@aol.com*

Cure for Amplifier Arcing

Tube-type linear power amplifiers often develop arcing somewhere in the plate tank circuit. This can be the result of improper operation, improper load impedance or circuit faults. Depending upon the cause, the arcing can occur at multiple locations in the plate tank circuit. When arcing occurs, immediately stop operating the amplifier

to determine and correct the cause, since extensive damage can occur if the arcing continues. This hint explains a form of permanent repair to the damaged capacitor plates that is effective once the problem causing the arcing is corrected.

When a tuning capacitor arc occurs in a high power tank circuit, huge currents flow through the arcs causing melting of the edges of the aluminum plates. Typically, this occurs when operating on the higher frequency bands, when the plates are mostly or completely unmeshed. Once the plates have melted, the edges become deformed and develop sharp spots instead of the factory smooth edge. The deformed shape decreases the spacing at that location in the capacitor. The sharp edges at the melted points increase the local electric field density and both of these changes further reduce the breakdown voltage of the capacitor well below its original rating. This is why once the capacitor arcs, it will continue to arc, even after the original fault is corrected. For normal operation, the capacitor must be restored to its original breakdown rating.

The usual method to repair a damaged capacitor involves using small jeweler's files to carefully file the raised edges and restore the edge to a smooth shape, restoring the spacing and eliminating sharp points. Carefully remove these filings since they can cause further arcing. Often it is nearly impossible to do an adequate job and the capacitor continues to arc.

To make a successful repair first perform the usual repair discussed above. Then, before the amplifier is returned to service, applying Kapton (polyimide) tape to the damaged capacitor edges can be effective in preventing further arcing. Kapton tape is available from toroid and balun kit manufacturers such as Amidon (**www.amidoncorp.**

com). It has an extremely high breakdown voltage with a single 2.7 mil thickness able to withstand 7000 V.

Use a piece of tape with a width at least double the spacing of the capacitor plates. Cut a length that is slightly longer than the radius of the damaged plates. Set the capacitor so it is completely unmeshed. Using tweezers, center the tape on the edge of the damaged plate so it covers the damaged areas all the way from the shaft extending past the outer edge of the plate (see **Figure 3**). Then fold the tape down between the neighboring plates and stick them to the damaged plate.

When properly applied, the tape should cover the entire damaged area and be stuck to the plate on both sides. Stick the tape to itself where it extends past the outer edge of the plate and cut it off with scissors, leaving about 0.1 inch extending past the edge of the plate. Check that this will clear the fixed parts of the capacitor throughout its range. It is not necessary to apply tape to every edge; only the damaged plates need to be covered.

The very high breakdown voltage of the tape will prevent any arcs from starting in the damaged areas and help restore the capacitor to its original breakdown voltage. I've used this method on an Amp Supply LK-500ZB amplifier, which did yeoman's duty through both the CW and phone sections of the ARRL® DX Contest without arcing. The addition of the Kapton dielectric does not appreciably change the capacitor tuning characteristics. — *73, Tony Brock-Fisher, K1KP, 15 Webster St, Andover, MA 01810-1109,* **k1kp@arrl.net**

Curing Erratic Tuning

Most ham gear made in the past 15-20 years tunes use some sort of rotary encoder. These encoders generate digital pulses that are fed either directly or through some sort of buffer to the rig's microprocessor, telling it to tune up or down.

I purchased an Alinco DR-610T at a hamfest and found that the tuning via the main tuning knob was very erratic. I would try to increase (or decrease) the frequency (or memory channel) and it would bounce up, then right back to the same frequency or, in some cases, it would decrement instead of incrementing. Playing with the tuning knob repeatedly might finally get the frequency set where it was supposed to be. I decided to investigate the problem and see if a fix was possible.

The Alinco uses a common type of rotary encoder. More information on how these work can be found at **www.ubasics.com/adam/electronics/doc/rotryenc.shtml**. I am not sure what failure mode causes the erratic operation, but, instead of replacing the control, I was able to devise a fix. Through experimentation, I found that a couple of 0.1 μF capacitors added across the

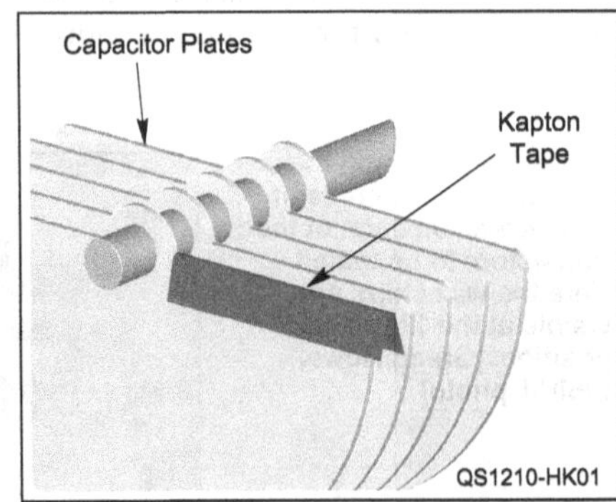

Figure 3 — Apply the tape over the damaged edge of the air-variable capacitor to help prevent further arcing. [Tony Brock-Fisher, K1KP]

Figure 4 — The capacitors, visible on the right, are mounted on the back of the control head circuit board, behind the rotary encoder. [Kirk Ellis, KI4RK, photo]

Figure 5 — This is a close-up view of the capacitors mounted to the back of the encoder. [Kirk Ellis, KI4RK, photo]

rotary controller's terminals corrected the problem. Too large a capacitor stopped the tuning altogether and too small had no effect on the problem.

I soldered the two 0.1µF capacitors between each output terminal and ground (see **Figures 4** and **5**) on the opposite side of the PC board from where the encoder is located. The encoder has an "up" and a "down" terminal, and a common terminal, which in this radio was connected to ground.

On your rig, consult the schematic and service documentation to determine the best way to access the terminals and install the capacitors. You may need to experiment with different values of capacitors, though 0.1µF should be a good starting point. After soldering the capacitors in place, I reassembled the radio and it operated like new. — *73, Kirk Ellis, KI4RK, 203 Edgebrook Dr, Pikeville NC 27863*, **ki4rk@arrl.net**

Rescue Plug Saves Battery

I pulled out my MFJ-259 antenna analyzer to do some testing only to discover the power switch had been left on, draining the batteries. In addition to the annoyance of finding a dead unit and the expense of new batteries, there is the danger that drained batteries may leak. This could potentially damage the unit.

A little investigation showed that switching between the batteries and an external power supply is managed by the 2.1 millimeter power jack. As with many similarly powered products, when a plug is inserted into the jack, the electrical path between the batteries and ground is interrupted. Knowing this, it is a simple matter to make a "dummy plug." Just dig a suitable plug out of your junk box or buy one through an electronics supplier. Insert the plug into the external dc socket before storage and you'll never have accidentally drained batteries again. I put the plug on a keeper string (using the plug ground/sleeve wire, not the power/tip wire) attached to one of the antenna jack screws as a simple reminder and to avoid losing the plug during use. **Figure 6** shows the unit with the dummy plug inserted.

This workaround has also been verified on an MFJ-269 (with the J3 Jumper in the off, or non-rechargeable, position). It is possible that this technique might work with other MFJ analyzers, but this hasn't been tested. I have also used this technique to protect the batteries of other products that use both internal and external power sources, although not all will work correctly (depending on how the power supplies are switched/controlled). — *73, Mat Breton, AB8VJ, 35229 Rosslyn St, Westland, MI 48185*, **ab8vj@arrl.net**

Homebrew Dip Coils for MFJ-259/269

Dip coils for the MFJ-259/269 are a very useful accessory for these ever-popular antenna analyzers. The homebrew coils I am describing here are inexpensive and easy to build. Plus, unlike the commercial versions, they are more compact and have PL-259 connectors, so no adaptor is needed.

I made the coil forms from a 1⅛-inch long piece of ⅜-inch diameter wooden dowel, which fits into the threaded end of a PL-259 connector. The coils are 16 AWG enameled wire. I drilled a ⅛-inch hole through the length of the dowels for the coil end going to the center pin. A lengthwise slot made with a Dremel or a hacksaw is just as good.

To assemble the coils, first scrape one wire end, insert it through the drilled dowel, and solder it to the tip of the PL-259. Fold the wire back down the side of the dowel and hand wind the necessary number of turns onto the dowel. Then cut, scrape, and solder the other end of the wire to the skirt of the PL-259. Lastly, cover the assembly with heat shrink tubing.

The coils do not need to be an exact value; they only provide coupling between

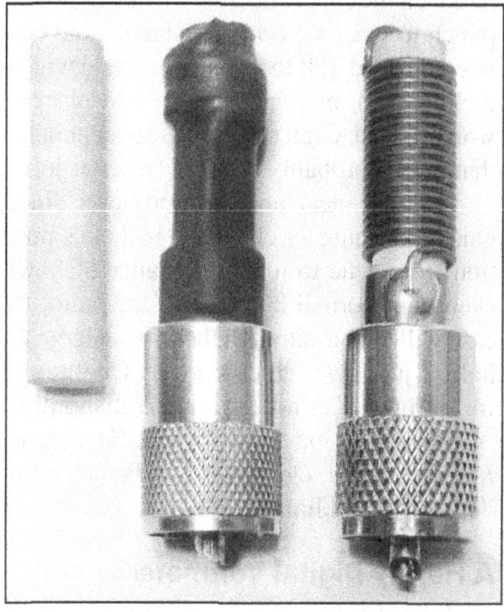

Figure 7 — This is a set of homebrew dip coils for the MFJ-259/269 analyzer. The completed high range coil is in the center; the low-range coil on the right is uninsulated to show construction details. [John Portune, W6NBC, photo]

Figure 6 — By inserting the appropriate dc plug into the external dc jack, you can disable the internal batteries preventing accidental discharge and possible corrosion damage. [Mat Breton, AB8VJ, photo]

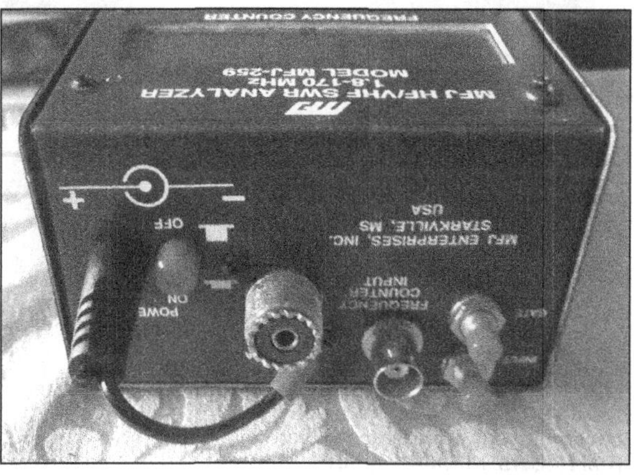

the resonant circuit being tested and the 50 Ω Wheatstone bridge in the analyzer. An inductance of 1 µH is satisfactory for the lower analyzer ranges and 0.25 µH for the higher. On the dowel form, use five turns for the high coil and 19 for the low coil (see **Figure 7**). — *73, John Portune, W6NBC, 519 W Taylor St, Spc 111, Santa Maria, CA 93458-1029,* **w6nbc@arrl.net**

Does the Case Fit?

I was about to give up on my handheld transceiver after many complaints that my signal was strong but my audio was "mushy." I take good care of my radio and even keep it in a protective leather case to prevent damage and keep debris out of all those little holes. One day, while admiring this little marvel, I noticed the protective vinyl window over the display was blocking more than dirt. Although the window had holes over the speaker, the pinhole for the microphone was behind the vinyl, which was sure to muffle my voice.

The solution was easy. Using a marking pen, I put a dot over the **MIC** pinhole. Next, I removed the cover and used a handheld paper punch to make a ³⁄₁₆ inch diameter hole centered on the dot. It took a little effort to wiggle the punch in position, but once in place, it worked great. Craft stores sell leather punches that would probably do an even better job.

I've had clear audio reports ever since. One other thing — check the pinhole's position from time to time. As **Figure 8** shows, even with normal handling of the radio, the case will shift around. Rather than enlarge the hole, I just give the case a tug. One thing's for certain, I'm no longer heard shouting, "Can you hear me now?!" — *73, Hal Rogers, K8CMD, 7811 Dogwood Ln, Parma, Ohio 44130,* **k8cmd.hal@arrl.net**.

A Handy Digital Voltmeter

Take a quick look around your shack. How much of your equipment runs on batteries or a dc power supply? If you're like most of us, the answer is most of it. You probably also have a multimeter for checking these power sources. So why would you need another meter? An extra voltmeter can be quite handy for monitoring the condition of those rechargeable batteries you keep around for emergency power or the performance of your solar energy system.

The EZ-Volt is an accurate digital meter you can construct in just a couple of hours. It's based on the Surmen V20D LED digital voltmeter module (**www.universal-radio.com/catalog/meters/0794.html**). This module uses only 10 – 20 mA and can be powered by the circuit it's measuring or a separate power source.

The EZ-Volt meter will directly measure any dc source from 4 – 30 V (four volts is the minimum required to operate the meter's electronics). The schematic is quite simple, as shown in **Figure 9**. The dc power from the device being measured is applied to the Vcc and GND terminals of the meter module through D1, a 1N914 switching diode, which protects against accidental polarity reversal. The meter module reads the dc voltage placed on the Sense terminal, which is connected to the dc input through R1, a 1000 Ω, ¼ W fixed resistor, which provides additional polarity reversal protection.

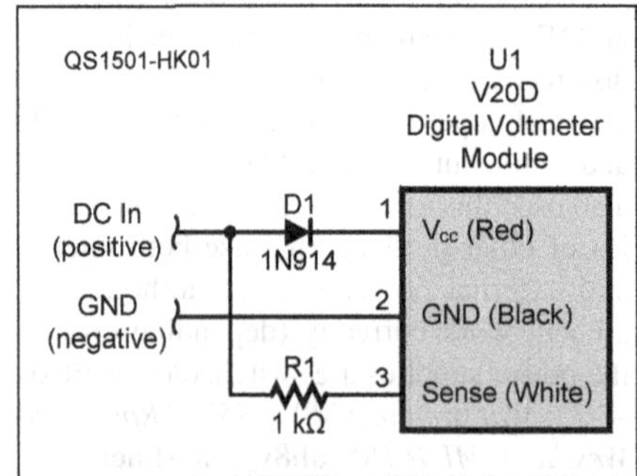

Figure 9 — Schematic of the EZ-Volt Digital Voltmeter.

Construction

I constructed this meter in about an hour's time. There's nothing critical here — just make sure to insulate your wiring with heat shrink tubing or electrical tape before final assembly. **Figure 10** is the inside view of the meter that I built in a small RadioShack plastic project box. This small DVM module lends itself to a variety of other mounting arrangements. The diode and resistor are within the black shrink-wrapped **Y** connection that feeds the red and white meter module wires. An Anderson Powerpole connector provides the dc input connection. All the components are simply hot-glued in place.

Operation

Simply connect the meter's leads to any dc source between 4 and 30 V and the display will show the voltage. The display only updates about twice a second, so this meter can't be used for checking rapidly changing voltages. — *73, Tom Wheeler, NØGSG, 10724 Horton, Overland Park, KS 66211,* **tom.n0gsg@gmail.com**.

DMM Field Strength Meter

This small project converts an inexpensive digital multimeter (DMM) into a field strength meter. I had received the DMM as a giveaway and found the rest of the parts in my junk box.

I soldered up the detector circuit (see **Figure 11**) on a piece of perforated board ³⁄₈ × 1¾ inches long and insulated it with electrical tape across the bottom. Once I had finished the field strength circuit board, I tested it and wired it inside the case of the multimeter (see **Figure 12**). I then drilled a hole in the center of the top of the case for the antenna socket and a hole in the right side of the case for a small toggle switch. The board's ground wire was soldered to the meter ground (common); the board's positive wire was soldered to the toggle switch and then to the meter's + dc side.

Figure 8 — Punching a "mic hole" in your handheld transceiver's protective vinyl cover may solve that low audio problem. [Hal Rogers, K8CMD, photo]

Figure 10 — The interior of the EZ-Volt. [Tom Wheeler, NØGSG, photo]

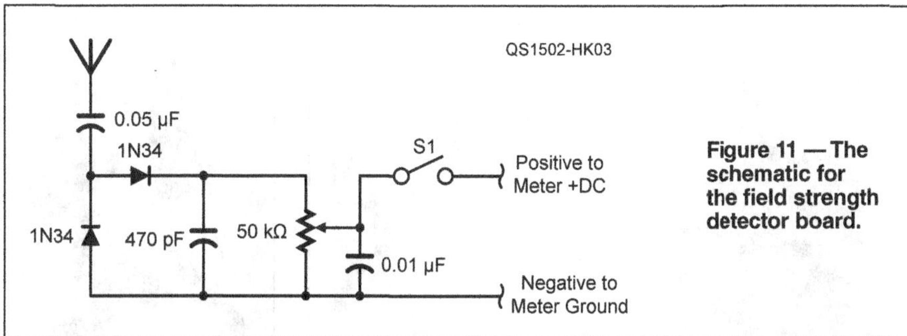

Figure 11 — The schematic for the field strength detector board.

Figure 12 — The detector board mounted inside the multimeter case. The bottom of the board is insulated with plastic tape. [Richard Russo, KB3VZL, photo]

Figure 13 — The original multimeter and its upgraded companion. [Richard Russo, KB3VZL, photo]

When the toggle switch is in the **OFF** position the meter works normally. When the toggle switch is in the **ON** position with the antenna connected, you can use the dc voltage scale to measure field strength (see **Figure 13**). When I tested the unit it worked perfectly. I now have a no-cost, scalable field strength meter with digital readout. — *73, Richard Russo, KB3VZL, 105 Colonial Ave, Norristown, PA 19403,* **kb3vzl@arrl.net**

Reverse Sawtooth Generator

Figure 14 is the schematic for a very simple- to-build reverse sawtooth generator, which is better known as a reverse linear ramp generator. It is made of discrete components and can be used in oscilloscopes, spectrum analyzers, or other projects as a sweep generator. In comparison to a standard sawtooth generator, this design can be used to trigger the oscilloscope and at the same time start the trace scan. Of course, the trace scan is reversed in this mode.

The final two transistors form a complementary Darlington current amplifier so that larger loads may be driven without overloading the generator. The frequency (sweep rate) of the generator can be altered by changing the 1 nF capacitor. You could also install a rotary switch at this point to switch in different value capacitors, which will increase the operating range of the generator.

The linearity of the generator is very good for such a simple circuit (see **Figure 15**). With a 12 V input voltage, the output is about 9 V.

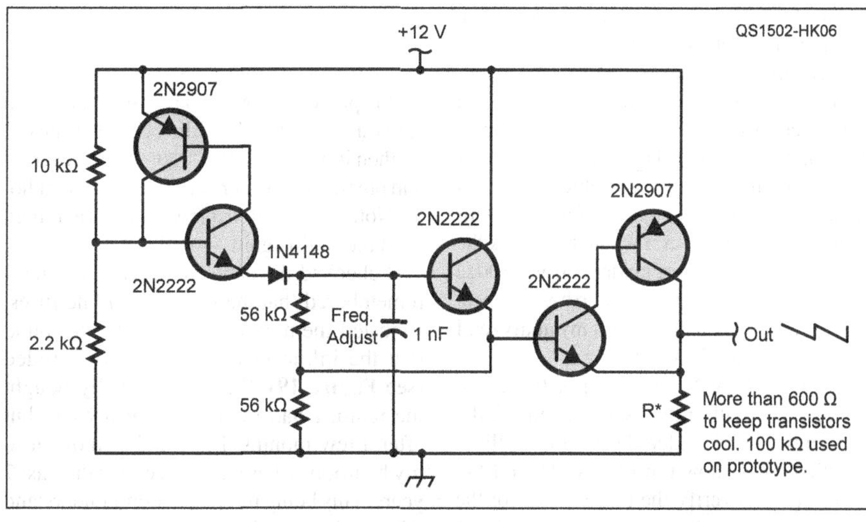

Figure 14— The schematic diagram for the sawtooth generator.

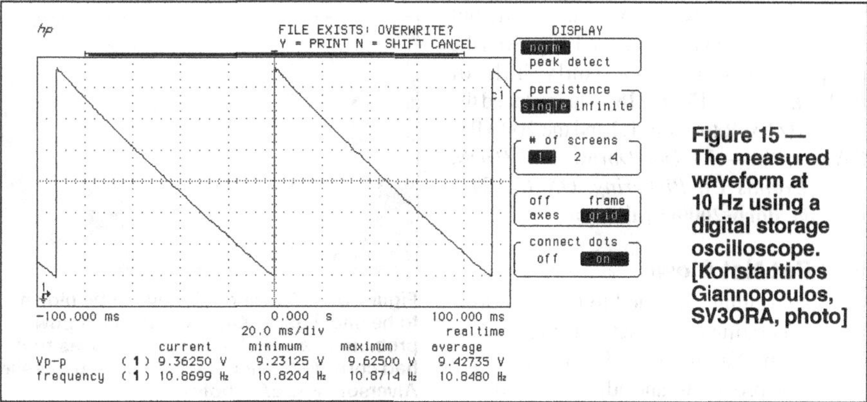

Figure 15 — The measured waveform at 10 Hz using a digital storage oscilloscope. [Konstantinos Giannopoulos, SV3ORA, photo]

The lower edge of the waveform shown is at about 1 V and the higher edge at about 10 V. — *73, Konstantinos Giannopoulos, SV3ORA, Ethnikis Antistasis 26, Arahovitika-Rio-Patra, 26504, Greece,* **sv3ora@qrp.gr**

Meter Sock

I have several digital volt-ohm meters (VOM) that acquired scratches or scuff marks on their meter faces through general use and storage. I have devised a solution that is inexpensive and very effective in protecting these easily scratched faces.

I take an old sock and cut off the bottom, slide it over the meter face, and then cut the top to fit. I simply slide this "meter sock," over the meter face when it's not in use. Using this method, I have eliminated all scratching and scuffing and my meters look pristine. If the "meter sock" is made longer to cover the complete VOM, the control buttons can also be protected. — *73, Michael Janis, KE3OQ, 149 Silo Cir, Nazareth, PA 18064,* **mjanis@ ptd.net**

Twinlead Analyzer Adapter

Today's antenna analyzers are quite versatile instruments. In addition to measuring antenna impedance and resonance, they can be extremely useful in tuning coaxial stubs, locating cable faults, and measuring cable characteristics such as velocity factor (VF). The manuals for most analyzers detail how to perform these measurements.

The analyzers are usually fitted with a Type N connector, to which one must connect the item being tested. This is usually fine for testing antennas, as it is possible to connect an adapter to the Type N plug for other connector types. However, I wanted to check the VF on some samples of ladder line and 300 Ω twinlead, but was unsure how to best connect these to the Type N female on my analyzer. I finally devised this adapter.

I took a Type N male plug of the crimp variety and, with a hacksaw, sawed off the back end. Then I soldered some mini-alligator clips on as shown in **Figures 16** and **17**.

In order to verify the performance of the clip connections, I connected a ½ W, 50 Ω carbon composition resistor to the clips and ran a SWR plot. The SWR at 150 MHz was about 1.5, which was not bad, but I wondered if I could do better. I improved the adapter by cutting off only half of the back end (see **Figure 18**). This reduced the 150 MHz SWR to about 1.2 and the 200 MHz SWR to 1.3! — *73, Don Dorward, VA3DDN, 1363 Brands Ct, Pickering, ON L1V2T2, Canada,* **dorbs2000@yahoo.ca**

Bad But Not Blown

One evening I decided to try some low-power operating on 20 meters using PSK31. I turned my Yaesu FT-857D transceiver on, started the programs, and adjusted my tuner. I

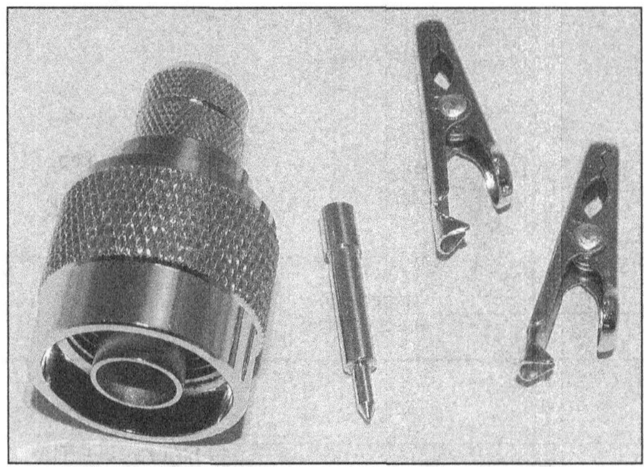

Figure 16 — The Type N connector, its center pin, and the two alligator clips that make up the adapter. [Don Dorward, VA3DDN, photo]

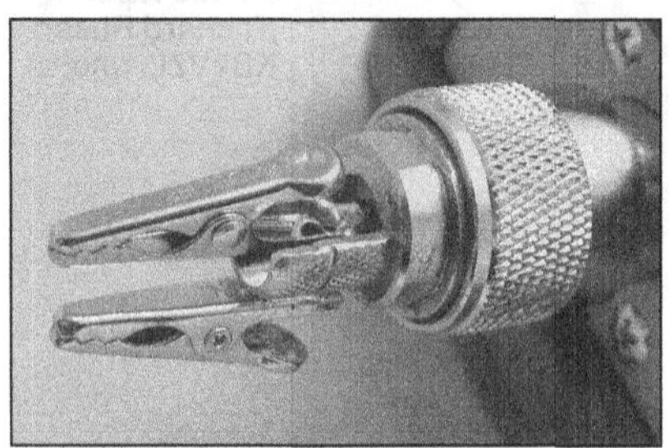

Figure 17 — Version 1 of the modified Type N connector converted into an adapter for checking twinlead-type lines. [Don Dorward, VA3DDN, photo]

Figure 18 — This is version 2 of the adapter, with only half of the back end cut away. This design produced an improved SWR. [Don Dorward, VA3DDN, photo]

set the power at 5 W, then pressed a button to start transmitting. The radio began to transmit — then it powered off. I stopped the program and pressed the power-on button for the radio.

Nothing. I waited a few minutes and then tried the radio again — still nothing.

I checked the cables and all were fine. I remembered that the radio had inline fuses. I located them, pulled them out, and found that the inline fuses were badly corroded (see **Figure 19**). When I originally bought the radio, I had mounted it in my car, but after a few months, I decided to move it to my bedroom where it has been for the past 7 years. This being the case, I don't understand where this corrosion came from.

As you can see, the fuses didn't appear to be blown. I sanded down the contacts, after which I did a continuity check. The check was "iffy." I looked at the fuse under a magnifying lens and saw that the corrosion/ rust had traveled up into the fuse.

I replaced both fuses and the radio powered up. Before I replaced the fuses, I did check the sockets, and I will probably need to replace them in the future. The lesson here is that, when checking your radio, it is a good idea to check the fuses themselves, too. — *73, Leslie Alverson, K4LEA, 111 Ascot Dr, Greer, SC 29651-1005,* **kd4sfd2@ gmail.com**

Dollar Continuity Checker

Looking for an inexpensive continuity tester? Look no further than your local discount store. For $1 you can purchase a device, known as a Window Alert or Intrud Alert, which buzzes if a door, window, or cabinet is opened. This same device can be modified to make a continuity tester.

The alarm works magnetically. A magnet is attached to one side of the door or window and the alarm body is attached to the other side adjacent to the magnet. A sensor in the alarm detects the presence of the magnet and silences the alarm. If you open the door, you move the magnet away from the sensor and the alarm goes off. Without the magnet

Figure 19 — Fuses don't have to be blown to be bad. Leslie, K4LEA, traced his power problem to a set of inline power fuses that had corroded inside the fuse casing. [Leslie Alverson, K4LEA, photo]

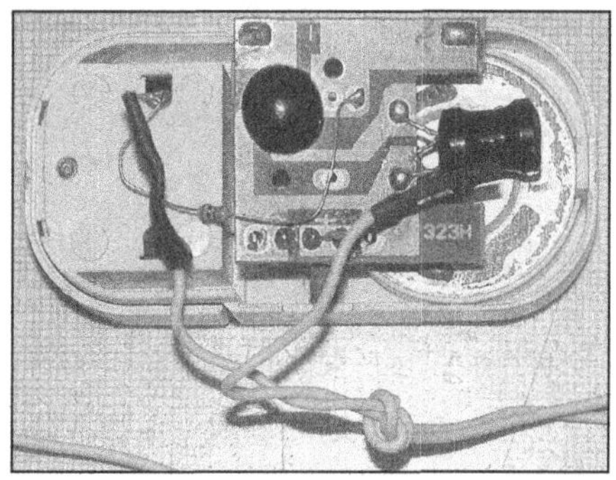

Figure 20 — Cutting the battery lead and extending the ends outside the case turns an inexpensive door alarm into a continuity checker. [Bill Kirk, NJ1X, photo]

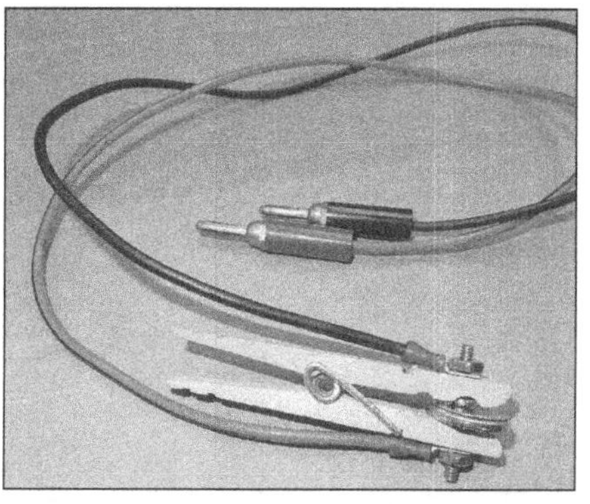

Figure 21 — A clothespin and a couple of screws will make a test fixture for checking button batteries. [Aaron Fineman, W3GIS, photo]

nearby, the alarm is on continuously and controlled only by the power switch.

To convert the alarm into a continuity tester, remove the batteries. Then remove the small screw that holds both sections together and open the case (see **Figure 20**). Cut one of the battery wires (in my unit, one battery wire was a resistor). Don't unsolder the conductors that go to the batteries, because soldering another conductor to either battery contacts is difficult.

Cut a notch in the case side opposite from the switch. This notch should be wide enough and deep enough for two #22 AWG wires. A small soldering iron could be used to make the notch. If an iron is used, remove any plastic from around the notch with a small blade, so the cover will fit securely.

Prepare the test leads by cutting two lengths of #22 AWG wire, 9 – 10 inches long, and strip insulation from all four ends. For ease of use, solder an alligator clip or "mini-grabber" onto one end of each wire. Then tie the two lengths together with a knot about 1½ inches from the other end of the wires and place a ¹⁄₁₆-inch piece of heat-shrink tubing on each side. Solder each knot-end wire to each end of the cut battery conductor. Cover the soldered wires with the heat-shrink tubing and place the knot on the inside of the notch.

Close the cover and replace the screw and batteries. Put the tester in the **ON** position. Without the magnet nearby, the tester is in an "armed" state. Touching the two wires together will supply power to the

alarm and sound the buzzer. Put the switch in the **OFF** position when not using the tester. The bar magnet that comes with the alarm is not used. — *73, Bill Kirk, NJ1X, 17 Bellevue Ave, Winchester, MA 01890,* **wk9879@ verizon.net**

Button Battery Test Helper

When testing small button batteries, such as the CR2032, it is a challenge to hold the test leads on the battery. Here is a very simple test jig for holding the battery while testing. I used a wooden clothespin as the basic fixture

as it provides insulation for each terminal. Take one apart, drill holes in each side, and insert a screw. Then attach a test lead using terminal lugs to each screw and then secure them to the clothespin with a nut. Reassemble the clothespin and the fixture is complete (see **Figure 21**). I put banana plugs on the leads coming off the fixture to plug directly into the VOM. If your VOM has alligator clips on the test leads, you can just leave two pigtails on the terminals and clip onto the pigtails. — *73, Aaron Fineman, W3GIS, 13910 Flint Rock Rd, Rockville, MD 20853,* **w3gis@verizon.net**

Indicator Lamps Help with Remote Troubleshooting

When constructing a device that will be mounted out of reach, it is very helpful to install one or more indicator lamps so you can see the device's status from a distance. This can be especially useful when the device is up on your tower. While building a coaxial antenna relay, I added an LED to show when power is applied to the relay coil (see **Figure 22**). If at some time in the future, I throw the switch and nothing happens, I can go outside and see if power is reaching the box, or if the problem is somewhere closer to the ground. — *73, Al Yerger, K2ATY, 1312 Union Ave, Newburgh, NY 12550-8907,* **k2aty@arrl.net**

Figure 22 — The LED mounted at the upper left serves as a power-on indicator for this tower mounted antenna relay. It is visible from the ground and might save you a trip up the tower. [Al Yerger, K2ATY, photo]

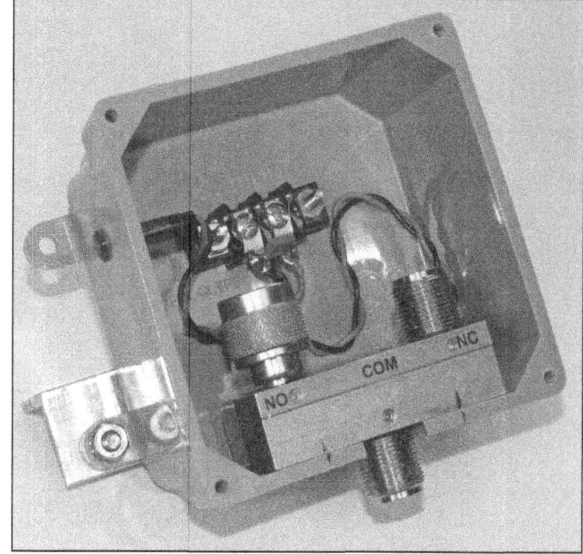

Appendix

RF Connectors and Transmission Lines

There are many different types of transmission lines and RF connectors for coaxial cable, but the three most common for amateur use are the UHF, Type N and BNC families. The type of connector used for a specific job depends on the size of the cable, the frequency of operation and the power levels involved. **Table 1** shows the characteristics of many popular transmission lines,

1 UHF Connectors

The so-called UHF connector (the series name is not related to frequency) is found on most HF and some VHF equipment. It is the only connector many hams will ever see on coaxial cable. PL-259 is another name for the UHF male, and the female is also known as the SO-239. These connectors are rated for full legal amateur power at HF. They are poor for UHF work because they do not present a constant impedance, so the UHF label is a misnomer. PL-259 connectors are designed to fit RG-8 and RG-11 size cable (0.405-inch OD). Adapters are available for use with smaller RG-58, RG-59 and RG-8X size cable. UHF connectors are not weatherproof. Most PL-259s are nickel plated, but silver-plated connectors are much easier to solder and only slightly more expensive.

Figure 1 shows how to install the solder type of PL-259 on RG-8 cable. Proper preparation of the cable end is the key to success. Follow these simple steps. Measure back

about ¾-inch from the cable end and slightly score the outer jacket around its circumference. With a sharp knife, cut through the outer jacket, through the braid and through the dielectric — almost to the center conductor. Be careful not to score the center conductor. Cutting through all outer layers at once keeps the braid from separating. (Using a coax stripping tool with preset blade depth makes this and subsequent trimming steps much easier.)

Pull the severed outer jacket, braid and dielectric off the end of the cable as one piece. Inspect the area around the cut, looking for any strands of braid hanging loose and snip them off. There won't be any if your knife was sharp enough. Next, score the outer jacket about ⁵⁄₁₆-inch back from the first cut. Cut through the jacket lightly; do not score the braid. This step takes practice. If you score the braid, start again. Remove the outer jacket.

Tin the exposed braid and center conductor, but apply the solder sparingly and avoid melting the dielectric. Slide the coupling ring onto the cable. Screw the connector body onto the

Figure 2 — Installing PL-259 plugs on RG-58 or RG-59 cable requires the use of UG-175 or UG-176 adapters, respectively. The adapter screws into the plug body using the threads of the connector that grip the jacket on larger cables. (*Courtesy Amphenol Electronic Components*)

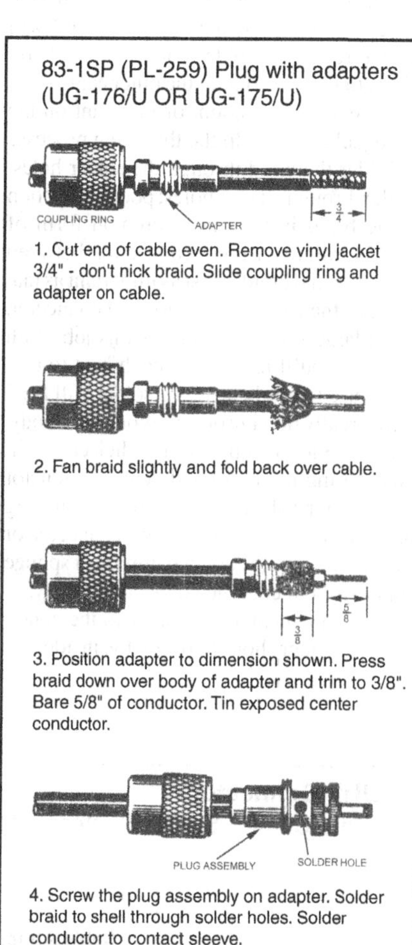

1. Cut end of cable even. Remove vinyl jacket 3/4" - don't nick braid. Slide coupling ring and adapter on cable.

2. Fan braid slightly and fold back over cable.

3. Position adapter to dimension shown. Press braid down over body of adapter and trim to 3/8". Bare 5/8" of conductor. Tin exposed center conductor.

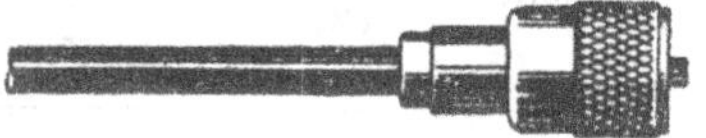

4. Screw the plug assembly on adapter. Solder braid to shell through solder holes. Solder conductor to contact sleeve.

5. Screw coupling ring on plug assembly.

HBK0460

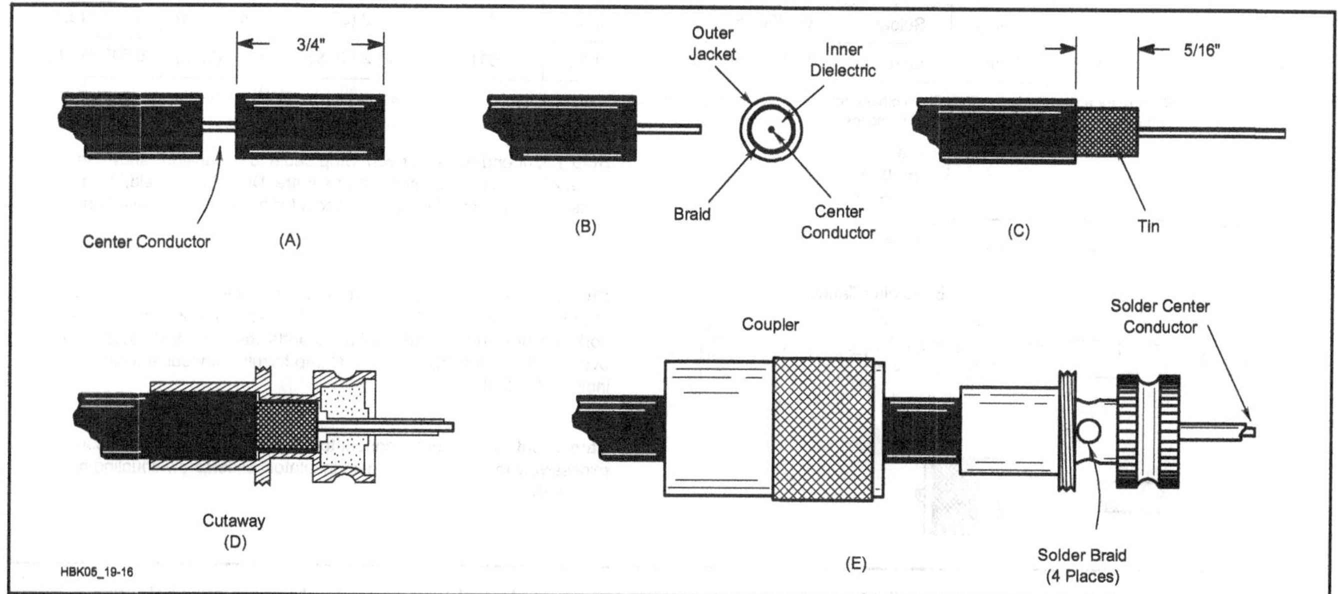

Figure 1 —The PL-259, or UHF, connector is almost universal for amateur HF work and is popular for equipment operating in the VHF range. Steps A through E are described in detail in the text.

cable. If you prepared the cable to the right dimensions, the center conductor will protrude through the center pin, the braid will show through the solder holes, and the body will actually thread onto the outer cable jacket. A very small amount of lubricant on the cable jacket will help the threading process.

Solder the braid through the solder holes. Solder through all four holes; poor connection to the braid is the most common form of PL-259 failure. A good connection between connector and braid is just as important as that between the center conductor and connector. Use a large soldering iron for this job. With practice, you'll learn how much heat to use. If you use too little heat, the solder will bead up, not really flowing onto the connector body. If you use too much heat, the dielectric will melt, letting the braid and center conductor touch. After soldering, to prevent distorting the dielectric do not move the connector or cable until both are cool. A damp rag or sponge can be used to cool the connector quickly.

Solder the center conductor to the center pin. The solder should flow on the inside, not the outside, of the center pin. If you wait until the connector body cools off from soldering the braid, you'll have less trouble with the dielectric melting. Trim the center conductor to be even with the end of the center pin. Use a small file to round the end, removing any solder that built up on the outer surface of the center pin. Use a sharp knife, very fine sandpaper or steel wool to remove any solder flux from the outer surface of the center pin. Screw the coupling ring onto the body, and you're finished.

Figure 2 shows how to install a PL-259 connector on RG-58 or RG-59 cable. An adapter is used for the smaller cable with standard RG-8 size PL-259s. (UG-175 for RG-58 and UG-176 for RG-59.) Prepare the cable as shown. Once the braid is prepared, screw the adapter into the PL-259 shell and finish the job as you would a PL-259 on RG-8 cable.

Figure 3 shows the instructions and dimensions for crimp-on UHF connectors that fit all common sizes of coaxial cable. While amateurs have been reluctant to adopt crimp-on connectors, the availability of good quality connectors and inexpensive crimping tools make crimp technology a good choice, even for connectors used outside. Soldering the center conductor to the connector tip is optional.

UHF connectors are not waterproof and must be waterproofed whether soldered or crimped as shown in the section of the **Safety** chapter on Antenna and Tower Safety.

2 BNC, N and F Connectors

The BNC connectors illustrated in **Figure 4** are popular for low power levels at VHF and UHF. They accept RG-58 and RG-59 cable, and are available for cable mounting in both male and female versions. Several different styles are available, so be sure to use the dimensions for the type you have. Follow the installation instructions carefully. If you prepare the cable to the wrong dimensions, the center pin will not seat properly with connectors of the opposite gender. Sharp scissors are a big help for trimming the braid evenly. Crimp-on BNC connectors are also available, with a large

---

UHF Connectors
Braid Crimp - Solder Center Contact

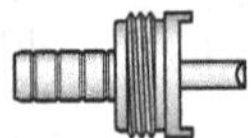

Ferrule Coupling Nut Body assembly

Amphenol	Cable RG-/U	Cable Attachment		Hex Crimp Data			Stripping Dims, inches (mm)		
		Outer	Inner	Cavity for Outer Ferrule	Die Set Tool 227-994	CTL Series Tool No.	a	b	c
83-58SP	58, 141	Crimp	Solder	0.213(5.4)	227-1221-11	CTL-1	1.14 (29.0)	0.780 (19.9)	0.250 (6.4)
83-58SP-1002	400	Crimp	Solder	0.213(5.4)	227-1221-11	CTL-1	1.14 (29.0)	0.780 (19.9)	0.250 (6.4)
83-59DCP-RFX	59	Crimp	Solder	0255(6.5)	227-1221-13	CTL-1	1.22 (30.9)	0.890* (22.6)	0.543 (13.8)
83-58SCP-RFX	58	Crimp	Solder	0.213(5.4)	227-1221-11	CTL-1	1.22 (30.9)	0.890* (22.6)	0.543 (13.8)
83-59SP	59	Crimp	Solder	0.255(6.5)	227-1221-13	CTL-1	1.22 (30.9)	0.890* (22.6)	0.543 (13.8)
83-8SP-RFX	8	Crimp	Solder	0.429(10.9)	227-1221-25	CTL-3	1.22 (30.9)	0.890* (22.6)	0.543 (13.8)

See www.AmphenolRF.com for assembly instructions for all other connectore types. These dimensions only apply to Amphenol connectors and may not be correct for other manufacturers.
* Manufacturer's assembly dimensions incorrectly show 0.574 inches.

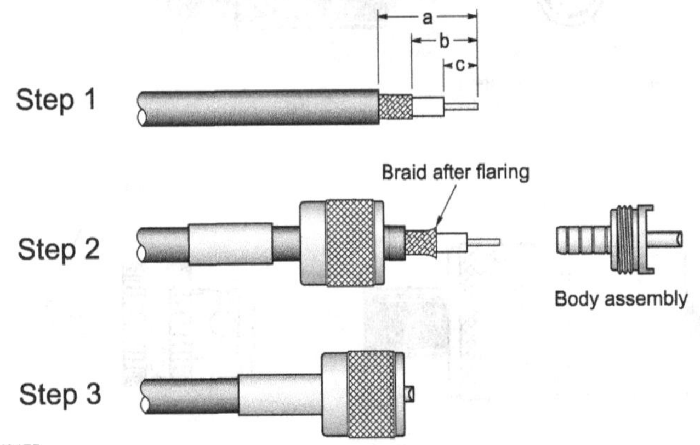

Step 1 Cut end of cable even. Strip cable to dimensions shown in table. All cuts are to be sharp and square. Do not nick braid, dielectric or center conductor. Tin center conductor avoiding excessive heat.

Step 2 Slide coupling nut and ferrule over cable jacket. Flair braid slightly as shown. Install cable into body assembly, so inner ferrule portion slides under braid, until braid butts shoulder. Slide outer ferrule over braid until it butts shoulder. Crimp ferrule with tool and die set indicated in table.

Step 3 Soft solder center conductor to contact. Avoid heating contact excessively to prevent damaging insulator. Slide/screw coupling nut over body.

Figure 3 — Crimp-on UHF connectors are available for all sizes of popular coaxial cable and save considerable time over soldered connectors. The performance and reliability of these connectors is equivalent to soldered connectors, if crimped properly. (*Courtesy Amphenol Electronic Components*)

Table 1

Nominal Characteristics of Commonly Used Transmission Lines

RG or Type	Part Number	Nom. Z_0 Ω	VF %	Cap. pF/ft	Cent. Cond. AWG	Diel. Type	Shield Type	Jacket Matl	OD inches	Max V (RMS)	Matched Loss (dB/100')			
											1 MHz	10	100	1000
RG-6	Belden 1694A	75	82	16.2	#18 Solid BC	FPE	FC	P1	0.275	300	0.3	.7	1.8	5.9
RG-6	Belden 8215	75	66	20.5	#21 Solid CCS	PE	D	PE	0.332	2700	0.4	0.8	2.7	9.8
RG-8	Belden 7810A	50	86	23.0	#10 Solid BC	FPE	FC	PE	0.405	300	0.1	0.4	1.2	4.0
RG-8	TMS LMR400	50	85	23.9	#10 Solid CCA	FPE	FC	PE	0.405	600	0.1	0.4	1.3	4.1
RG-8	Belden 9913	50	84	24.6	#10 Solid BC	ASPE	FC	P1	0.405	300	0.1	0.4	1.3	4.5
RG-8	CXP1318FX	50	84	24.0	#10 Flex BC	FPE	FC	P2N	0.405	600	0.1	0.4	1.3	4.5
RG-8	Belden 9913F	50	83	24.6	#11 Flex BC	FPE	FC	P1	0.405	300	0.1	0.4	1.3	4.5
RG-8	Belden 9914	50	82	24.8	#10 Solid BC	FPE	FC	P1	0.405	300	0.2	0.6	1.5	4.8
RG-8	TMS LMR400UF	50	85	23.9	#10 Flex BC	FPE	FC	PE	0.405	600	0.1	0.4	1.4	4.9
RG-8	DRF-BF	50	84	24.5	#9.5 Flex BC	FPE	FC	PE	0.405	600	0.1	0.5	1.6	5.2
RG-8	WM CQ106	50	84	24.5	#9.5 Flex BC	FPE	FC	P2N	0.405	600	0.2	0.6	1.8	5.3
RG-8	CXP008	50	78	26.0	#13 Flex BC	FPE	S	P1	0.405	600	0.1	0.5	1.8	7.1
RG-8	Belden 8237	52	66	29.5	#13 Flex BC	PE	S	P1	0.405	3700	0.2	0.6	1.9	7.4
RG-8X	Belden 7808A	50	86	23.5	#15 Solid BC	FPE	FC	PE	0.240	300	0.2	0.7	2.3	7.4
RG-8X	TMS LMR240	50	84	24.2	#15 Solid BC	FPE	FC	PE	0.242	300	0.2	0.8	2.5	8.0
RG-8X	WM CQ118	50	82	25.0	#16 Flex BC	FPE	FC	P2N	0.242	300	0.3	0.9	2.8	8.4
RG-8X	TMS LMR240UF	50	84	24.2	#15 Flex BC	FPE	FC	PE	0.242	300	0.2	0.8	2.8	9.6
RG-8X	Belden 9258	50	82	24.8	#16 Flex BC	FPE	S	P1	0.242	300	0.3	0.9	3.2	11.2
RG-8X	CXP08XB	50	80	25.3	#16 Flex BC	FPE	S	P1	0.242	300	0.3	1.0	3.1	14.0
RG-9	Belden 8242	51	66	30.0	#13 Flex SPC	PE	SCBC	P2N	0.420	5000	0.2	0.6	2.1	8.2
RG-11	Belden 8213	75	84	16.1	#14 Solid BC	FPE	S	PE	0.405	300	0.1	0.4	1.3	5.2
RG-11	Belden 8238	75	66	20.5	#18 Flex TC	PE	S	P1	0.405	300	0.2	0.7	2.0	7.1
RG-58	Belden 7807A	50	85	23.7	#18 Solid BC	FPE	FC	PE	0.195	300	0.3	1.0	3.0	9.7
RG-58	TMS LMR200	50	83	24.5	#17 Solid BC	FPE	FC	PE	0.195	300	0.3	1.0	3.2	10.5
RG-58	WM CQ124	52	66	28.5	#20 Solid BC	PE	S	PE	0.195	1400	0.4	1.3	4.3	14.3
RG-58	Belden 8240	52	66	29.9	#20 Solid BC	PE	S	P1	0.193	1400	0.3	1.1	3.8	14.5
RG-58A	Belden 8219	53	73	26.5	#20 Flex TC	FPE	S	P1	0.195	300	0.4	1.3	4.5	18.1
RG-58C	Belden 8262	50	66	30.8	#20 Flex TC	PE	S	P2N	0.195	1400	0.4	1.4	4.9	21.5
RG-58A	Belden 8259	50	66	30.8	#20 Flex TC	PE	S	P1	0.192	1400	0.5	1.5	5.4	22.8
RG-59	Belden 1426A	75	83	16.3	#20 Solid BC	FPE	S	P1	0.242	300	0.3	0.9	2.6	8.5
RG-59	CXP 0815	75	82	16.2	#20 Solid BC	FPE	S	P1	0.232	300	0.5	0.9	2.2	9.1
RG-59	Belden 8212	75	78	17.3	#20 Solid CCS	FPE	S	P1	0.242	300	0.2	1.0	3.0	10.9
RG-59	Belden 8241	75	66	20.4	#23 Solid CCS	PE	S	P1	0.242	1700	0.6	1.1	3.4	12.0
RG-62A	Belden 9269	93	84	13.5	#22 Solid CCS	ASPE	S	P1	0.240	750	0.3	0.9	2.7	8.7
RG-62B	Belden 8255	93	84	13.5	#24 Flex CCS	ASPE	S	P2N	0.242	750	0.3	0.9	2.9	11.0
RG-63B	Belden 9857	125	84	9.7	#22 Solid CCS	ASPE	S	P2N	0.405	750	0.2	0.5	1.5	5.8
RG-83	WM165	35	66	44.0	#10 Solid BC	PE	S	P2	0.405	2000	0.23	0.8	2.8	9.6
RG-142	CXP 183242	50	69.5	29.4	#19 Solid SCCS	TFE	D	FEP	0.195	1900	0.3	1.1	3.8	12.8
RG-142B	Belden 83242	50	69.5	29.0	#19 Solid SCCS	TFE	D	TFE	0.195	1400	0.3	1.1	3.9	13.5
RG-174	Belden 7805R	50	73.5	26.2	#25 Solid BC	FPE	FC	P1	0.110	300	0.6	2.0	6.5	21.3
RG-174	Belden 8216	50	66	30.8	#26 Flex CCS	PE	S	P1	0.110	1100	0.8	2.5	8.6	33.7
RG-213	Belden 8267	50	66	30.8	#13 Flex BC	PE	S	P2N	0.405	3700	0.2	0.6	2.1	8.0
RG-213	CXP213	50	66	30.8	#13 Flex BC	PE	S	P2N	0.405	600	0.2	0.6	2.0	8.2
RG-214	Belden 8268	50	66	30.8	#13 Flex SPC	PE	D	P2N	0.425	3700	0.2	0.7	2.2	8.0
RG-216	Belden 9850	75	66	20.5	#18 Flex TC	PE	D	P2N	0.425	3700	0.2	0.7	2.0	7.1
RG-217	WM CQ217F	50	66	30.8	#10 Flex BC	PE	D	PE	0.545	7000	0.1	0.4	1.4	5.2
RG-217	M17/78-RG217	50	66	30.8	#10 Solid BC	PE	D	P2N	0.545	7000	0.1	0.4	1.4	5.2
RG-218	M17/79-RG218	50	66	29.5	#4.5 Solid BC	PE	S	P2N	0.870	11000	0.1	0.2	0.8	3.4
RG-223	Belden 9273	50	66	30.8	#19 Solid SPC	PE	D	P2N	0.212	1400	0.4	1.2	4.1	14.5
RG-303	Belden 84303	50	69.5	29.0	#18 Solid SCCS	TFE	S	TFE	0.170	1400	0.3	1.1	3.9	13.5
RG-316	CXP TJ1316	50	69.5	29.4	#26 Flex BC	TFE	S	FEP	0.098	1200	1.2	2.7	8.0	26.1
RG-316	Belden 84316	50	69.5	29.0	#26 Flex SCCS	TFE	S	FEP	0.096	900	0.8	2.5	8.3	26.0
RG-393	M17/127-RG393	50	69.5	29.4	#12 Flex SPC	TFE	D	FEP	0.390	5000	0.2	0.5	1.7	6.1
RG-400	M17/128-RG400	50	69.5	29.4	#20 Flex SPC	TFE	D	FEP	0.195	1400	0.4	1.3	4.3	15.0
LMR500	TMS LMR500UF	50	85	23.9	#7 Flex BC	FPE	FC	PE	0.500	2500	0.1	0.4	1.2	4.0
LMR500	TMS LMR500	50	85	23.9	#7 Solid CCA	FPE	FC	PE	0.500	2500	0.1	0.3	0.9	3.3
LMR600	TMS LMR600	50	86	23.4	#5.5 Solid CCA	FPE	FC	PE	0.590	4000	0.1	0.2	0.8	2.7
LMR600	TMS LMR600UF	50	86	23.4	#5.5 Flex BC	FPE	FC	PE	0.590	4000	0.1	0.2	0.8	2.7
LMR1200	TMS LMR1200	50	88	23.1	#0 Copper Tube	FPE	FC	PE	1.200	4500	0.04	0.1	0.4	1.3
Hardline														
1/2"	CATV Hardline	50	81	25.0	#5.5 BC	FPE	SM	none	0.500	2500	0.05	0.2	0.8	3.2
1/2"	CATV Hardline	75	81	16.7	#11.5 BC	FPE	SM	none	0.500	2500	0.1	0.2	0.8	3.2
7/8"	CATV Hardline	50	81	25.0	#1 BC	FPE	SM	none	0.875	4000	0.03	0.1	0.6	2.9
7/8"	CATV Hardline	75	81	16.7	#5.5 BC	FPE	SM	none	0.875	4000	0.03	0.1	0.6	2.9
LDF4-50A	Heliax – ½"	50	88	25.9	#5 Solid BC	FPE	CC	PE	0.630	1400	0.02	0.2	0.6	2.4
LDF5-50A	Heliax – ⅞"	50	88	25.9	0.355" BC	FPE	CC	PE	1.090	2100	0.03	0.10	0.4	1.3
LDF6-50A	Heliax – 1¼"	50	88	25.9	0.516" BC	FPE	CC	PE	1.550	3200	0.02	0.08	0.3	1.1
Parallel Lines														
TV Twinlead (Belden 9085)		300	80	4.5	#22 Flex CCS	PE	none	P1	0.400	**	0.1	0.3	1.4	5.9
Twinlead (Belden 8225)		300	80	4.4	#20 Flex BC	PE	none	P1	0.400	8000	0.1	0.2	1.1	4.8
Generic Window Line		450	91	2.5	#18 Solid CCS	PE	none	P1	1.000	10000	0.02	0.08	0.3	1.1
WM CQ 554		440	91	2.7	#14 Flex CCS	PE	none	P1	1.000	10000	0.04	0.01	0.6	3.0
WM CQ 552		440	91	2.5	#16 Flex CCS	PE	none	P1	1.000	10000	0.05	0.2	0.6	2.6
WM CQ 553		450	91	2.5	#18 Flex CCS	PE	none	P1	1.000	10000	0.06	0.2	0.7	2.9
WM CQ 551		450	91	2.5	#18 Solid CCS	PE	none	P1	1.000	10000	0.05	0.02	0.6	2.8
Open-Wire Line		600	0.95-99***	1.7	#12 BC	none	none	none	**	12000	0.02	0.06	0.2	—

(continued on next page)

Approximate Power Handling Capability (1:1 SWR, 40°C Ambient):

	1.8 MHz	7	14	30	50	150	220	450	1 GHz
RG-58 Style	1350	700	500	350	250	150	120	100	50
RG-59 Style	2300	1100	800	550	400	250	200	130	90
RG-8X Style	1830	840	560	360	270	145	115	80	50
RG-8/213 Style	5900	3000	2000	1500	1000	600	500	350	250
RG-217 Style	20000	9200	6100	3900	2900	1500	1200	800	500
LDF4-50A	38000	18000	13000	8200	6200	3400	2800	1900	1200
LDF5-50A	67000	32000	22000	14000	11000	5900	4800	3200	2100
LMR500	18000	9200	6500	4400	3400	1900	1600	1100	700
LMR1200	52000	26000	19000	13000	10000	5500	4500	3000	2000

Legend:

**	Not Available or varies		N	Non-Contaminating
***	Varies with spacer material and spacing		P1	PVC, Class 1
ASPE	Air Spaced Polyethylene		P2	PVC, Class 2
BC	Bare Copper		PE	Polyethylene
CC	Corrugated Copper		S	Single Braided Shield
CCA	Copper Cover Aluminum		SC	Silver Coated Braid
CCS	Copper Covered Steel		SCCS	Silver Plated Copper Coated Steel
CXP	Cable X-Perts, Inc.		SM	Smooth Aluminum
D	Double Copper Braids		SPC	Silver Plated Copper
DRF	Davis RF		TC	Tinned Copper
FC	Foil + Tinned Copper Braid		TFE	Teflon®
FEP	Teflon ® Type IX		TMS	Times Microwave Systems
Flex	Flexible Stranded Wire		UF	Ultra Flex
FPE	Foamed Polyethylene		WM	Wireman
Heliax	Andrew Corp Heliax			

Figure 4 (below) — BNC connectors are common on VHF and UHF equipment at low power levels. (*Courtesy Amphenol Electronic Components*)

BNC CONNECTORS

Standard Clamp

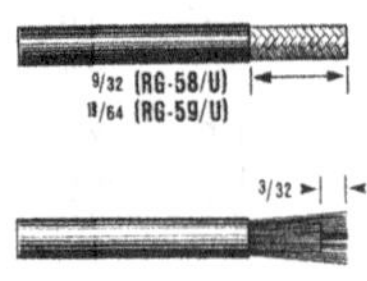

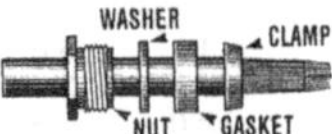

1. Cut cable even. Strip jacket. Fray braid and strip dielectric. **Don't nick braid or center conductor.** Tin center conductor.

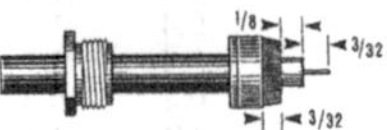

2. Taper braid. Slide nut, washer, gasket and clamp over braid. Clamp inner shoulder should fit squarely against end of jacket.

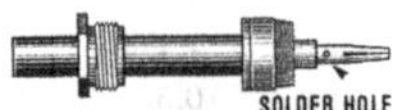

3. With clamp in place, comb out braid, fold back smooth as shown. Trim center conductor.

4. Solder contact on conductor through solder hole. Contact should butt against dielectric. Remove excess solder from outside of contact. Avoid excess heat to prevent swollen dielectric which would interfere with connector body.

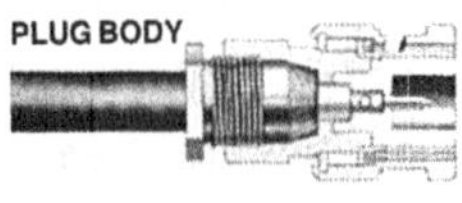

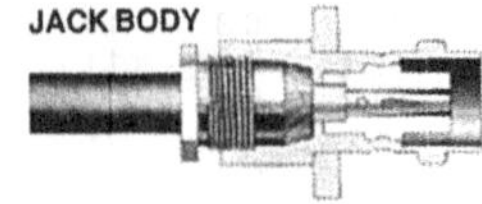

5. Push assembly into body. Screw nut into body with wrench until tight. **Don't rotate body on cable to tighten.**

Improved Clamp

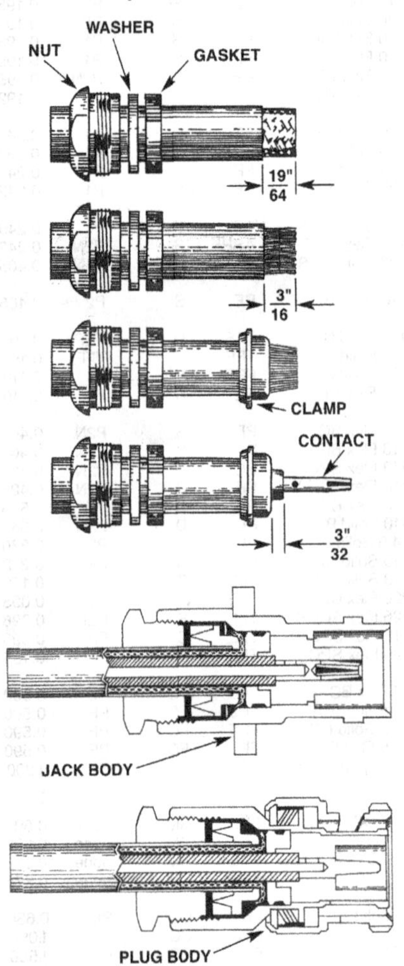

Follow 1, 2, 3 and 4 in BNC connectors (standard clamp) except as noted. Strip cable as shown. Slide gasket on cable *with groove facing clamp.* Slide clamp *with sharp edge facing gasket.* Clamp *should* cut gasket to seal properly.

HBK05_19-18

C. C. Clamp

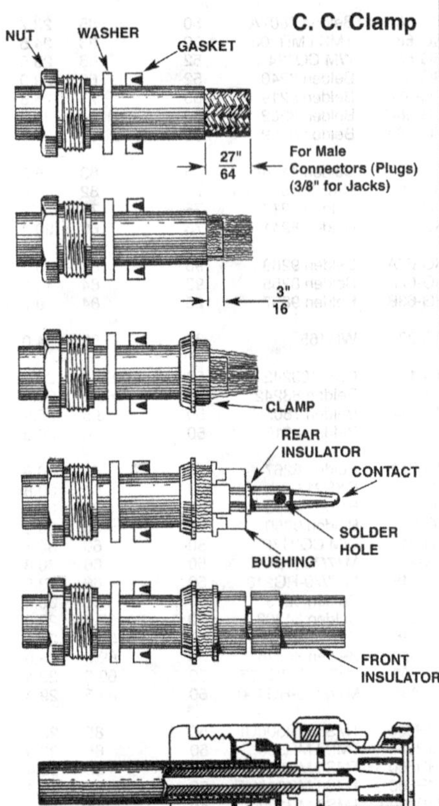

1. Follow steps 1, 2, and 3 as outlined for the standard-clamp BNC connector.

2. Slide on bushing, rear insulator and contact. The parts must butt securely against each other, as shown.

3. Solder the center conductor to the contact. Remove flux and excess solder.

4. Slide the front insulator over the contact, making sure it butts against the contact shoulder.

5. Insert the prepared cable end into the connector body and tighten the nut. Make sure the sharp edge of the clamp seats properly in the gasket.

Table 2
Coaxial Cable Connectors

UHF Connectors

Military No.	Style	Cable RG- or Description
PL-259	Str (m)	8, 9, 11, 13, 63, 87, 149, 213, 214, 216, 225
UG-111	Str (m)	59, 62, 71, 140, 210
SO-239	Pnl (f)	Std, mica/phenolic insulation
UG-266	Blkhd (f)	Rear mount, pressurized, copolymer of styrene ins.

Adapters

PL-258	Str (f/f)	Polystyrene ins.
UG-224,363	Blkhd (f/f)	Polystyrene ins.
UG-646	Ang (f/m)	Polystyrene ins.
M-359A	Ang (m/f)	Polystyrene ins.
M-358	T (f/m/f)	Polystyrene ins.

Reducers

UG-175	55, 58, 141, 142 (except 55A)
UG-176	59, 62, 71, 140, 210

Family Characteristics:

All are nonweatherproof and have a nonconstant impedance. Frequency range: 0-500 MHz. Maximum voltage rating: 500 V (peak).

N Connectors

Military No.	Style	Cable RG-	Notes
UG-21	Str (m)	8, 9, 213, 214	50 Ω
UG-94A	Str (m)	11, 13, 149, 216	70 Ω
UG-536	Str (m)	58, 141, 142	50 Ω
UG-603	Str (m)	59, 62, 71, 140, 210	50 Ω
UG-23, B-E	Str (f)	8, 9, 87, 213, 214, 225	50 Ω
UG-602	Str (f)	59, 62, 71, 140, 210	—
UG-228B, D, E	Pnl (f)	8, 9, 87, 213, 214, 225	—
UG-1052	Pnl (f)	58, 141, 142	50 Ω
UG-593	Pnl (f)	59, 62, 71, 140, 210	50 Ω
UG-160A, B, D	Blkhd (f)	8, 9, 87, 213, 214, 225	50 Ω
UG-556	Blkhd (f)	58, 141, 142	50 Ω
UG-58, A	Pnl (f)		50 Ω
UG-997A	Ang (f)		50 Ω

Panel mount (f) with clearance above panel

M39012/04-	Blkhd (f)	Front mount hermetically sealed
UG-680	Blkhd (f)	Front mount pressurized

N Adapters

Military No.	Style	Notes
UG-29,A,B	Str (f/f)	50 Ω, TFE ins.
UG-57A.B	Str (m/m)	50 Ω, TFE ins.
UG-27A,B	Ang (f/m)	Mitre body
UG-212A	Ang (f/m)	Mitre body
UG-107A	T (f/m/f)	—
UG-28A	T (f/f/f)	—
UG-107B	T (f/m/f)	—

Family Characteristics:

N connectors with gaskets are weatherproof. RF leakage: –90 dB min @ 3 GHz. Temperature limits: TFE: –67° to 390°F (–55° to 199°C). Insertion loss 0.15 dB max @ 10 GHz. Copolymer of styrene: –67° to 185°F (–55° to 85°C). Frequency range: 0-11 GHz. Maximum voltage rating: 1500 V P-P. Dielectric withstanding voltage 2500 V RMS. SWR (MIL-C-39012 cable connectors) 1.3 max 0-11 GHz.

BNC Connectors

Military No.	Style	Cable RG-	Notes
UG-88C	Str (m)	55, 58, 141, 142, 223, 400	

Military No.	Style	Cable RG-	Notes
UG-959	Str (m)	8, 9	
UG-260,A	Str (m)	59, 62, 71, 140, 210	Rexolite ins.
UG-262	Pnl (f)	59, 62, 71, 140, 210	Rexolite ins.
UG-262A	Pnl (f)	59, 62, 71, 140, 210	nwx, Rexolite ins.
UG-291	Pnl (f)	55, 58, 141, 142, 223, 400	
UG-291A	Pnl (f)	55, 58, 141, 142, 223, 400	nwx
UG-624	Blkhd (f)	59, 62, 71, 140, 210	Front mount Rexolite ins.
UG-1094A	Blkhd		Standard
UG-625B	Receptacle		
UG-625			

BNC Adapters

Military No.	Style	Notes
UG-491,A	Str (m/m)	
UG-491B	Str (m/m)	Berylium, outer contact
UG-914	Str (f/f)	
UG-306	Ang (f/m)	
UG-306A,B	Ang (f/m)	Berylium outer contact
UG-414,A	Pnl (f/f)	# 3-56 tapped flange holes
UG-306	Ang (f/m)	
UG-306A,B	Ang (f/m)	Berylium outer contact
UG-274	T (f/m/f)	
UG-274A,B	T (f/m/f)	Berylium outer contact

Family Characteristics:

Z = 50 Ω. Frequency range: 0-4 GHz w/low reflection; usable to 11 GHz. Voltage rating: 500 V P-P. Dielectric withstanding voltage 500 V RMS. SWR: 1.3 max 0-4 GHz. RF leakage –55 dB min @ 3 GHz. Insertion loss: 0.2 dB max @ 3 GHz. Temperature limits: TFE: –67° to 390°F (–55° to 199°C); Rexolite insulators: –67° to 185°F (–55° to 85°C). "Nwx" = not weatherproof.

HN Connectors

Military No.	Style	Cable RG-	Notes
UG-59A	Str (m)	8, 9, 213, 214	
UG-1214	Str (f)	8, 9, 87, 213, 214, 225	Captivated contact
UG-60A	Str (f)	8, 9, 213, 214	Copolymer of styrene ins.
UG-1215	Pnl (f)	8, 9, 87, 213, 214, 225	Captivated contact
UG-560	Pnl (f)		
UG-496	Pnl (f)		
UG-212C	Ang (f/m)		Berylium outer contact

Family Characteristics:

Connector Styles: Str = straight; Pnl = panel; Ang = Angle; Blkhd = bulkhead. Z = 50 Ω. Frequency range = 0-4 GHz. Maximum voltage rating = 1500 V P-P. Dielectric withstanding voltage = 5000 V RMS SWR = 1.3. All HN series are weatherproof. Temperature limits: TFE: –67° to 390°F (–55° to 199°C); copolymer of styrene: –67° to 185°F (–55° to 85°C).

Cross-Family Adapters

Families	Description	Military No.
HN to BNC	HN-m/BNC-f	UG-309
N to BNC	N-m/BNC-f	UG-201,A
	N-f/BNC-m	UG-349,A
	N-m/BNC-m	UG-1034
N to UHF	N-m/UHF-f	UG-146
	N-f/UHF-m	UG-83,B
	N-m/UHF-m	UG-318
UHF to BNC	UHF-m/BNC-f	UG-273
	UHF-f/BNC-m	UG-255

Type N assembly instructions

CLAMP TYPES

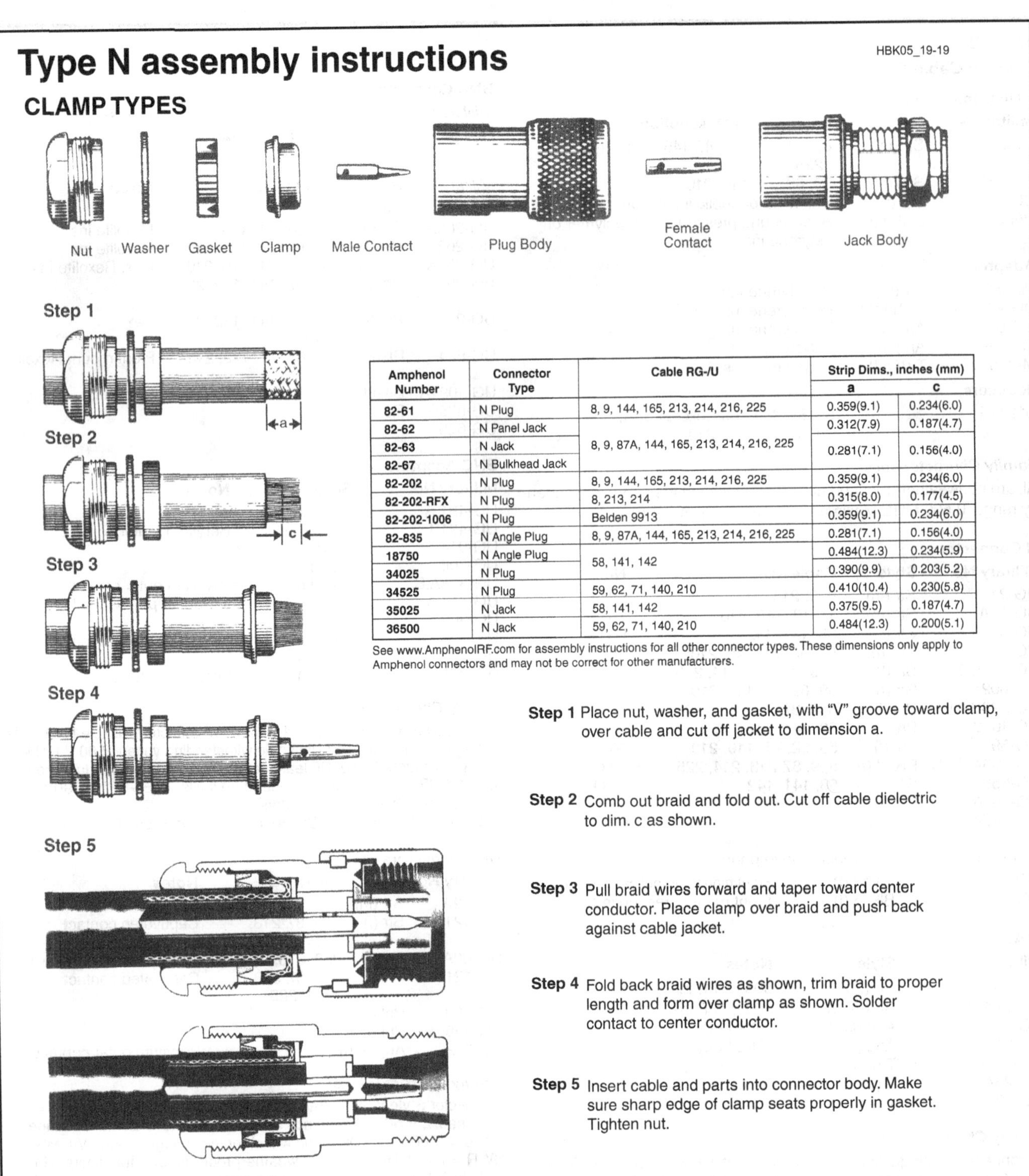

Amphenol Number	Connector Type	Cable RG-/U	Strip Dims., inches (mm)	
			a	c
82-61	N Plug	8, 9, 144, 165, 213, 214, 216, 225	0.359(9.1)	0.234(6.0)
82-62	N Panel Jack		0.312(7.9)	0.187(4.7)
82-63	N Jack	8, 9, 87A, 144, 165, 213, 214, 216, 225	0.281(7.1)	0.156(4.0)
82-67	N Bulkhead Jack			
82-202	N Plug	8, 9, 144, 165, 213, 214, 216, 225	0.359(9.1)	0.234(6.0)
82-202-RFX	N Plug	8, 213, 214	0.315(8.0)	0.177(4.5)
82-202-1006	N Plug	Belden 9913	0.359(9.1)	0.234(6.0)
82-835	N Angle Plug	8, 9, 87A, 144, 165, 213, 214, 216, 225	0.281(7.1)	0.156(4.0)
18750	N Angle Plug	58, 141, 142	0.484(12.3)	0.234(5.9)
34025	N Plug		0.390(9.9)	0.203(5.2)
34525	N Plug	59, 62, 71, 140, 210	0.410(10.4)	0.230(5.8)
35025	N Jack	58, 141, 142	0.375(9.5)	0.187(4.7)
36500	N Jack	59, 62, 71, 140, 210	0.484(12.3)	0.200(5.1)

See www.AmphenolRF.com for assembly instructions for all other connector types. These dimensions only apply to Amphenol connectors and may not be correct for other manufacturers.

Step 1 Place nut, washer, and gasket, with "V" groove toward clamp, over cable and cut off jacket to dimension a.

Step 2 Comb out braid and fold out. Cut off cable dielectric to dim. c as shown.

Step 3 Pull braid wires forward and taper toward center conductor. Place clamp over braid and push back against cable jacket.

Step 4 Fold back braid wires as shown, trim braid to proper length and form over clamp as shown. Solder contact to center conductor.

Step 5 Insert cable and parts into connector body. Make sure sharp edge of clamp seats properly in gasket. Tighten nut.

Figure 5 — Type N connectors are a must for high-power VHF and UHF operation. (*Courtesy Amphenol Electronic Components*)

number of variations, including a twist-on version. A guide to installing these connectors is available with the downloadable supplemental content.

The Type N connector, illustrated in **Figure 5**, is a must for high-power VHF and UHF operation. N connectors are available in male and female versions for cable mounting and are designed for RG-8 size cable. Unlike UHF connectors, they are designed to maintain a constant impedance at cable joints. Like BNC connectors, it is important to prepare the cable to the right dimensions. The center pin must be positioned correctly to mate with the center pin of connectors of the opposite gender. Use the right dimensions for the connector style you have. Crimp-on N connectors are also available, again with a large number of variations. A guide to installing these connectors is available with the downloadable supplemental content.

Type F connectors, used primarily on cable TV connections, are also popular for receive-only antennas and can be used with RG-59

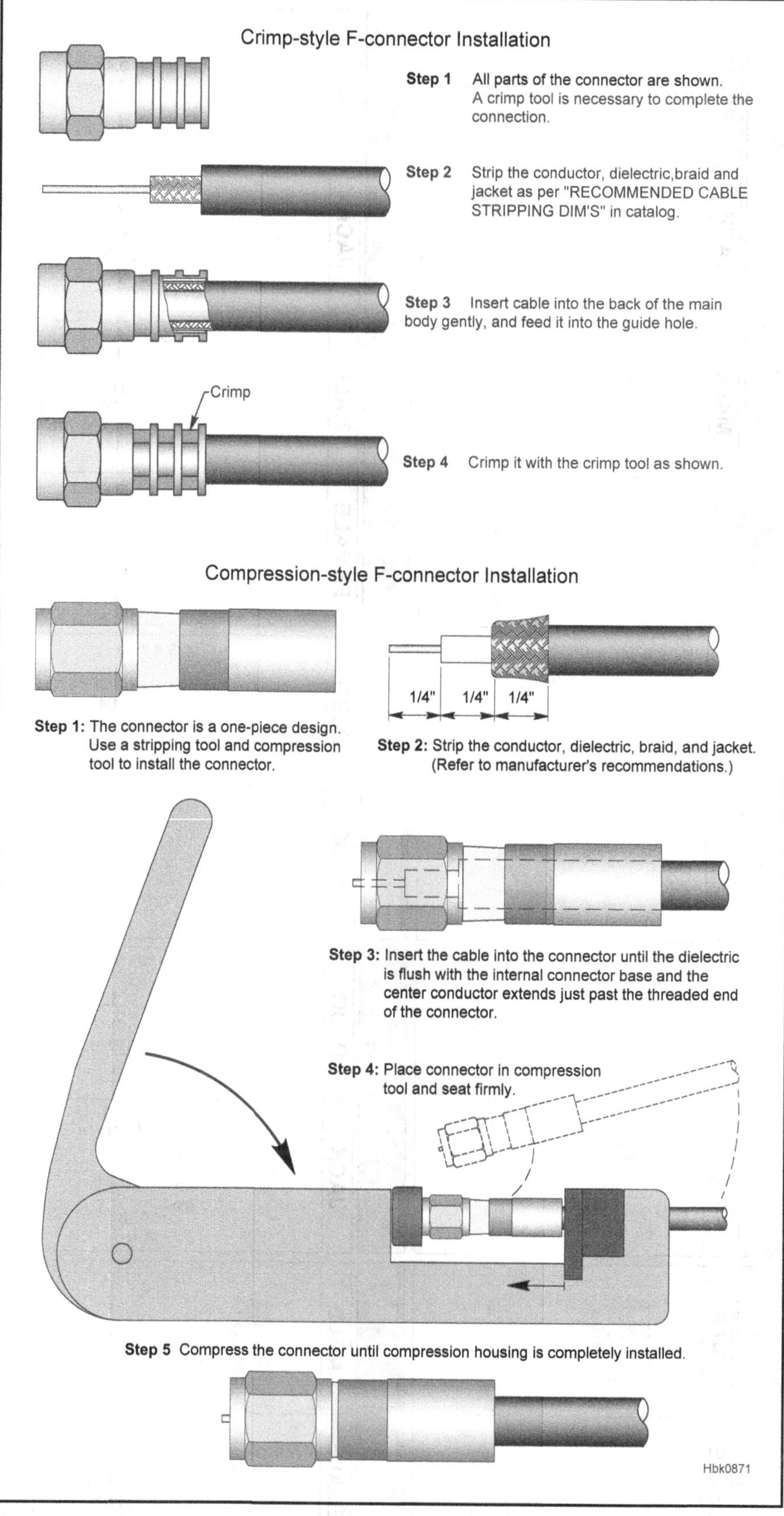

Figure 6 — Type F connectors, commonly used for cable TV connections, can be used for receive-only antennas with inexpensive RG-59 and RG-6 cable. The older crimp-style connectors have largely been replaced by the more reliable and weather-resistant compression-style connectors. Follow the manufacturer's instruction for the cable and connector selected. [Courtesy Amphenol Electronic Components and Technetix, Inc.]

or the increasingly popular RG-6 cable available at low cost. Crimp-on connectors are the only option for these connectors and **Figure 6** shows a general guide for installing them. The exact dimensions vary between connector styles and manufacturers — information on crimping is generally provided with the connectors. There are two styles of crimp; ferrule and compression. The ferrule crimp method is similar to that for UHF, BNC, and N connectors in which a metal ring is compressed around the exposed coax shield. The compression crimp forces a bushing into the back of the connector, clamping the shield against the connector body. In all cases, the exposed center conductor of the cable — a solid wire — must end flush with the end of the connector. A center conductor that is too short may not make a good connection.

3 Connector Identifier and Range Chart

The following pages of figures provide dimensions and side views to help identify the different types of connectors used for RF through microwave frequencies. Dimensions are provided in both imperial and metric units as appropriate. For mm-wave and microwave connectors, calipers or a micrometer may be required to provide an accurate measurement capable of distinguishing between similar connectors. These specifications are intended for connector identification only and should not be the sole dimensions used when laying out a circuit board or drilling a mounting hole.

These figures and chart were provided by Pasternack (**www.pasternack.com**), a major distributor of coaxial connectors, cable, tools, and other RF materials and supplies.

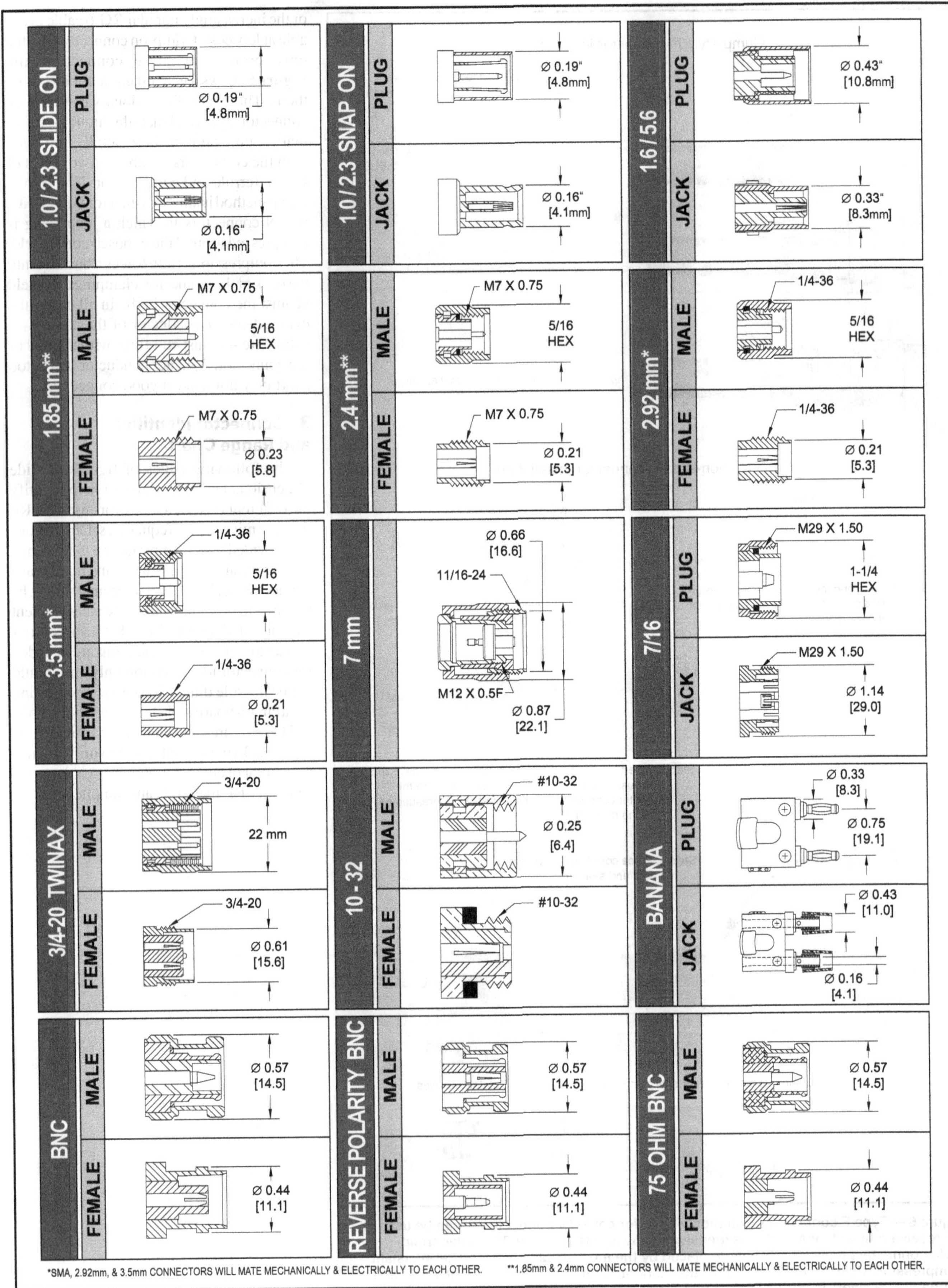

Figure 7A — Connector side views, set 1 of 4. [Courtesy of Pasternack]

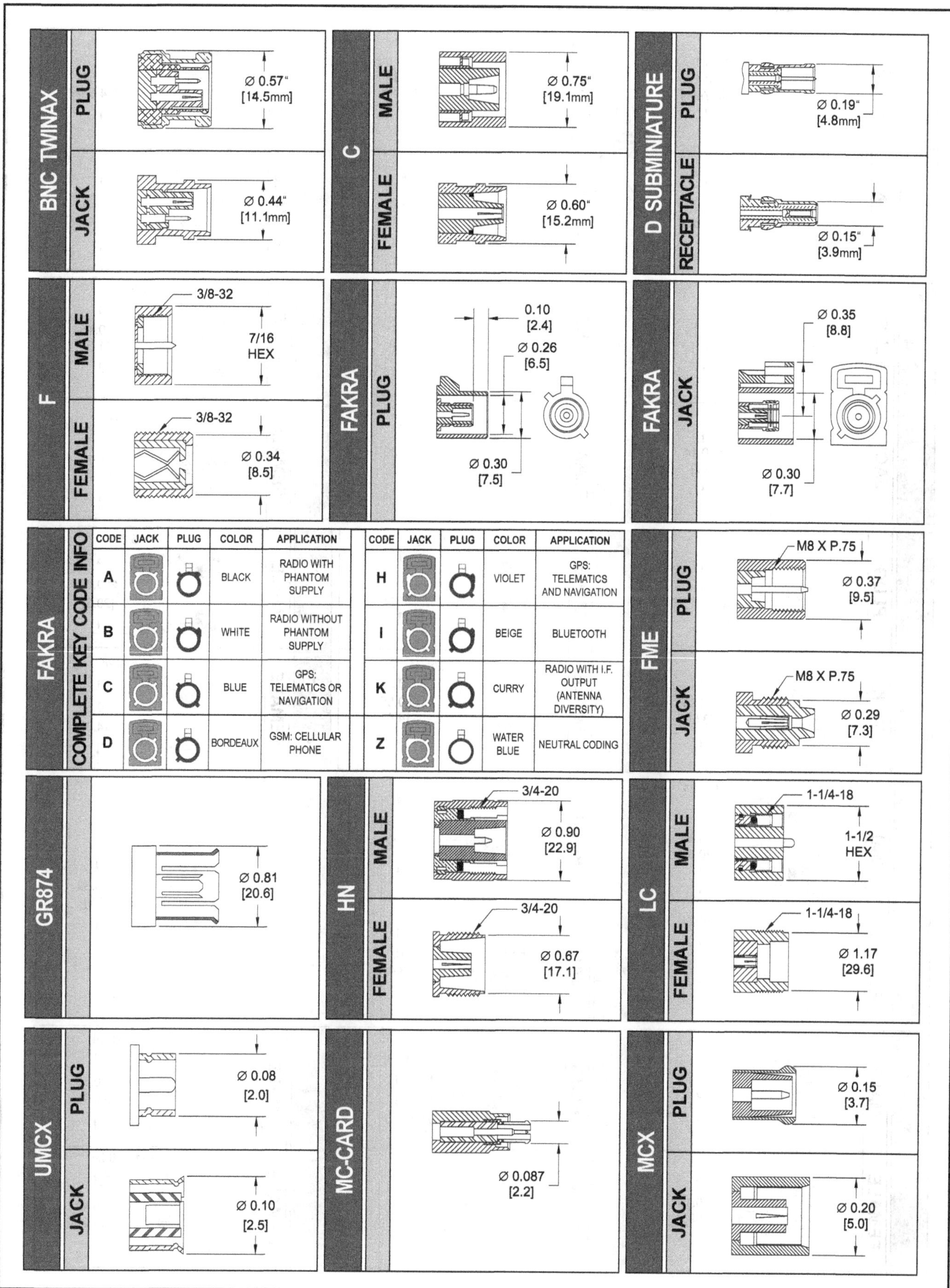

FAKRA — COMPLETE KEY CODE INFO

CODE	JACK	PLUG	COLOR	APPLICATION	CODE	JACK	PLUG	COLOR	APPLICATION
A			BLACK	RADIO WITH PHANTOM SUPPLY	H			VIOLET	GPS: TELEMATICS AND NAVIGATION
B			WHITE	RADIO WITHOUT PHANTOM SUPPLY	I			BEIGE	BLUETOOTH
C			BLUE	GPS: TELEMATICS OR NAVIGATION	K			CURRY	RADIO WITH I.F. OUTPUT (ANTENNA DIVERSITY)
D			BORDEAUX	GSM: CELLULAR PHONE	Z			WATER BLUE	NEUTRAL CODING

Figure 7B — Connector side views, set 2 of 4. [Courtesy of Pasternack]

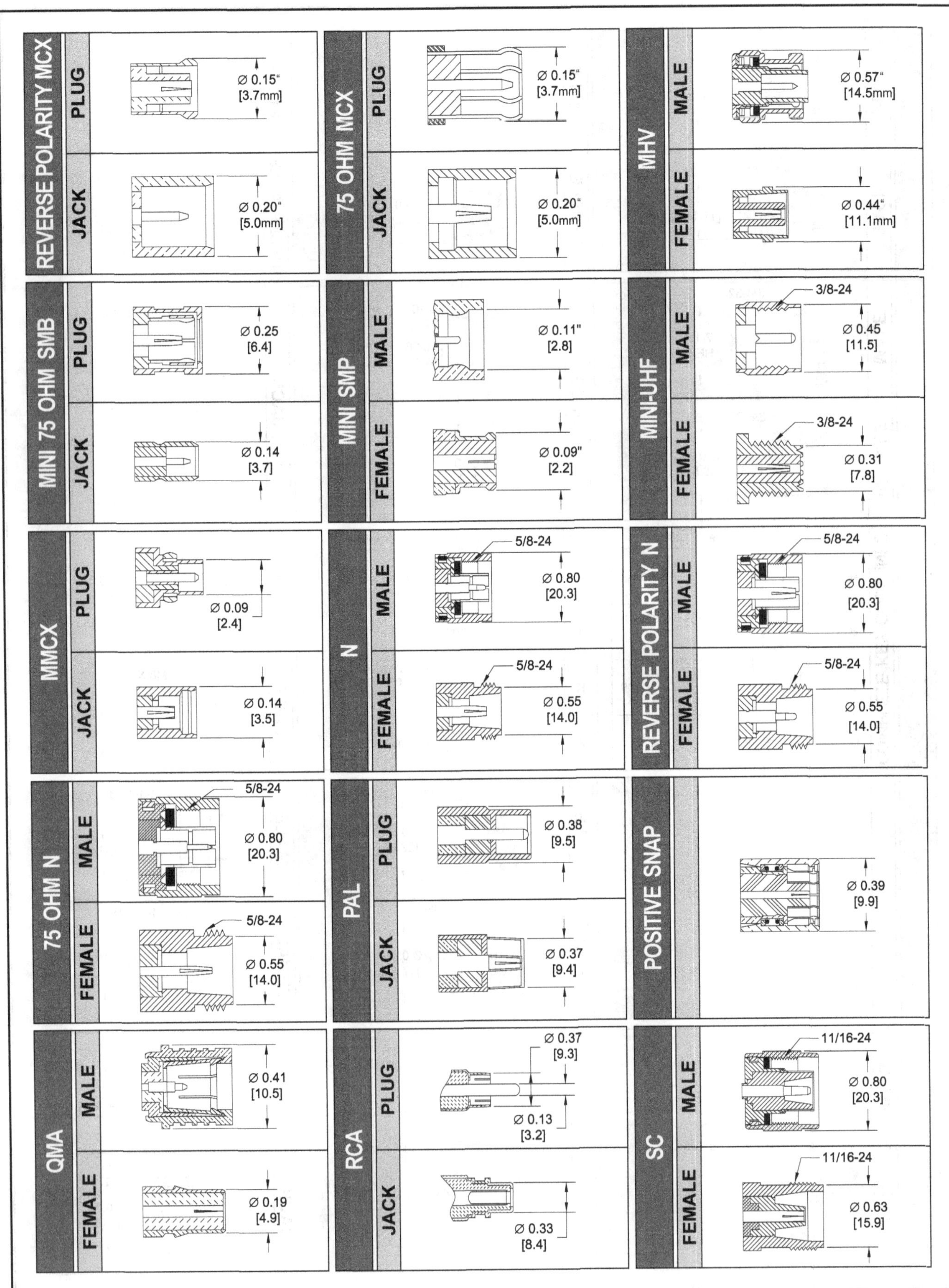

Figure 7C — Connector side views, set 3 of 4. [Courtesy of Pasternack]

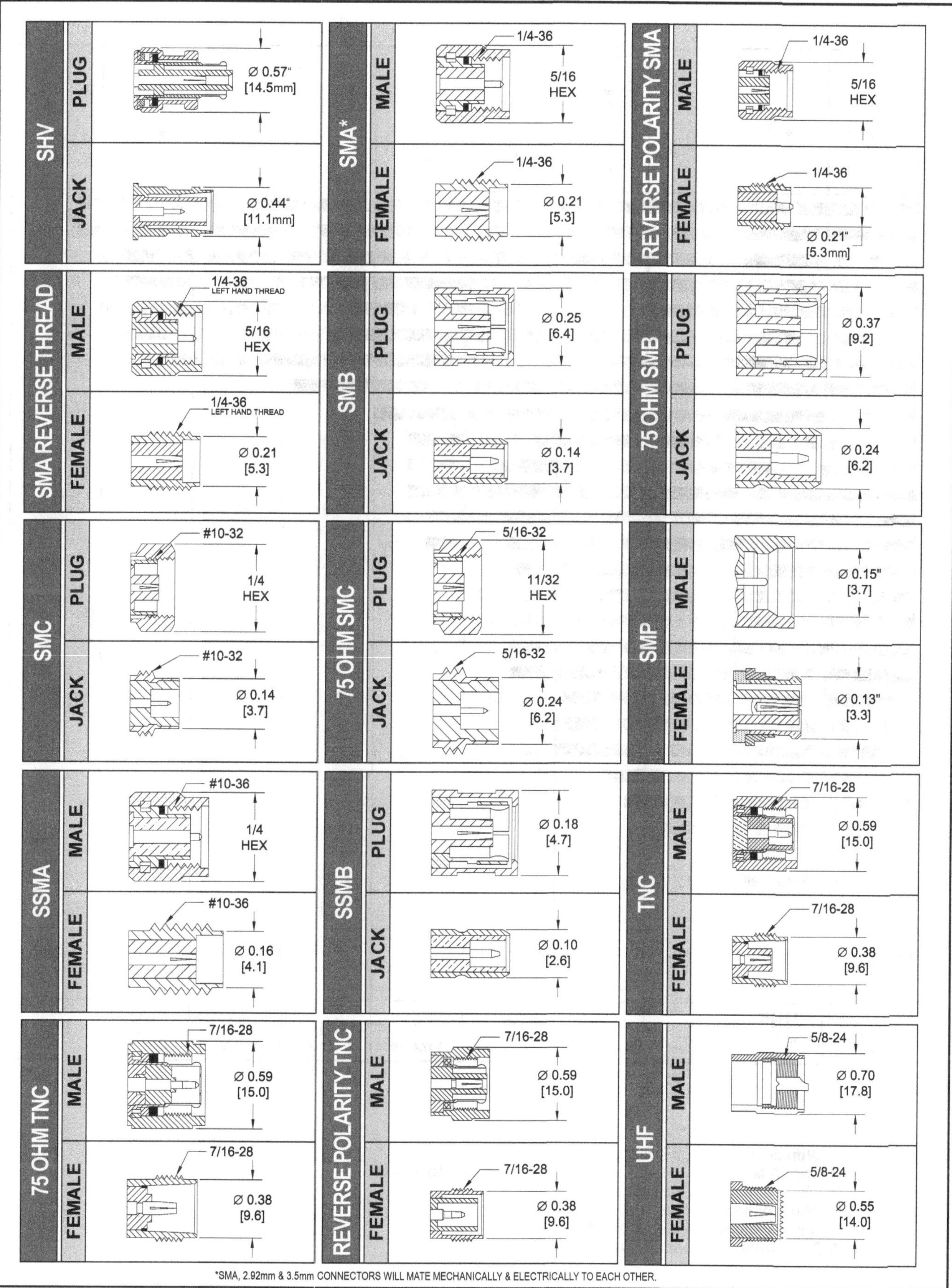

Figure 7D — Connector side views, set 4 of 4. [Courtesy of Pasternack]

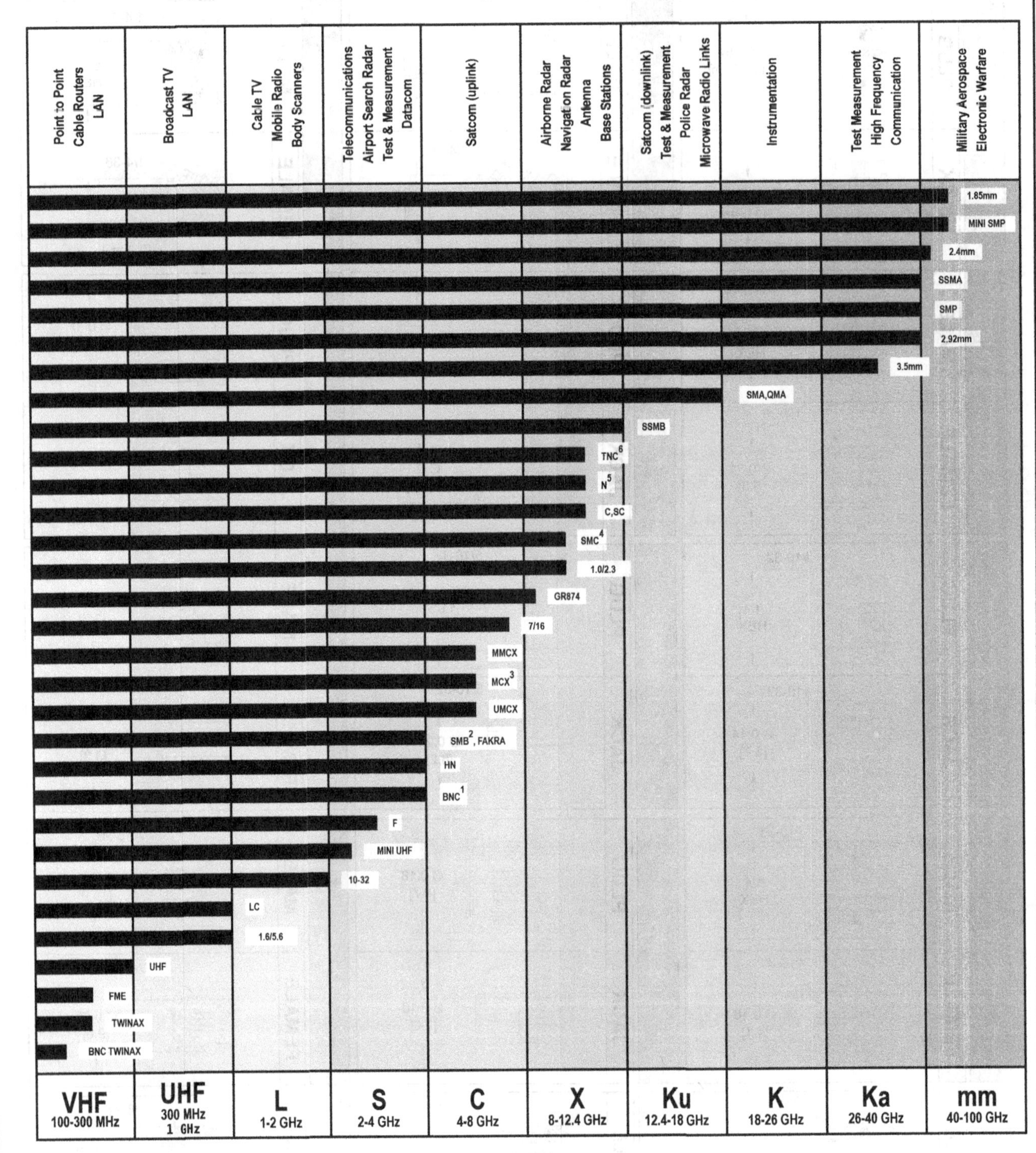

Notes:

1: BNC-75 Ohm connectors operate up to 1 GHz
2: SMB-75 Ohm & Mini SMB-75 Ohm connectors operate up to 4 GHz
3: MCX-75 Ohm connectors operate up to 6 GHz
4: SMC-75 Ohm connectors operate up to 10 GHz
5: N-75 Ohm connectors operate up to 1.5 GHz
6: TNC-75 Ohm connectors operate up to 1 GHz

Figure 8 — Recommended frequency ranges by connector type. [Courtesy of Pasternack]

Avoiding Poor Quality UHF Connectors

*[This guidance is provided by Hal Kennedy, N4GG, and is repeated in the **Station Construction** chapter. Many easily-avoided station problems are caused by poor-quality or poorly attached connectors or adapters — Ed.]*

PL-259s have four parts: the outer sleeve called the "knurled nut," the connector body, the insulator/dielectric and the center pin. All four components can be compromised to the point of making a bargain connector useless.

Problems frequently encountered:

• **Finish:** Bargain connectors sometimes have a finish you can't solder to! They may have a chrome-like appearance, but the plating may not take solder well and has to be filed down for a good connection.

• **Threading:** The internal threads at the rear of the body are there to accept a UG-style insert that narrows the connector barrel to accept smaller diameter coax such as RG-8X or RG-58. The threads may be metric! UG inserts also sometimes appear in the US market with metric threads. Either way, the insert will not screw into the body.

• **Dielectric:** Good connectors use quality phenolic or Teflon insulation between the center pin and the body. Bargain connectors might use anything, including materials such as polystyrene, which will melt when the center pin is soldered.

• **Center pin diameter:** This is one of the most common and insidious problems in mystery PL-259s. The center pin outer diameter (OD) is almost always slightly smaller than it should be and it's hard to notice. The center pin connection between a PL-259 and an SO-239 or barrel connector depends on the male side pin OD being correct and the matching fingers on the female side being the correct diameter and made of the proper spring material.

• **Center socket spring tension:** If the SO-239 socket metal relaxes over time and/or temperature, an intermittent connection will be created that can be very hard to track down.

• **Mating indentions:** The indentations on the end of the SO-239 that mate to a PL-259 (the annulus flange) may only have four indentations to match up with the short prongs on the body of the male connector. A quality SO-239 or barrel connector has indentations all the way around. If the PL-259 and SO-239 don't seat completely, an intermittent connection is likely to develop.

• **Tee and right-angle (elbow) UHF adapters:** The center conductor has to make a right-angle turn inside the shell. In poor-quality adapters the right-angle connection is done with a spring contact — these do not hold up. Quality tee and right-angle adapters are reliable because the internal conductors are tapped and threaded — the conductors are screwed together within the body at the right angle junction.

How can we tell the good connectors? If the price is too good to be true — well, it is. PL-259s with good silver plating have a dull appearance. Good connectors have a part number and manufacturer's name stamped into them. You can look up the connector's specifications if it's marked. An example is the connectors made by Amphenol — all of which have parts numbers such as 83-1SP (PL-259) or 83-1R (SO-239) stamped into or onto the connector body.

Table 3
Powdered-Iron Toroidal Cores: Magnetic Properties

There are differing conventions for referring to the type of core material: #, mix and type are all used. For example, all of the following designate the same material: #12, Mix 12, 12-Mix, Type 12 and 12-Type.

Inductance and Turns Formula
The turns required for a given inductance or inductance for a given number of turns can be calculated from:

$$N = 100\sqrt{\frac{L}{A_L}} \qquad L = A_L\left(\frac{N^2}{10{,}000}\right) \quad \text{(Amidon cores)} \qquad\qquad N = \sqrt{\frac{L}{A_L}} \qquad L = A_L \times N^2 \quad \text{(Non-Amidon cores)}$$

where N = number of turns; L = desired inductance; A_L = inductance index (μH per 100 turns-squared for Amidon cores; nH or μH per turns-squared for non-Amidon cores; see core data sheet)

Amidon Associates literature gives the value of A_L as inductance per 100 turns but the correct units are inductance per 100 turns-squared. The units of inductance are generally in nH but may also be mH. Make sure you understand which units apply and use the A_L value and formula provided by the manufacturer of the core to calculate number of turns or inductance.

Toroid diameter is indicated by the number following "T" — T-200 is 2.00 in. dia; T-68 is 0.68 in. diameter, etc.

AL Values

Size	26*	3	15	1	2	7	6	10	12	17	0
T-12	na	60	50	48	20	18	17	12	7.5	7.5	3.0
T-16	145	61	55	44	22	na	19	13	8.0	8.0	3.0
T-20	180	76	65	52	27	24	22	16	10.0	10.0	3.5
T-25	235	100	85	70	34	29	27	19	12.0	12.0	4.5
T-30	325	140	93	85	43	37	36	25	16.0	16.0	6.0
T-37	275	120	90	80	40	32	30	25	15.0	15.0	4.9
T-44	360	180	160	105	52	46	42	33	18.5	18.5	6.5
T-50	320	175	135	100	49	43	40	31	18.0	18.0	6.4
T-68	420	195	180	115	57	52	47	32	21.0	21.0	7.5
T-80	450	180	170	115	55	50	45	32	22.0	22.0	8.5
T-94	590	248	200	160	84	na	70	58	32.0	na	10.6
T-106	900	450	345	325	135	133	116	na	na	na	19.0
T-130	785	350	250	200	110	103	96	na	na	na	15.0
T-157	870	420	360	320	140	na	115	na	na	na	na
T-184	1640	720	na	500	240	na	195	na	na	na	na
T-200	895	425	na	250	120	105	100	na	na	na	na

*Mix-26 is similar to the older Mix-41, but can provide an extended frequency range.

Magnetic Properties Iron Powder Cores

Mix	Color	Material	μ	Temp stability (ppm/°C)	f (MHz)	Notes
26	Yellow/white	Hydrogen reduced	75	825	dc - 1	Used for EMI filters and dc chokes
3	Gray	Carbonyl HP	35	370	0.05 - 0.50	Excellent stability, good Q for lower frequencies
15	Red/white	Carbonyl GS6	25	190	0.10 - 2	Excellent stability, good Q
1	Blue	Carbonyl C	20	280	0.50 - 5	Similar to Mix-3, but better stability
2	Red	Carbonyl E	10	95	2 - 30	High Q material
7	White	Carbonyl TH	9	30	3 - 35	Similar to Mix-2 and Mix-6, but better temperature stability
6	Yellow	Carbonyl SF	8	35	10 - 50	Very good Q and temperature stability for 20-50 MHz
10	Black	Powdered iron W	6	150	30 - 100	Good Q and stability for 40 - 100 MHz
12	Green/white	Synthetic oxide	4	170	50 - 200	Good Q, moderate temperature stability
17	Blue/yellow	Carbonyl	4	50	40 - 180	Similar to Mix-12, better temperature stability, Q drops about 10% above 50 MHz, 20% above 100 MHz
0	Tan	phenolic	1	0	100 - 300	Inductance may vary greatly with winding technique

Courtesy of Amidon Assoc and Micrometals
Note: Color codes hold only for cores manufactured by Micrometals, which makes the cores sold by most Amateur Radio distributors.

Table 4

Powdered-Iron Toroidal Cores: Dimensions

Toroid diameter is indicated by the number following "T" — T-200 is 2.00 in. dia; T-68 is 0.68 in. diameter, etc.
See Table 3 for a core sizing guide.

Red E Cores—500 kHz to 30 MHz (μ = 10)

No.	OD (in)	ID (in)	H (in)
T-200-2	2.00	1.25	0.55
T-94-2	0.94	0.56	0.31
T-80-2	0.80	0.50	0.25
T-68-2	0.68	0.37	0.19
T-50-2	0.50	0.30	0.19
T-37-2	0.37	0.21	0.12
T-25-2	0.25	0.12	0.09
T-12-2	0.125	0.06	0.05

Black W Cores—30 MHz to 200 MHz (μ=6)

No.	OD (In)	ID (In)	H (In)
T-50-10	0.50	0.30	0.19
T-37-10	0.37	0.21	0.12
T-25-10	0.25	0.12	0.09
T-12-10	0.125	0.06	0.05

Yellow SF Cores—10 MHz to 90 MHz (μ=8)

No.	OD (In)	ID (In)	H (In)
T-94-6	0.94	0.56	0.31
T-80-6	0.80	0.50	0.25
T-68-6	0.68	0.37	0.19
T-50-6	0.50	0.30	0.19
T-26-6	0.25	0.12	0.09
T-12-6	0.125	0.06	0.05

Number of Turns vs Wire Size and Core Size

Approximate maximum number of turns—single layer wound—enameled wire.

Wire Size	T-200	T-130	T-106	T-94	T-80	T-68	T-50	T-37	T-25	T-12
10	33	20	12	12	10	6	4	1		
12	43	25	16	16	14	9	6	3		
14	54	32	21	21	18	13	8	5	1	
16	69	41	28	28	24	17	13	7	2	
18	88	53	37	37	32	23	18	10	4	1
20	111	67	47	47	41	29	23	14	6	1
22	140	86	60	60	53	38	30	19	9	2
24	177	109	77	77	67	49	39	25	13	4
26	223	137	97	97	85	63	50	33	17	7
28	281	173	123	123	108	80	64	42	23	9
30	355	217	154	154	136	101	81	54	29	13
32	439	272	194	194	171	127	103	68	38	17
34	557	346	247	247	218	162	132	88	49	23
36	683	424	304	304	268	199	162	108	62	30
38	875	544	389	389	344	256	209	140	80	39
40	1103	687	492	492	434	324	264	178	102	51

Actual number of turns may differ from above figures according to winding techniques, especially when using the larger size wires. Chart prepared by Michel J. Gordon, Jr, WB9FHC.
Courtesy of Amidon Assoc.

W5SWL Electronics

www.W5SWL.com

Premium Quality RF Connectors Order Direct !

— Wide Selection of Connectors —

- **UHF & N**
- **BNC & SMA**
- **Mini-UHF & FME**
- **TNC & C**
- **MC MCX & MMCX**

- **QMA SMB & SMC**
- **DIN & Low PIM**
- **Reverse Polarity**
- **RF Adapters**
- **Bulkheads**

— And Much More! —

- **Dave's Hobby Shop by W5SWL**
- **Ham Radio Gadgets**
- **RF Technical Parts**
- **New & Surplus Materials**

- **Order at www.W5SWL.com**

Ships Fast From The Arkansas River Valley

ChattRadio.com
CHATTANOOGA, TENNESSEE
1.833.456.4673
We Are Your One Stop
Radio Shop